# Editorial Board

Mathematics Monograph Series   41

# Algebraic Theory of Generalized Inverses

(广义逆的代数理论)

Jianlong Chen (陈建龙)　　Xiaoxiang Zhang (张小向)

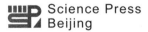
Science Press
Beijing

Springer

Responsible Editors: Xin Li, Yueting Li

Jianlong Chen  
School of Mathematics  
Southeast University  
Nanjing, Jiangsu, China

Xiaoxiang Zhang  
School of Mathematics  
Southeast University  
Nanjing, Jiangsu, China

ISBN 978-7-03-076570-3  
Science Press, Beijing, China

Jointly published with Springer Nature Singapore Pte Ltd.

The print edition is only for sale in Chinese mainland. Customers from outside of Chinese mainland please order the print book from Springer Nature Singapore Pte Ltd.

Published by Science Press  
16 Donghuangchenggen North Street  
Beijing 100717,P.R.China  
Printed in Beijing

# Preface to Mathematics Monograph Series

Science Press asked me to write a preface for their series of books called "Mathematics Monograph Series". They told me that the Press had published nearly 30 mathematical monographs in this series since 2006. This reminded me that, also in 2006, I received an email message from the Editor in Chief of "Sugaku Tushin" ("Mathematical Communications", the membership magazine of the Mathematical Society of Japan). The Editor in Chief told me that they were planning to have a special section on "Recent development in Chinese mathematical community", and invited me to write an article for the special section. As a result, I published in their magazine an article called "Some Aspects of Mathematical Community in China". Among other things, in the article I demonstrated that, with the favorable environment, the Chinese mathematical community had made great progress since the late 1970s (when China started to implement "Reform and Opening-up Policy"): A large number of publications (including articles and monographs) have been written by Chinese mathematicians, there have been always Chinese mathematicians presenting their speeches at various international academic conferences or workshops, many Chinese mathematicians have served as editors of international academic journals, or as members in various academic organizations. All these reveal that the Chinese mathematical community which has been growing rapidly has exerted more and more influence in the world.

Indeed, the series of mathematical monographs published by Science Press reflects partly more and more influence of Chinese mathematical community in the world. Chinese people are good at mathematics. In the past, Science Press published many high level mathematical monographs and textbooks. Among them some were written in Chinese and some were written in English. Some monographs which appeared originally in Chinese have been purchased by international publishers who then re-published them abroad in English, and this has gained influence in corresponding areas of the international mathematical community.

In recent years most Chinese mathematicians have mastered good English. In accordance with this situation, Science Press has decided to publish "Mathematics Monograph Series" —— a series in which high level mathematical monographs and textbooks are written directly in English. The goal of this series is to provide further good service for Chinese mathematicians and to enhance further the influence of the mathematics study in China in the international mathematical community.

I would like to conclude this short preface with the following wish which I expressed also at the end of the afore mentioned article "Some Aspects of Mathematical Community in China":

The Chinese mathematical community will continuously make its effort to work hard, and to strengthen its international exchanges and collaborations, so as to make more contributions to the study and development of mathematics in the world.

Zhiming Ma

March 15, 2015

# Preface

It has been nearly 120 years since the idea of generalized inverses was first proposed by Ivar Fredholm [43] in the process of studying integral equations. Over the past hundred years, the theory of generalized inverses has been substantially developed; many different types of generalized inverses have been introduced and studied whether at the level of matrices and operators, or at the level of elements in rings and semigroups. It has been proved that the theory of generalized inverses not only has important theoretical significance itself, but also plays an important and extensive role in numerical analysis, differential equations, probability statistics, cryptography, optimization, and so on.

This book focuses on Moore-Penrose inverses, group inverses, Drazin inverses, core inverses, and pseudo core inverses in several algebraic systems such as rings, semigroups, and categories. The authors wrote this book for two main purposes: on one hand, considering that the current textbooks and monographs on generalized inverses mostly focus on the properties, computations, and applications of related generalized inverses of complex matrices and operators, it seems helpful to write a book to collect the theory of generalized inverses in several more general algebraic systems; on the other hand, we wish to provide an appropriate textbook or reference book on the algebraic theory of generalized inverses for graduate students in related fields (having been acquainted with basic linear algebra and abstract algebra at the undergraduate level). Therefore, the contents of this book are organized step by step: we start with decompositions of matrices, introduce the basic properties of generalized inverses of matrices, and then discuss generalized inverses of elements in rings and semigroups, as well as morphisms in categories. The algebraic nature of generalized inverses is presented, and the behaviors of generalized inverses are related to the properties of the algebraic system.

There are six chapters in this book.

Chapter 1 is devoted to setting the stage for the whole discussion. We first briefly review some commonly used decompositions of complex matrices and constructions of some special matrices. Then we collect some basic notions and properties of rings such as regularity, finiteness conditions, properties related to involutions, as well as some element-wise techniques.

In Chaps. 2–6, we discuss the Moore-Penrose inverse, group inverse, Drazin inverse, core inverse, and pseudo core inverse in turn, and each chapter focuses on one of these five generalized inverses. In summary, each chapter shall begin with a specific generalized inverse of complex matrices. We give concrete examples of such a generalized inverse, give its basic construction through various matrix decompositions, and list some of its typical properties. After these, we turn to abstract algebraic setting. First, for an arbitrary element, we give the existence criteria of its generalized inverse in terms of inclusion relations of certain sets, direct sum decompositions of a ring, cancellability of some elements with respect to the multiplication, and existence of some specific idempotents. Some elementary properties of the generalized inverse are presented, and various equivalent characterizations of the generalized inverse are given by using one-sided principal ideals and one-sided annihilators. In particular, for von Neumann regular elements, we establish the connection between the existence of their generalized inverses and of some units, and give the expressions of their generalized inverses by using these units. Then, we discuss the generalized inverses of some special elements such as triple products $paq$, sums and differences of elements with appropriate natures in a ring, block matrices and companion matrices over a ring, as well as morphisms with factorizations and sums of morphisms in an additive category. Various expressions and computing methods of their generalized inverses are obtained. In addition, some chapters contain specific research topics. For example, the classical Jacobson's lemma states that if $1 - ab$ is invertible then so is $1 - ba$, in which case $(1 - ba)^{-1} = 1 + b(1 - ab)^{-1}a$. Section 4.4 discusses the corresponding version of Jacobson's lemma for Drazin inverses. In Sect. 6.4, we study equivalent conditions for $1 - ba$ to have a pseudo core inverse when $1 - ab$ has a pseudo core inverse, while it is shown that Jacobson's lemma for pseudo core inverses is not valid in general. In a $C^*$-algebra, note that regularity of an element amounts to Moore-Penrose invertibility; and rings with this property are called GN rings. Section 2.5 investigates the relationship between the (generalized) GN property of a ring and Jacobson's lemma for Moore-Penrose inverses of matrices over that ring.

This book is based on our original lecture notes for a course on the generalized inverse theory. The excellent work of many researchers constitutes the cornerstone of this book. In the process of writing, we have received strong support from the School of Mathematics and the Graduate School of Southeast University, as well as funding from the National Natural Science Foundation of China (Nos. 11371089, 11771076, 11871145, and 12171083), the Qing Lan Project of Jiangsu Province, and the construction funds for excellent textbooks of postgraduates of Southeast University. Since the first draft of this book was completed, it has been used many times for graduate students in School of Mathematics, Southeast University. Doctoral students Cang Wu, Guiqi Shi, Yukun Zhou, Wende Li, Xiaofeng Chen, and Mengmeng Zhou have carefully proofread this book and put forward many valuable suggestions. We take this opportunity to express our sincere thanks to them all.

We would like to dedicate this book to the 120th anniversary of Southeast University.

Nanjing, China                                                                                          Jianlong Chen
February 26, 2023                                                                                   Xiaoxiang Zhang

# Contents

# Chapter 1
# Algebraic Basic Knowledge

In this beginning chapter, we shall review some of the basic concepts and set up some notations for the subsequent chapters. The readers are assumed to be familiar with most of the basic knowledge on sets, groups, rings, fields and vector spaces.

## 1.1 Complex Matrices

Throughout the text, the set of all $m \times n$ complex matrices is denoted by $\mathbb{C}^{m \times n}$, the symbol $I_n$ will stand for the $n \times n$ identity matrix. For any $A \in \mathbb{C}^{m \times n}$, the symbols $A^*, \mathcal{R}(A), \mathcal{N}(A)$ and $\text{rank}(A)$ will denote the conjugate transpose, range, null space and rank of $A$, respectively.

### 1.1.1 Some Decompositions of Complex Matrices

The matrix decomposition is a highlight of linear algebra. Many times, the existence and structure of a generalized inverse can be easily deduced from a particular matrix decomposition. We now review some of the matrix decompositions that are commonly used in the following discussion.

- The standard decomposition:

  Let $A \in \mathbb{C}^{m \times n}$ be of rank $r$. Then there exist invertible matrices $P \in \mathbb{C}^{m \times m}$, $Q \in \mathbb{C}^{n \times n}$ such that

$$A = P \begin{bmatrix} I_r & 0 \\ 0 & 0 \end{bmatrix} Q. \tag{1.1.1}$$

© The Author(s), under exclusive license to Springer Nature Singapore Pte Ltd. 2024
J. Chen, X. Zhang, *Algebraic Theory of Generalized Inverses*,
https://doi.org/10.1007/978-981-99-8285-1_1

- The full rank decomposition:

Let $A \in \mathbb{C}^{m \times n}$ be of rank $r$. Then there exist $G \in \mathbb{C}^{m \times r}$, $H \in \mathbb{C}^{r \times n}$ such that $\mathrm{rank}(G) = \mathrm{rank}(H) = r$ and

$$A = GH. \tag{1.1.2}$$

- The Jordan normal form:

Let $A \in \mathbb{C}^{n \times n}$. Then there exists an invertible matrix $P \in \mathbb{C}^{n \times n}$ such that

$$A = P^{-1} \begin{bmatrix} J_1 \\ & \ddots \\ & & J_h \end{bmatrix} P, \tag{1.1.3}$$

where $J_i = \begin{bmatrix} \lambda_i & 1 \\ & \ddots & \ddots \\ & & \ddots & 1 \\ & & & \lambda_i \end{bmatrix}$ for any $i \in \{1, \cdots, h\}$, $\lambda_1, \cdots, \lambda_s \neq 0$,

$\lambda_{s+1}, \cdots, \lambda_h = 0$. Moreover, take

$$D = \begin{bmatrix} J_1 \\ & \ddots \\ & & J_s \end{bmatrix}, \quad N = \begin{bmatrix} J_{s+1} \\ & \ddots \\ & & J_h \end{bmatrix}.$$

Then $D$ is invertible, $N$ is nilpotent, and

$$A = P^{-1} \begin{bmatrix} D & 0 \\ 0 & N \end{bmatrix} P. \tag{1.1.4}$$

- The core-nilpotent decomposition:

**Theorem 1.1.1** *Let $A \in \mathbb{C}^{n \times n}$. Then A can be written as the sum of matrices $A_1$ and $A_2$, i.e., $A = A_1 + A_2$, where:*

(1) $\mathrm{rank}(A_1) = \mathrm{rank}(A_1^2)$.
(2) $A_2$ *is nilpotent.*
(3) $A_1 A_2 = A_2 A_1 = 0$.

***Proof*** Let $A$ be the form of (1.1.4), and let $A_1 = P^{-1} \begin{bmatrix} D & 0 \\ 0 & 0 \end{bmatrix} P$ and $A_2 = P^{-1} \begin{bmatrix} 0 & 0 \\ 0 & N \end{bmatrix} P$. Then $A_1$, $A_2$ meet the above conditions.                    □

- The singular value decomposition:

Let $A \in \mathbb{C}^{m \times n}$ be of rank $r$. Then there exist unitary matrices $U \in \mathbb{C}^{m \times m}$ and $V \in \mathbb{C}^{n \times n}$ such that

$$A = U \begin{bmatrix} \Sigma & 0 \\ 0 & 0 \end{bmatrix} V^*, \qquad (1.1.5)$$

where $\Sigma = \mathrm{diag}(\sigma_1 I_{r_1}, \ldots, \sigma_t I_{r_t})$ is the diagonal matrix of singular values of $A$, $\sigma_1 > \sigma_2 > \cdots > \sigma_t > 0, r_1 + r_2 + \cdots + r_t = r$.

- The Hartwig-Spindelböck decomposition and its dual:

**Theorem 1.1.2 ([60, Corollary 6])** *Let $A \in \mathbb{C}^{n \times n}$ be of rank $r$. Then there exist unitary matrices $U, V$ such that*

$$A = U \begin{bmatrix} \Sigma K & \Sigma L \\ 0 & 0 \end{bmatrix} U^* \qquad (1.1.6)$$

$$= V \begin{bmatrix} K\Sigma & 0 \\ M\Sigma & 0 \end{bmatrix} V^*, \qquad (1.1.7)$$

*where $\Sigma = \mathrm{diag}(\sigma_1 I_{r_1}, \ldots, \sigma_t I_{r_t})$ is the diagonal matrix of singular values of $A$, $\sigma_1 > \sigma_2 > \cdots > \sigma_t > 0, r_1 + r_2 + \cdots + r_t = r$, and $K \in \mathbb{C}^{r \times r}, L \in \mathbb{C}^{r \times (n-r)}, M \in \mathbb{C}^{(n-r) \times r}$ satisfy*

$$KK^* + LL^* = K^*K + M^*M = I_r.$$

*Proof* Let $A$ be the form of (1.1.5). Then

$$A = U \begin{bmatrix} \Sigma & 0 \\ 0 & 0 \end{bmatrix} V^* U U^*$$

$$= U \begin{bmatrix} \Sigma & 0 \\ 0 & 0 \end{bmatrix} \begin{bmatrix} K & L \\ M & N \end{bmatrix} U^*$$

$$= U \begin{bmatrix} \Sigma K & \Sigma L \\ 0 & 0 \end{bmatrix} U^*.$$

Moreover, since $U, V$ are unitary, it follows that $V^*U = \begin{bmatrix} K & L \\ M & N \end{bmatrix}$ is also unitary.

Hence $KK^* + LL^* = I_r$. Similarly, we can show that $A = V \begin{bmatrix} K\Sigma & 0 \\ M\Sigma & 0 \end{bmatrix} V^*$ and $K^*K + M^*M = I_r$. $\square$

Equations (1.1.6) and (1.1.7) above are known as the Hartwig-Spindelböck decomposition and dual Hartwig-Spindelböck decomposition of $A$, respectively.

The following result was given in [114]. Here we use the Hartwig-Spindelböck decomposition to give another proof.

**Theorem 1.1.3** *Let $A \in \mathbb{C}^{n \times n}$. Then there exists a unitary matrix $U$ such that*

$$A = U \begin{bmatrix} D & L \\ 0 & N \end{bmatrix} U^*, \tag{1.1.8}$$

*where $D$ is invertible and $N$ is nilpotent.*

**Proof** Let $A = U_1 \begin{bmatrix} K_1 & L_1 \\ 0 & 0 \end{bmatrix} U_1^*$ be the Hartwig-Spindelböck decomposition of $A$,

and let $K_1 = U_{11} \begin{bmatrix} K_2 & L_2 \\ 0 & 0 \end{bmatrix} U_{11}^*$ be the Hartwig-Spindelböck decomposition of $K_1$.

Then

$$A = U_1 \begin{bmatrix} U_{11} \begin{bmatrix} K_2 & L_2 \\ 0 & 0 \end{bmatrix} U_{11}^* & L_1 \\ 0 & 0 \end{bmatrix} U_1^*$$

$$= U_1 \begin{bmatrix} U_{11} & 0 \\ 0 & I \end{bmatrix} \begin{bmatrix} K_2 & L_2 & L_{11} \\ 0 & 0 & L_{12} \\ 0 & 0 & 0 \end{bmatrix} \begin{bmatrix} U_{11} & 0 \\ 0 & I \end{bmatrix}^* U_1^*$$

$$= U_2 \begin{bmatrix} K_2 & L_{21} \\ 0 & N_1 \end{bmatrix} U_2^*,$$

where $N_1 = \begin{bmatrix} 0 & L_{12} \\ 0 & 0 \end{bmatrix}$ is nilpotent, $U_2 = U_1 \begin{bmatrix} U_{11} & 0 \\ 0 & I \end{bmatrix}$ is unitary. Continuing this process, we can obtain

$$A = U \begin{bmatrix} D & L \\ 0 & N \end{bmatrix} U^*$$

eventually.                                                                              □

The next two decompositions can be achieved by applying Theorem 1.1.3.

- The core-EP decomposition:

**Theorem 1.1.4 ([114, Theorem 2.1])** *Let $A \in \mathbb{C}^{n \times n}$. Then $A$ can be written as the sum of matrices $A_1$ and $A_2$, i.e., $A = A_1 + A_2$, where:*

(1) $\text{rank}(A_1) = \text{rank}(A_1^2)$.
(2) $A_2$ *is nilpotent.*
(3) $A_1^* A_2 = A_2 A_1 = 0$.

**Proof** Let $A$ be the form of Eq. (1.1.8). Then it suffices to set $A_1 = U \begin{bmatrix} D & L \\ 0 & 0 \end{bmatrix} U^*$

and $A_2 = U \begin{bmatrix} 0 & 0 \\ 0 & N \end{bmatrix} U^*$. □

- The EP-nilpotent decomposition:

**Theorem 1.1.5 ([116, Theorem 2.1])** *Let $A \in \mathbb{C}^{n \times n}$. Then $A$ can be written as the sum of matrices $A_1$ and $A_2$, i.e., $A = A_1 + A_2$, where:*

(1) $\mathcal{R}(A_1) = \mathcal{R}(A_1^*)$.
(2) $A_2$ *is nilpotent.*
(3) $A_2 A_1 = 0$.

**Proof** Let $A$ be the form of (1.1.8). Then it suffices to set $A_1 = U \begin{bmatrix} D & 0 \\ 0 & 0 \end{bmatrix} U^*$ and

$A_2 = U \begin{bmatrix} 0 & L \\ 0 & N \end{bmatrix} U^*$. □

### *1.1.2  The Index of a Square Complex Matrix*

For any $A \in \mathbb{C}^{n \times n}$, the smallest nonnegative integer $k$ such that $\text{rank}(A^k) = \text{rank}(A^{k+1})$ is called the *index* of $A$ and denoted by $\text{ind}(A)$.

Consider the following two chains,

$$\mathcal{R}(A) \supseteq \mathcal{R}(A^2) \supseteq \mathcal{R}(A^3) \cdots ,$$

$$\mathcal{N}(A) \subseteq \mathcal{N}(A^2) \subseteq \mathcal{N}(A^3) \cdots .$$

By comparing the rank, we know that there exist $s, t \in \mathbb{N} \, (\leq n)$ such that $\mathcal{R}(A^s) = \mathcal{R}(A^{s+1})$, $\mathcal{N}(A^t) = \mathcal{N}(A^{t+1})$.

**Lemma 1.1.6** *Let $A \in \mathbb{C}^{n \times n}$. Then*

$$\text{ind}(A) = \min\{s \in \mathbb{N} : \mathcal{R}(A^s) = \mathcal{R}(A^{s+1})\}$$

$$= \min\{t \in \mathbb{N} : \mathcal{N}(A^t) = \mathcal{N}(A^{t+1})\}.$$

***Proof*** When $A$ is nonsingular, the result is clear. When $A$ is singular, let $A = P^{-1} \begin{bmatrix} D & 0 \\ 0 & N \end{bmatrix} P$ be the Jordan normal form of $A$, where $P, D$ are invertible and $N$ is nilpotent. Then clearly

$$\text{ind}(A) = \min\{s \in \mathbb{N} : \mathcal{R}(A^s) = \mathcal{R}(A^{s+1})\}$$
$$= \min\{t \in \mathbb{N} : \mathcal{N}(A^t) = \mathcal{N}(A^{t+1})\}$$

is the index of nilpotency of $N$ (i.e., the smallest integer $k > 0$ such that $N^k = 0$).

$\square$

We next consider the matrices that are of index 1; they are also known as core matrices or group matrices.

**Theorem 1.1.7** *Let* $A \in \mathbb{C}^{n \times n}$. *Then the following statements are equivalent:*

(1) $\text{ind}(A) = 1$.
(2) $\mathcal{R}(A) = \mathcal{R}(A^2)$.
(3) $\mathcal{N}(A) = \mathcal{N}(A^2)$.
(4) *There exist invertible matrices* $P, D$ *such that* $A = P^{-1} \begin{bmatrix} D & 0 \\ 0 & 0 \end{bmatrix} P$.
(5) *If* $A = GH$ *is the full rank decomposition of* $A$, *then* $HG$ *is invertible.*

***Proof*** The proof is left to the readers. $\square$

### 1.1.3   Idempotents, Projections and EP Matrices

Recall that $A \in \mathbb{C}^{n \times n}$ is called

- an idempotent matrix if $A^2 = A$;
- a projection matrix if $A^2 = A = A^*$;
- an EP matrix if $\mathcal{R}(A) = \mathcal{R}(A^*)$.

From the Jordan normal form of $A$, it is easy to get the next result.

**Theorem 1.1.8** *Let* $A \in \mathbb{C}^{n \times n}$ *be of rank* $r$. *Then the following statements are equivalent:*

(1) *$A$ is an idempotent matrix.*
(2) *There exists an invertible matrix* $P \in \mathbb{C}^{n \times n}$ *such that* $A = P^{-1} \begin{bmatrix} I_r & 0 \\ 0 & 0 \end{bmatrix} P$.
(3) $\text{rank}(A) + \text{rank}(I_n - A) = n$.

**Theorem 1.1.9** *Let $A \in \mathbb{C}^{n \times n}$ be of rank $r$. Then the following statements are equivalent:*

(1) *$A$ is a projection matrix.*

(2) *There is a unitary matrix $U \in \mathbb{C}^{n \times n}$ such that $A = U \begin{bmatrix} I_r & 0 \\ 0 & 0 \end{bmatrix} U^*$.*

**Proof** $(1) \Rightarrow (2)$. By Theorem 1.1.3, there exists a unitary matrix $U$ such that

$$A = U \begin{bmatrix} D & L \\ 0 & N \end{bmatrix} U^*,$$

where $D$ is invertible, $N$ is nilpotent. Since $A$ is Hermitian, we obtain $A^* = U \begin{bmatrix} D^* & 0 \\ L^* & N^* \end{bmatrix} U^* = A$. Hence $L = 0$. Also, since $A$ is idempotent, we obtain $A^2 = U \begin{bmatrix} D^2 & 0 \\ 0 & N^2 \end{bmatrix} U^* = A$. It follows that $D^2 = D$ and $N^2 = N$. Hence $D = I_r$, $N = 0$, as needed.

$(2) \Rightarrow (1)$. It is clear. □

Take advantage of the core-EP decomposition, we also have the following characterization of EP matrices.

**Theorem 1.1.10** *Let $A \in \mathbb{C}^{n \times n}$ be of rank $r$. Then the following statements are equivalent:*

(1) *$A$ is an EP matrix.*

(2) *There is a unitary matrix $U \in \mathbb{C}^{n \times n}$ such that $A = U \begin{bmatrix} D & 0 \\ 0 & 0 \end{bmatrix} U^*$, where $D \in \mathbb{C}^{r \times r}$ is invertible.*

## 1.2 Definitions and Examples of Rings

### 1.2.1 Basic Concepts and Examples

**Definition 1.2.1** A *ring* $(R, +, \cdot, 0, 1)$ is a nonempty set $R$ equipped with two binary operations, addition $(+)$ and multiplication $(\cdot)$, such that $(R, +, 0)$ is an abelian group, $(R, \cdot, 1)$ is a monoid (i.e., a semigroup with identity 1) and the multiplication is distributive over the addition, that is,

$$a \cdot (b + c) = a \cdot b + a \cdot c \text{ and } (a + b) \cdot c = a \cdot c + b \cdot c$$

for all $a, b, c \in R$. Usually, the underlying set $R$ is used to refer to the ring. To emphasise that the multiplication is associative, and that the existence of identity 1, we shall say that $R$ is an associative ring with identity.

In addition,

- a ring $R$ is called a *commutative ring* if $ab = ba$ for any $a, b \in R$.
- a ring $R$ is called a *domain* if $ab = 0$ implies $a = 0$ or $b = 0$ for each $a, b \in R$.
- a ring $R$ is called a *division ring* if for any $a \neq 0 \in R$, there is $b \in R$ such that $ab = ba = 1$. A commutative division ring is called a field.

**Example 1.2.2** Let $\mathbb{Z}$, $\mathbb{Q}$, $\mathbb{R}$ and $\mathbb{C}$ be the sets of integers, rational numbers, real numbers, and complex numbers with addition and multiplication defined as usual. Then we get the integer ring $\mathbb{Z}$, the rational number field $\mathbb{Q}$, the real number field $\mathbb{R}$ and the complex number field $\mathbb{C}$.

**Example 1.2.3** Let $V_F$ be a vector space over a number field $F$ and let $\text{End}(V_F)$ be the set of all linear transforms in $V_F$. Then $\text{End}(V_F)$ is a ring with addition and multiplication defined as usual.

If $S$ is a nonempty subset of a ring $R$ such that $S$ becomes a ring with the operations and identity inherited from $R$, then the ring $S$ is called a *subring* of $R$.

**Example 1.2.4 (Center)** An element $a \in R$ is said to be *central* if $ab = ba$ for each $b \in R$. The set $Z(R)$ of all central elements of $R$ is called the *center* of $R$. Clearly, $Z(R)$ has at least two elements 0 and 1, and $Z(R)$ is a (commutative) subring of $R$.

If $I$ is a nonempty subset of $R$ such that $(I, +, 0)$ is a subgroup of $(R, +, 0)$ and $ra \in I$ (resp., $ar \in I$) for all $r \in R$, then $I$ is called a left (resp., right) *ideal* of $R$. Moreover, $I$ is called an ideal of $R$ if it is both a left ideal and a right ideal of $R$.

A left (resp., right) ideal $M$ of $R$ is called *maximal* if there is no other proper left (resp., right) ideal of $R$ which contains $M$. According to Zorn's Lemma, maximal left (resp., right) ideals always exist.

**Example 1.2.5 (Jacobson Radical, See also [77, Chapter 2, p. 50])** The *Jacobson radical* of a ring $R$, denoted by $J(R)$, is defined to be the intersection of all the maximal left ideals of $R$. As it turned out, $J(R)$ is also the intersection of all the maximal right ideals of $R$, and hence a two-sided ideal.

**Example 1.2.6 (Annihilators)** Given any nonempty set $X$ of $R$, the set of all left *annihilators* of $X$ in $R$ is

$$\mathbf{l}_R(X) = \{r \in R : rx = 0 \text{ for all } x \in X\}.$$

It is easy to determine that $\mathbf{l}_R(X)$ is a left ideal of $R$. For convenience, $\mathbf{l}_R(X)$ is also written as $\mathbf{l}(X)$, and in case $X = \{x\}$ we usually mark $\mathbf{l}_R(\{x\})$ simply as $\mathbf{l}(x)$ or $^\circ x$.

The definition and notation of right annihilators are similar to that of left annihilators.

Let $I$ be an ideal of a ring $R$. Then there is a natural ring structure consisting of the factor set $R/I = \{r + I : r \in R\}$ with addition and multiplication defined as follows:

$$(a + I) + (b + I) = (a + b) + I \text{ and } (a + I) \cdot (b + I) = (ab) + I$$

for each $a, b \in R$. Such ring $R/I$ is called the *factor ring of $R$ modulo $I$*.

**Example 1.2.7** Let $\mathbb{Z}$ be the integer ring and $n\mathbb{Z} = \{na : a \in \mathbb{Z}\}$ be the ideal of $\mathbb{Z}$ generated by an integer $n$. The factor ring of $\mathbb{Z}$ modulo $n\mathbb{Z}$ is denoted by $\mathbb{Z}_n$.

**Definition 1.2.8** An *algebra over a field $K$* is a $K$-vector space $A$ equipped with a multiplication $(\cdot)$ such that $(A, +, \cdot, 0, 1)$ is a ring and

$$k(ab) = (ka)b = a(kb)$$

for any $k \in K$ and $a, b \in A$.

**Example 1.2.9 (The Hamilton's Algebra (Ring) of Quaternions)** Let

$$\mathbb{H} = \{a + bi + cj + dk : a, b, c, d \in \mathbb{R}\}$$

be a 4-dimensional $\mathbb{R}$-vector space. Let the multiplication in $\mathbb{H}$ be determined by the distributive law and the product of the base elements as follows: $i^2 = j^2 = k^2 = -1$, $ij = k = -ji$, $jk = i = -kj$ and $ki = j = -ik$. Then $\mathbb{H}$ is a 4-dimensional $\mathbb{R}$-algebra.

## 1.2.2 Some Extensions of Rings

For any ring $R$ and positive integers $m, n$,

$$A = \begin{bmatrix} a_{11} & a_{12} & \cdots & a_{1n} \\ a_{21} & a_{22} & \cdots & a_{2n} \\ \vdots & \vdots & & \vdots \\ a_{m1} & a_{m2} & \cdots & a_{mn} \end{bmatrix}$$

is called an $m \times n$ matrix over $R$ if $a_{ij} \in R$ for all $i \in \{1, 2, \cdots, m\}$ and $j \in \{1, 2, \cdots, n\}$. The element $a_{ij}$ is referred to as the $(i, j)$-entry of $A$, and $A$ is denoted by $A = (a_{ij})$ for simplicity. The *transpose* of $A$ is the $n \times m$ matrix $A^{\mathrm{T}}$ whose $(j, i)$-entry equals $a_{ij}$.

Let $R^{m \times n}$ be the set of all $m \times n$ matrices over $R$. Then

**Example 1.2.10** $R^{n \times n}$ is a ring (named as the $n \times n$ *full matrices ring over* $R$) with addition and multiplication defined as follows: for any $A = (a_{ij})$, $B = (b_{ij}) \in R^{n \times n}$,

$$(a_{ij}) + (b_{ij}) := (c_{ij}), \quad (a_{ij})(b_{ij}) := (d_{ij}),$$

where $c_{ij} = a_{ij} + b_{ij}$ and $d_{ij} = \sum_{k=1}^{n} a_{ik} b_{kj}$.

**Example 1.2.11** The subring of $R^{n \times n}$ consisting of all upper triangular matrices (i.e., matrices whose $(i, j)$-entry is zero whenever $i > j$) with addition and multiplication inherited from $R^{n \times n}$ is called the $n \times n$ *upper triangular matrices ring over* $R$ and denoted by $T R^{n \times n}$.

Let $\Gamma = \{1, \cdots, n\}$. Then every matrix $A = (a_{ij}) \in R^{n \times n}$ corresponds to a map

$$f_A : \Gamma \times \Gamma \to R; (i, j) \longmapsto a_{ij}.$$

More generally, if $\Gamma$ is an infinite set, then the map

$$f : \Gamma \times \Gamma \to R; f(\alpha, \beta) = a_{\alpha\beta}$$

prompts us to consider the infinite matrix $(a_{\alpha\beta})$ over $R$. We shall call $(a_{\alpha\beta})$ a $|\Gamma| \times |\Gamma|$ matrix over $R$ and denote the set of all $|\Gamma| \times |\Gamma|$ matrices over $R$ by $R^{|\Gamma| \times |\Gamma|}$. Clearly, $R^{|\Gamma| \times |\Gamma|}$ is an abelian group with addition defined as the usual matrices sum. But $R^{|\Gamma| \times |\Gamma|}$ fails to be a ring with the multiplication defined as usual matrices product. In fact, let

$$A = \begin{bmatrix} 1 & 1 & 1 & \cdots \\ 0 & 1 & 1 & \cdots \\ 0 & 0 & 1 & \cdots \\ \vdots & \vdots & \vdots & \ddots \end{bmatrix}, \quad B = \begin{bmatrix} 1 & -1 & 0 & 0 & \cdots \\ 0 & 1 & -1 & 0 & \cdots \\ 0 & 0 & 1 & -1 & \cdots \\ \vdots & \vdots & \vdots & \vdots & \ddots \end{bmatrix}, \quad C = \begin{bmatrix} 0 & 0 & 0 & \cdots \\ -1 & 0 & 0 & \cdots \\ -1 & -1 & 0 & \cdots \\ \vdots & \vdots & \ddots & \ddots \end{bmatrix} \in R^{|\Gamma| \times |\Gamma|},$$

and adopt the usual product of matrices. Then $(AB)C = C \neq A = A(BC)$ (see [9, Example 3]), which means that the associative law of "multiplication" fails.

However,

**Example 1.2.12**

(1) The subset

$$RFM(R) = \{(a_{\alpha\beta}) : a_{\alpha\beta} \in R, \text{ for every } \alpha \in \Gamma,$$

$$\text{the set } \{\beta \in \Gamma : a_{\alpha\beta} \neq 0\} \text{ is a finite set}\}$$

of $R^{|\Gamma| \times |\Gamma|}$ can be a ring (called the *row finite matrix ring over R*) with the addition and multiplication defined as usual matrices sum and product.

(2) The subset

$$CFM(R) = \{(a_{\alpha\beta}) : a_{\alpha\beta} \in R, \text{ for every } \beta \in \Gamma,$$

$$\text{the set } \{\alpha \in \Gamma : a_{\alpha\beta} \neq 0\} \text{ is a finite set}\}$$

of $R^{|\Gamma| \times |\Gamma|}$ can be a ring (called the *column finite matrix ring over R*) with the addition and multiplication defined as usual matrices sum and product.

(3) The subring $BFM(R) = RFM(R) \cap CFM(R)$ of $RFM(R)$ and $CFM(R)$ is called the *double finite matrix ring over R*.

For any ring $R$,

$$\sum_{i=0}^{\infty} a_i x^i = a_0 + a_1 x + \cdots + a_n x^n + \cdots \quad (\text{where } a_i \in R)$$

is called a *formal power series over R* in an indeterminant $x$.

**Example 1.2.13 (Formal Power Series Rings)** Let $R[[x]]$ be the set of all formal power series over $R$ in an indeterminant $x$. Then $R[[x]]$ is a ring with addition defined by

$$\sum_{i=0}^{\infty} a_i x^i + \sum_{i=0}^{\infty} b_i x^i = \sum_{i=0}^{\infty} (a_i + b_i) x^i,$$

and multiplication defined by

$$\sum_{i=0}^{\infty} a_i x^i \cdot \sum_{i=0}^{\infty} b_i x^i = \sum_{i=0}^{\infty} (\sum_{j+k=i} a_j b_k) x^i,$$

where $\sum_{i=0}^{\infty} a_i x^i, \sum_{i=0}^{\infty} b_i x^i \in R[[x]]$.

**Example 1.2.14 (Polynomial Rings)** The subring of $R[[x]]$ consisting of formal power series which have only a finite number of non-zero coefficients with addition and multiplication inherited from $R[[x]]$ is called the *polynomial ring over R* and denoted by $R[x]$.

**Example 1.2.15 (Group Rings)** Let $R$ be a ring and $G$ be a group. Let $RG$ be the set of all formal linear combinations of the form

$$\alpha = \sum_{g \in G} a_g g$$

where $a_g \in R$ and $a_g = 0$ almost everywhere, that is, only a finite number of coefficients are different from 0 in each of these sums. For any $\alpha = \sum_{g \in G} a_g g$, $\beta = \sum_{g \in G} b_g g \in RG$, define

$$\alpha + \beta = \sum_{g \in G} (a_g + b_g)g,$$

and

$$\alpha \cdot \beta = \sum_{u \in G} \left( \sum_{gh=u} a_g b_h \right) u.$$

It is easy to verify that, with the two operations above, $RG$ is a ring. We shall call $RG$ the *group ring of G over R*.

### 1.2.3  Idempotents, Units and Regular Elements

An element $a$ in a ring $R$ is said to be left (resp., right) invertible if there is $b \in R$ such that $ba = 1$ (resp., $ab = 1$) in which case such $b$ is called a left (resp., right) inverse of $a$. If $a$ is both left and right invertible, then it is said to be invertible, or simply called a unit. The set of all units of $R$ is denoted by $R^{-1}$ in this text.

An element $e \in R$ is an *idempotent* if $e^2 = e$. Note that if $e$ is an idempotent, then $1 - e$, $e + er(1 - e)$ and $e + (1 - e)re$ (for each $r \in R$) are also idempotents, and $1 - 2e$ is a unit since $(1 - 2e)^2 = 1$.

The next result reveals that idempotents can be uniquely determined by their one-sided ideals.

**Lemma 1.2.16 ([108, Lemma 2.9])** *If $e_1$ and $e_2$ are idempotents such that $Re_1 \subseteq Re_2$ and $e_2 R \subseteq e_1 R$, then $e_1 = e_2$.*

**Proof** If $Re_1 \subseteq Re_2$, then $e_1 = ue_2$ for some $u \in R$, so $e_1 e_2 = ue_2^2 = ue_2 = e_1$. Similarly, $e_2 R \subseteq e_1 R$ implies $e_1 e_2 = e_2$. Thus, $e_1 = e_2$. □

**Lemma 1.2.17** *Let $e \in R$ be an idempotent. Then:*

$$^\circ e = R(1 - e), \quad e^\circ = (1 - e)R. \tag{1.2.1}$$

**Corollary 1.2.18** *Let $e, f \in R$ be idempotents. Then:*

(1) $eR = fR \Leftrightarrow {}^\circ e = {}^\circ f \Leftrightarrow fe = e, ef = f$.
(2) $Re = Rf \Leftrightarrow e^\circ = f^\circ \Leftrightarrow ef = e, fe = f$.

An element $a \in R$ is *nilpotent* if there exists a positive integer $n$ such that $a^n = 0$; the smallest such $n$ is called the *index of nilpotency* of $a$. If $a$ is nilpotent (say with index $n$), then $1 - a$ is a unit since

$$(1 - a)(1 + a + \cdots + a^{n-1}) = (1 + a + \cdots + a^{n-1})(1 - a) = 1.$$

Note that $(R^{-1}, \cdot, 1)$ is a group, and that for any $a, b \in R^{-1}$, we have the *reverse order law*

$$(ab)^{-1} = b^{-1}a^{-1} \tag{1.2.2}$$

and the *absorption law*

$$a^{-1}(a + b)b^{-1} = a^{-1} + b^{-1}. \tag{1.2.3}$$

The next result is known as Jacobson's lemma.

**Theorem 1.2.19 (Jacobson's Lemma)** *Let $a, b \in R$.*

(1) *If $1 - ab$ is left invertible, then so is $1 - ba$. In this case, if $x \in R$ is a left inverse of $1 - ab$, then $1 + bxa$ is a left inverse of $1 - ba$.*
(2) *If $1 - ab$ is right invertible, then so is $1 - ba$. In this case, if $x \in R$ is a right inverse of $1 - ab$, then $1 + bxa$ is a right inverse of $1 - ba$.*
(3) *If $1 - ab$ is invertible, then so is $1 - ba$ and*

$$(1 - ba)^{-1} = 1 + b(1 - ab)^{-1}a. \tag{1.2.4}$$

***Proof*** It can be verified directly.                                    □

An element $a \in R$ is said to be *(von Neumann) regular* if $a = aba$ for some $b \in R$, in which case $ab, ba$ are idempotents. And $a$ is said to be *unit-regular* if $a = ava$ for some $v \in R^{-1}$. It turns out that $a$ is unit-regular if and only if $a = ue$ (or, $a = fu$) for some idempotents $e, f$ and a unit $u$. Moreover, if there exist an idempotent $e$ and a unit $u$ such that $a = ue = eu$, then $a$ is said to be *strongly regular*.

**Lemma 1.2.20 ([71, Lemma 2.1])** *Let $a, b \in R$. Then:*

(1) *$a$ is left invertible if and only if $a$ is regular with $a° = \{0\}$.*
(2) *$a$ is right invertible if and only if $a$ is regular with $°a = \{0\}$.*
(3) *If $a - aba$ is regular, then so is $a$.*

**Lemma 1.2.21 ([108, Lemmas 2.5, 2.6])** *Let $a, b \in R$.*

(1) *If $aR \subseteq bR$, then $°b \subseteq °a$.*
(2) *If $Ra \subseteq Rb$, then $b° \subseteq a°$.*
(3) *If $b$ is regular and $°b \subseteq °a$, then $aR \subseteq bR$.*
(4) *If $b$ is regular and $b° \subseteq a°$, then $Ra \subseteq Rb$.*

**Proof** (1) Since $aR \subseteq bR$, there exists $x \in R$ such that $a = bx$. Now for any $u \in R$ satisfying $ub = 0$, one can get $ua = ubx = 0$.

(3) Suppose that $b$ is regular with $bb^-b = b$, and $^\circ b \subseteq {}^\circ a$. Since $(1 - bb^-)b = 0$, it follows that $(1 - bb^-)a = 0$. Thus we have $a = bb^-a$, which implies that $aR \subseteq bR$.

Similarly, we have (2) and (4).                                                                         □

### 1.2.4  One-Sided Invertibility and Invertibility

We begin with the following observation.

**Proposition 1.2.22** *Let $V$ be a finite dimensional vector space over a number field $F$. Then for any $f, g \in \text{End}(V_F)$, if $fg = 1_V$ then $gf = 1_V$.*

**Proof** Let $f, g \in \text{End}(V_F)$ such that $fg = 1_V$. We have $\dim \mathcal{N}(g) + \dim \mathcal{R}(g) = n$. Since $fg = 1_V$, we obtain $\mathcal{N}(g) = 0$. It follows that $\dim \mathcal{R}(g) = n$, and hence $\mathcal{R}(g) = V$. So, $g$ is bijective. Therefore $gf = gf(gg^{-1}) = g(fg)g^{-1} = gg^{-1} = 1_V$.                                                                         □

Nevertheless, such a fact is not valid for infinite dimensional vector spaces.

**Example 1.2.23**

(1) ([65]) Let

$$
A = \begin{bmatrix} 0 & 1 & & \\ & 0 & 1 & \\ & & 0 & 1 \\ & & & \ddots & \ddots \end{bmatrix}, \quad B = \begin{bmatrix} 0 & & \\ 1 & 0 & \\ & 1 & 0 \\ & & \ddots & \ddots \end{bmatrix} \in CFM(\mathbb{C}).
$$

Then $AB = I$, $BA \neq I$.

(2) ([128, Example 2.2.4 (3)]) Let $V$ be the $\mathbb{R}$-vector space of all differentiable functions over the interval $[0, 1]$, and let $R = \text{End}(V)$. Let

$$a : V \to V, \ f(x) \longmapsto f'(x),$$

$$b : V \to V, \ f(x) \longmapsto \int_0^x f(t)dt.$$

Then $a, b \in R$, $ab = 1$, but $ba \neq 1$.

(3) Let $V = \mathbb{R}^\infty$ and $R = \text{End}(V)$. Let

$$a : V \to V, \ (a_1, a_2, a_3, \cdots) \longmapsto (a_2, a_3, \cdots),$$

$$b : V \to V, \ (a_1, a_2, a_3, \cdots) \longmapsto (0, a_1, a_2, \cdots).$$

Then $a, b \in R$, $ab = 1$, but $ba \neq 1$.

**Definition 1.2.24** ([76, Chapter 1, p. 5]) A ring $R$ is said to be *Dedekind-finite* if $ab = 1 \Rightarrow ba = 1$ for all $a, b \in R$. Moreover, $R$ is said to be *stably finite* if the matrix ring $R^{n \times n}$ is Dedekind-finite for every positive integer $n$.

**Example 1.2.25**

(1) ([76, Proposition 1.12]) Any commutative ring $R$ is stably finite.
(2) ([76, Proposition 1.13]) Any right noetherian ring $R$ (i.e., $R$ satisfies the *ascending chain condition* on right ideals) is stably finite. In particular, every semisimple ring $R$ (i.e., $R \cong D_1^{n_1 \times n_1} \times \cdots \times D_s^{n_s \times n_s}$ for some division rings $D_1, \cdots, D_s$; see also Wedderburn-Artin theorem [77]) is stably finite.
(3) ([76, p. 18, (11)]) Let $R$ be a ring for which there exists a positive integer $n$ such that $c^n = 0$ for any nilpotent element $c \in R$. Then $R$ is Dedekind-finite.
(4) ([76, p. 19, (18)]) There exists a ring $R$ such that $R$ is Dedekind-finite while $R^{2 \times 2}$ is not: Let R be the $K$-algebra generated over a field $K$ by $\{s, t, u, v; w, x, y, z\}$ with relations dictated by the matrix equation $AB = I_2$, where $A = \begin{bmatrix} s & u \\ t & v \end{bmatrix}$, $B = \begin{bmatrix} x & y \\ z & w \end{bmatrix}$. Then $R$ is a domain (hence Dedekind-finite), but $R^{2 \times 2}$ is not since $BA \neq I_2$.
(5) Recall that a ring $R$ is right (resp., left) *self-injective* if, for every right (resp., left) ideal $I$ of $R$, every right (resp., left) $R$-homomorphism $I \to R$ can be extended to $R \to R$. $R$ is called self-injective if it is both left and right self-injective. By Lam [76, Corollary 6.49], any self-injective ring is Dedekind-finite.
(6) ([77, Proposition 20.8]) A semilocal ring $R$ (i.e., $R/J(R)$ is semisimple) is Dedekind-finite.
(7) Every unit-regular ring $R$ is Dedekind-finite.

**Proof** Let $a, b \in R$ with $ab = 1$. By unit-regularity, there exists a unit $u$ such that $aua = a$. So, $au = au(ab) = (aua)b = ab = 1$. It follows that $a = u^{-1}$ is a unit, and hence $b = a^{-1}$. □

Note that if $R$ is a Dedekind-finite ring then, for any $A = \begin{bmatrix} a_1 & a_3 \\ 0 & a_2 \end{bmatrix} \in R^{2 \times 2}$, $A$ is invertible if and only if $a_1, a_2$ are invertible. However, this fact fails for non-Dedekind-finite rings.

**Example 1.2.26** Let $R$ be a non-Dedekind-finite ring, let $a, b \in R$ with $ab = 1, ba \neq 1$, and let $A = \begin{bmatrix} b & 1 \\ 0 & a \end{bmatrix} \in R^{2 \times 2}$. It is straightforward to check that $A$ is invertible with $A^{-1} = \begin{bmatrix} a & -1 \\ 1 - ba & b \end{bmatrix}$. But $a, b$ are not invertible.

## 1.3   Semigroups, Rings and Categories with Involution

### 1.3.1   Definitions and Examples

**Definition 1.3.1**  A unary operation $*$ in a semigroup $S$ (sends $s$ into $s^*$ for each element $s$) is called an *involution* if $(a^*)^* = a$ and $(ab)^* = b^*a^*$ for all $a, b \in S$. If $S$ is endowed with such an involution $*$, then $S$ is called a $*$-semigroup.

**Definition 1.3.2**  A unary operation $*$ in a ring $R$ is called an *involution* if $*$ is an involution in both $(R, +, 0)$ and $(R, \cdot, 1)$, or, equivalently, if $(a^*)^* = a$, $(a+b)^* = a^* + b^*$, $(ab)^* = b^*a^*$ for any $a, b \in R$. In this case, $R$ is called a $*$-ring.

An element $a$ in a $*$-semigroup (resp., $*$-ring) is said to be *Hermitian* (or *self-adjoint*, or *symmetric*) with respect to $*$ if $a^* = a$. An Hermitian idempotent is called a *projection*.

**Lemma 1.3.3 ([108, Lemma 2.10])**  *If $p_1$ and $p_2$ are projections such that $Rp_1 = Rp_2$ or $p_1 R = p_2 R$, then $p_1 = p_2$.*

**Proof**  If $Rp_1 = Rp_2$, then we have $p_1 = p_1 p_2$ and $p_2 = p_2 p_1$. But $p_2 = p_2^* = p_1^* p_2^* = p_1 p_2 = p_1$. Similarly, $p_1 R = p_2 R$ implies $p_1 = p_2$.                       □

Given a $*$-ring $R$, there is a natural involution $\bar{*}$ induced by $*$ in the matrix ring $R^{n \times n}$: for any $A = (a_{ij}) \in R^{n \times n}$, $A^{\bar{*}}$ is defined as $(a_{ij}^*)^{\mathrm{T}}$. If there is no risk of causing confusion, $\bar{*}$ is also simply written as $*$.

**Proposition 1.3.4**

(1) *The identity map of a semigroup $S$ is an involution if and only if $S$ is commutative.*
(2) *The identity map of a ring $R$ is an involution if and only if $R$ is commutative.*

**Example 1.3.5**  Let $RG$ be the group ring of $G$ over a commutative ring $R$ (see Example 1.2.15). Then there is a classical involution in $RG$ defined as follows: for any $\alpha = \sum_{g \in G} a_g g \in RG$, set

$$\alpha^* = \sum_{g \in G} a_g g^{-1}.$$

**Example 1.3.6 ([135, Example 2.15])**  Let $M$ be the transformation semigroup over the set $A = \{1, 2, 3\}$, and let $S$ be the subsemigroup of $M$ generated by

$$x = \begin{pmatrix} 1 & 2 & 3 \\ 2 & 3 & 3 \end{pmatrix}, \quad y = \begin{pmatrix} 1 & 2 & 3 \\ 1 & 1 & 3 \end{pmatrix}.$$

Then $S = \{x, x^2, y, xy, yx\}$. Set

$$x^* = x, \ (x^2)^* = x^2, \ y^* = y, \ (xy)^* = yx, \ (yx)^* = xy.$$

Then $*$ is an involution in $S$.

**Example 1.3.7** Let $M$ be the transformation semigroup over the set $A = \{1, 2, 3\}$, and let $T$ be the subsemigroup of $M$ generated by

$$x = \begin{pmatrix} 1\ 2\ 3 \\ 2\ 3\ 3 \end{pmatrix}, \quad y = \begin{pmatrix} 1\ 2\ 3 \\ 3\ 1\ 3 \end{pmatrix}.$$

Then $x^2 = y^2$, $x^2 y = x^3 = yx^2 = x^2$, $xyx = x$, $yxy = y$, and thus $T = \{x, x^2, y, xy, yx\}$. Set

$$x^* = y, \quad (x^2)^* = x^2, \quad y^* = x, \quad (xy)^* = xy, \quad (yx)^* = yx.$$

Then it is straightforward to verify that $*$ is an involution in $T$, and we have the relations

$$xyx = x, \quad yxy = y, \quad (xy)^* = xy, \quad (yx)^* = yx.$$

**Example 1.3.8** Let $R = \mathbb{C}^{n \times n}$. Then:

(1) $R$ is a $*$-ring if we take $*$ as the transpose map of complex matrices.
(2) $R$ is a $*$-ring if we take $*$ as the conjugate transpose map of complex matrices.

**Example 1.3.9** Let $\mathbb{P}$ be a number field and let $R = \mathbb{P}^{2 \times 2}$. For any $A = \begin{bmatrix} a & b \\ c & d \end{bmatrix} \in \mathbb{P}^{2 \times 2}$, set $A^* = \begin{bmatrix} d & -b \\ -c & a \end{bmatrix}$ (i.e., $A^*$ is the adjoint matrix of $A$). Then $*$ is an involution in $R$.

**Example 1.3.10** Let $R = \left\{ \begin{bmatrix} a & b \\ 0 & a \end{bmatrix} : a, b \in \mathbb{P} \right\}$ be the subring of $\mathbb{P}^{2 \times 2}$. For any $\begin{bmatrix} a & b \\ 0 & a \end{bmatrix} \in R$, set $\begin{bmatrix} a & b \\ 0 & a \end{bmatrix}^* = \begin{bmatrix} a & -b \\ 0 & a \end{bmatrix}$. Then $*$ is an involution in $R$.

**Example 1.3.11** ([130, Example 3(5)]) Let $\mathbb{Z}_2 \langle x, y \rangle$ be the polynomial algebra over $\mathbb{Z}_2$ in the non-commuting indeterminants $x$, $y$, and let

$$R = \mathbb{Z}_2 \langle x, y : x^2 = x, y^2 = y, xyx = 0 \rangle$$

be the factor algebra of $\mathbb{Z}_2 \langle x, y \rangle$ modulo the relations $x^2 = x$, $y^2 = y$, $xyx = 0$. Define $1^* = 1$, $x^* = x$, $y^* = y$, $(xy)^* = yx$, $(yx)^* = xy$ and $(yxy)^* = yxy$. Then $*$ is an involution in $R$.

**Example 1.3.12** ([121]) Let $K \langle x, y \rangle$ be the polynomial algebra over a field $K$ in the non-commuting indeterminants $x$, $y$, and let

$$R = K \langle x, y : x^2 y = x, xy^2 = y \rangle$$

be the factor algebra of $K\langle x, y\rangle$ modulo the relations $x^2 y = x$, $xy^2 = y$. Then the set $\mathcal{B} = \{y^l (xy)^m x^n : l, m, n \in \mathbb{N}\}$ forms a basis of $R$. For any element $r \in R$, there exist $p \in \mathbb{N}$, $l_i, m_i, n_i \in \mathbb{N}$ and $k_i \in K$ (for any integer $0 \le i \le p$) such that

$$r = \sum_{i=0}^{p} k_i y^{l_i} (xy)^{m_i} x^{n_i}.$$

Define

$$r^* = \sum_{i=0}^{p} k_i y^{n_i} (xy)^{m_i} x^{l_i},$$

then we have

$$(r^*)^* = r,$$

$$(r + s)^* = r^* + s^* \text{ for any } s \in R,$$

$$x^* = [y^0 (xy)^0 x^1]^* = y^1 (xy)^0 x^0 = y,$$

$$\text{and } y^* = (x^*)^* = x.$$

For any $\alpha_1 = y^{l_1} (xy)^{m_1} x^{n_1}$, $\alpha_2 = y^{l_2} (xy)^{m_2} x^{n_2} \in \mathcal{B}$, if we show $(\alpha_1 \alpha_2)^* = \alpha_2^* \alpha_1^*$, then it is enough to prove $(rs)^* = s^* r^*$. In fact, if $n_1 > l_2$, then

$$(\alpha_1 \alpha_2)^* = [y^{l_1} (xy)^{m_1} x^{n_1} y^{l_2} (xy)^{m_2} x^{n_2}]^*$$

$$= [y^{l_1} (xy)^{m_1} x^{n_1 - l_2} (xy)^{m_2} x^{n_2}]^*$$

$$= [y^{l_1} (xy)^{m_1} x^{n_1 - l_2 + n_2}]^*$$

$$= y^{n_1 - l_2 + n_2} (xy)^{m_1} x^{l_1},$$

and

$$\alpha_2^* \alpha_1^* = [y^{n_2} (xy)^{m_2} x^{l_2}][y^{n_1} (xy)^{m_1} x^{l_1}]$$

$$= y^{n_2} (xy)^{m_2} y^{n_1 - l_2} (xy)^{m_1} x^{l_1}$$

$$= y^{n_2 + n_1 - l_2} (xy)^{m_1} x^{l_1}$$

$$= (\alpha_1 \alpha_2)^*.$$

Similarly, we can prove $(\alpha_1 \alpha_2)^* = \alpha_2^* \alpha_1^*$ when $n_1 \le l_2$. Therefore, we conclude that $*$ is an involution in $R$.

**Example 1.3.13** ([121]) Let $K\langle x, y\rangle$ be the polynomial algebra over a field $K$ in the non-commuting indeterminants $x$, $y$, and let

$$R = K\langle x, y : x^2 y = x, xy^2 = y, x = xyx, yxy = y\rangle$$

be the factor algebra of $K\langle x, y\rangle$ modulo the relations $x^2 y = x$, $xy^2 = y$, $x = xyx$, $yxy = y$. Then the set $\mathcal{B} = \{xy, y^l x^n : l, n \in \mathbb{N}\}$ forms a basis of $R$. For any element $r \in R$, there exist $p \in \mathbb{N}$, $k_i, k_{p+1} \in K$, $l_i, n_i \in \mathbb{N}$ (for any integer $0 \le i \le p$) such that

$$r = \sum_{i=0}^{p} k_i y^{l_i} x^{n_i} + k_{p+1} xy.$$

Define

$$r^* = \sum_{i=0}^{p} k_i y^{n_i} x^{l_i} + k_{p+1} xy,$$

then it can be seen from the proof of Example 1.3.12 that $*$ is an involution.

**Definition 1.3.14** A *category* $\mathcal{C}$ consists of three ingredients:

(1) A class obj($\mathcal{C}$) of *objects*.
(2) A set $\mathrm{Hom}_{\mathcal{C}}(A, B)$ of *morphisms* (form $A$ to $B$) for each ordered pair of objects $(A, B)$.
(3) A map $(f, g) \mapsto gf$ of the product set $\mathrm{Hom}_{\mathcal{C}}(A, B) \times \mathrm{Hom}_{\mathcal{C}}(B, C)$ into $\mathrm{Hom}_{\mathcal{C}}(A, C)$ for each ordered triple of objects $(A, B, C)$ and morphisms $f \in \mathrm{Hom}_{\mathcal{C}}(A, B)$, $g \in \mathrm{Hom}_{\mathcal{C}}(B, C)$.

It is assumed that the objects and morphisms satisfy the following conditions:

(i) If $(A, B) \ne (C, D)$, then $\mathrm{Hom}_{\mathcal{C}}(A, B)$ and $\mathrm{Hom}_{\mathcal{C}}(C, D)$ are disjoint.
(ii) (Associativity) If $f \in \mathrm{Hom}_{\mathcal{C}}(A, B)$, $g \in \mathrm{Hom}_{\mathcal{C}}(B, C)$, and $h \in \mathrm{Hom}_{\mathcal{C}}(C, D)$, then $(hg)f = h(gf)$.
(iii) For every object $A$, there is an *identity morphism* $1_A \in \mathrm{Hom}_{\mathcal{C}}(A, A)$ such that $f1_A = f$ and $1_A g = g$ for all $f \in \mathrm{Hom}_{\mathcal{C}}(A, B)$ and $g \in \mathrm{Hom}_{\mathcal{C}}(B, A)$.

**Definition 1.3.15** A category $\mathcal{C}$ is said to have an *involution* $*$ provided that $*$ is a unary operation on the morphisms such that $f \in \mathrm{Hom}_{\mathcal{C}}(A, B)$ implies $f^* \in \mathrm{Hom}_{\mathcal{C}}(B, A)$ and that $(f^*)^* = f$, $(fg)^* = g^* f^*$ for all objects $A$, $B$ and morphisms $f$, $g$ of $\mathcal{C}$.

**Example 1.3.16** For any ring $R$ and positive integers $m, n$, there exists a category $\mathcal{C}$ consisting of:

(1) two objects $R^n$ and $R^m$;

(2) the sets $\text{Hom}_C(R^m, R^m) = R^{m \times m}$, $\text{Hom}_C(R^n, R^n) = R^{n \times n}$, $\text{Hom}_C(R^n, R^m)$
$= R^{m \times n}$, $\text{Hom}_C(R^m, R^n) = R^{n \times m}$ of morphisms;
(3) and composition of matrices.

Moreover, if $R$ has an involution $*$, then $C$ has the induced involution $\bar{*}$ of matrices as its involution.

### 1.3.2  Proper Involutions

**Definition 1.3.17**  An involution $*$ in a ring $R$ is said to be $k$-*proper* for some positive integer $k$ if $\sum_{i=1}^{k} a_i^* a_i = 0$ implies $a_i = 0$ for any $i \in \{1, \cdots, k\}$ and $a_i \in R$. Customarily, a 1-proper involution $*$ is directly said to be *proper*, in which case the ring $R$ is also called a $*$-*proper ring*.

**Proposition 1.3.18**  *An involution $*$ in a ring $R$ is $k$-proper if and only if the induced involution $*$ in $R^{k \times k}$ is proper.*

**Proof**  It is clear.                                                                                    □

**Example 1.3.19**

(1) Let $R = \mathbb{C}^{n \times n}$ and $*$ be the conjugate transpose map of complex matrices. Then $*$ is $k$-proper for any positive integer $k$.
(2) Let $R = \mathbb{C}^{n \times n}$ and $*$ be the transpose map of complex matrices. Then $*$ is not proper as

$$\begin{bmatrix} 1 & 0 \\ i & 0 \end{bmatrix}^* \begin{bmatrix} 1 & 0 \\ i & 0 \end{bmatrix} = 0 \text{ while } \begin{bmatrix} 1 & 0 \\ i & 0 \end{bmatrix} \neq 0.$$

(3) Let $R = \mathbb{R}^{n \times n}$ and $*$ be the transpose map of real matrices. Then $*$ is $k$-proper for any positive integer $k$.
(4) Let $R = \mathbb{Z}^{n \times n}$ and $*$ be the transpose map of integral matrices. Then $*$ is proper.
(5) Let $R = \left\{ \begin{bmatrix} a & b \\ b & a \end{bmatrix} : a, b \in \mathbb{R} \right\}$ be the commutative subring of $\mathbb{R}^{2 \times 2}$ and $*$ be identity map of $R$. Then $*$ is $k$-proper for any positive integer $k$.

**Example 1.3.20**  Let $R$ be a commutative ring and $*$ be the identity map of $R$. Then $*$ is proper if and only if $R$ has no non-zero nilpotent elements. In particular, if $R = \mathbb{Z}_{p^n}$ for some prime number $p$, then $*$ (the identity map of $R$) is proper if and only if $n = 1$.

### 1.3.3  The Gelfand-Naimark Property

**Definition 1.3.21 ([75, Definition 4])** Let $R$ be a ∗-ring. Then $R$ is said to have the *Gelfand-Naimark property* (GN-property for short) if $1 + aa^*$ is invertible for each $a \in R$.

**Example 1.3.22**

(1) Let $R = \mathbb{C}^{n \times n}$ and ∗ be the conjugate transpose map of complex matrices. Then $R$ has the GN-property.
(2) Every $C^*$-algebra possesses the GN-property.
(3) ([130, Example 1]) Let $n = p_1^{s_1} p_2^{s_2} \cdots p_m^{s_m}$, where $p_1, p_2, \cdots, p_m$ are pairwise different prime numbers and $s_1, s_2, \cdots, s_m$ are positive integers. Let $R = \mathbb{Z}_n$ and $* = 1_R$. Then $R$ has the GN-property if and only if each $p_i \equiv -1 \pmod 4$.
(4) ([130, Example 2]) Let $n$ be a positive integer and consider the cyclic group $C_n$. Then the group ring $\mathbb{R}C_n$ has the GN-property with respect to the classical involution.

More generally,

**Definition 1.3.23 ([27, Definition 3.6])** A ∗-ring $R$ is said to have the $k$-GN property for a positive integer $k$, if $1 + x_1 x_1^* + \cdots + x_k x_k^*$ is invertible for all $x_1, \cdots, x_k \in R$.

Clearly, if $R$ has the $k$-GN property then it has the $j$-GN property for any integer $0 < j < k$.

## 1.4  Regularity and ∗-Regularity of Rings

### 1.4.1  Regularity and FP-Injectivity

**Definition 1.4.1** A ring $R$ is said to be (*von Neumann*) *regular* if every element in $R$ is regular. Analogously, $R$ is said to be *unit-regular* (resp., *strongly regular*) if every element in $R$ is unit-regular (resp., strongly regular).

**Example 1.4.2** Let $D$ be a division ring. Then $D^{n \times n}$ is unit-regular.

**Proof** For any $A \in D^{n \times n}$, there are invertible $P, Q \in D^{n \times n}$ such that

$$A = P \begin{bmatrix} I_r & 0 \\ 0 & 0 \end{bmatrix} Q.$$

Let $U = Q^{-1}P^{-1}$. Then

$$AUA = P\begin{bmatrix} I_r & 0 \\ 0 & 0 \end{bmatrix}QQ^{-1}P^{-1}P\begin{bmatrix} I_r & 0 \\ 0 & 0 \end{bmatrix}Q$$

$$= P\begin{bmatrix} I_r & 0 \\ 0 & 0 \end{bmatrix}Q = A.$$

If follows that $A$ is unit-regular, and thus $D^{n \times n}$ is a unit-regular ring.    □

It can be seen from the above proof that every semisimple ring is unit-regular.

**Example 1.4.3** Let $D$ be a division ring. Then $CFM(D)$ is regular.

**Proof** Let $V_D$ be an infinite dimensional vector space over $D$. Since $CFM(D) \cong$ End$(V_D)$, it suffices to show that End$(V_D)$ is regular. Let $f \in$ End$(V_D)$. Then

$$V_D = \text{Im} f \oplus V_1 = \text{Ker} f \oplus V_2$$

for some subspaces $V_1$, $V_2$ of $V_D$. Let $\pi$ be the projection of $V_D$ on Im$f$ along $V_1$, $\pi_1$ be the projection of $V_D$ on $V_2$ along Ker$f$, $i$ be the injection of $V_2$ to $V_D$, and set $\alpha : \text{Im} f \to V_2$, $f(a) \longmapsto \pi_1(a)$. Then $\alpha$ is well-defined. Write $g = i\alpha\pi$. Then $g \in$ End$(V_D)$ and $fgf = f$. It follows that End$(V_D)$ is regular, and thus $CFM(D)$ is regular.    □

**Proposition 1.4.4** *Let n be a positive integer. Then a ring $R$ is regular (resp., unit-regular) if and only if $R^{n \times n}$ is regular (resp., unit-regular).*

**Proof** It follows by Goodearl [47, Corollary 4.7].    □

**Definition 1.4.5 ([18, Theorems 1.4.6, 3.2.2])** A ring $R$ is said to be *right P-injective* if $\mathbf{l}(\mathbf{r}(a)) = Ra$ for each $a \in R$. In particular, $R$ is said to be *right FP-injective* if the matrix ring $R^{n \times n}$ is right P-injective for every positive integer $n$. The left version of P-injectivity (resp., FP-injectivity) can be defined similarly. In addition, $R$ is said to be P-injective (resp., FP-injective) if it is both left and right P-injective (resp., FP-injective).

Note that if an element $a \in R$ is regular (with $aba = a$), then $\mathbf{l}(\mathbf{r}(a)) = \mathbf{l}((1 - ba)R) = Ra$. So, from Proposition 1.4.4 we can derive the following implication:

$$\text{regular rings} \Rightarrow \text{FP-injective rings} \Rightarrow \text{P-injective rings.} \tag{1.4.1}$$

Given any matrix $A \in R^{m \times n}$, take the following notations:

$$\mathcal{R}(A) = \{Ax : x \in R^{n \times 1}\}, \qquad \mathcal{R}_l(A) = \{xA : x \in R^{1 \times m}\},$$

$$\mathcal{N}(A) = \{x \in R^{n \times 1} : Ax = 0\}, \qquad \mathcal{N}_l(A) = \{x \in R^{1 \times m} : xA = 0\}.$$

**Theorem 1.4.6 ([68, Theorem 2.8])** *The following statements are equivalent for any ring $R$:*

(1) *$R$ is right FP-injective.*
(2) *$\mathcal{N}(A) \subseteq \mathcal{N}(B)$ if and only if $\mathcal{R}_l(B) \subseteq \mathcal{R}_l(A)$ for any positive integers $m, n$ and any $A, B \in R^{m \times n}$.*
(3) *$\mathcal{N}(A) \subseteq \mathcal{N}(B)$ if and only if $\mathcal{R}_l(B) \subseteq \mathcal{R}_l(A)$ for any positive integer $n$ and any $A, B \in R^{n \times n}$.*

**Proof** (1)⇒(2). Let $A, B \in R^{m \times n}$. If $\mathcal{R}_l(B) \subseteq \mathcal{R}_l(A)$, then clearly $\mathcal{N}(A) \subseteq \mathcal{N}(B)$. Conversely, if $\mathcal{N}(A) \subseteq \mathcal{N}(B)$, then taking $\bar{A} = \begin{bmatrix} A & 0 \\ 0 & 0 \end{bmatrix}, \bar{B} = \begin{bmatrix} B & 0 \\ 0 & 0 \end{bmatrix} \in R^{k \times k}$, we have $\mathcal{N}(\bar{A}) \subseteq \mathcal{N}(\bar{B})$. It follows that $\mathbf{r}(\bar{A}) \subseteq \mathbf{r}(\bar{B})$, and hence $\mathbf{l}(\mathbf{r}(\bar{B})) \subseteq \mathbf{l}(\mathbf{r}(\bar{A}))$. By the hypothesis that $R$ is right FP-injective, we obtain $R^{k \times k} \bar{B} \subseteq R^{k \times k} \bar{A}$. So, there exists $C \in R^{m \times m}$ such that $B = CA$. Hence $\mathcal{R}_l(B) \subseteq \mathcal{R}_l(A)$.

(2)⇒(3). It is trivial.

(3)⇒(1). For any positive integer $n$ and any $A \in R^{n \times n}$, let $B \in \mathbf{l}(\mathbf{r}(A))$. Then we have $\mathcal{N}(A) \subseteq \mathcal{N}(B)$, and hence $\mathcal{R}_l(B) \subseteq \mathcal{R}_l(A)$ by (3). So there exists $C \in R^{n \times n}$ such that $B = CA$, which follows that $\mathbf{l}(\mathbf{r}(A)) \subseteq R^{n \times n} A$. Since $R^{n \times n} A \subseteq \mathbf{l}(\mathbf{r}(A))$ is clear, we conclude that $R^{n \times n} A = \mathbf{l}(\mathbf{r}(A))$. Therefore, $R$ is right FP-injective. □

**Corollary 1.4.7 ([68, Corollary 2.9])** *If $R$ is a right FP-injective ring, then $\mathcal{N}(A) = \mathcal{N}(B)$ if and only if $\mathcal{R}_l(B) = \mathcal{R}_l(A)$ for any positive integers $m, n$ and any $A, B \in R^{m \times n}$.*

For any $m \times n$ matrix $A = (a_{ij})$ over a ∗-ring $R$, unless otherwise stated, $A^*$ represents the $n \times m$ matrix whose $(i, j)$-entry equals to $a_{ji}^*$; for any $X \subseteq R^{n \times 1}$ and $Y \subseteq R^{1 \times n}$, let

$$^{\perp}X = \{z \in R^{n \times 1} : z^* x = 0 \text{ for all } x \in X\},$$

$$Y^{\perp} = \{z \in R^{1 \times n} : yz^* = 0 \text{ for all } y \in Y\}.$$

**Theorem 1.4.8 ([18, Theorem 3.2.7])** *The following statements are equivalent for any ∗-ring $R$:*

(1) *$R$ is right FP-injective.*
(2) *$\mathcal{R}(A^*) = {}^{\perp}\mathcal{N}(A)$ for any positive integers $m, n$ and any $A \in R^{m \times n}$.*
(3) *$\mathcal{R}(A^*) = {}^{\perp}\mathcal{N}(A)$ for any positive integer $n$ and any $A \in R^{n \times n}$.*
(4) *$\mathcal{R}(A) = {}^{\perp}[{}^{\perp}\mathcal{R}(A)]$ for any positive integers $m, n$ and any $A \in R^{m \times n}$.*
(5) *$\mathcal{R}(A) = {}^{\perp}[{}^{\perp}\mathcal{R}(A)]$ for any positive integer $n$ and any $A \in R^{n \times n}$.*

**Proof** (1)⇒(2). It is obvious that $\mathcal{R}(A^*) \subseteq {}^{\perp}\mathcal{N}(A)$. Let $x \in {}^{\perp}\mathcal{N}(A)$. Write $\bar{A} = \begin{bmatrix} A & 0 \\ 0 & 0 \end{bmatrix} \in R^{k \times k}, \bar{x} = \begin{bmatrix} x \\ 0 \end{bmatrix} \in R^{k \times 1}$, then for any $B \in \mathbf{r}(\bar{A})$, we have $(\bar{x})^* B = 0$. Since $R$ is right FP-injective, there exists $y = [y_1, y_2] \in R^{1 \times k}$ such that $\bar{x}^* = y\bar{A} = [y_1 A, 0]$. Hence $x = A^* y_1^* \in \mathcal{R}(A^*)$.

(2)⇒(3). It is trivial.

(3)⇒(1). Let $B = \begin{bmatrix} b_1 \\ \vdots \\ b_n \end{bmatrix} \in \mathbf{l}(\mathbf{r}(A))$, where $b_i \in R^{1 \times n}$ for every $i = 1, \cdots, n$.

Then for any $x \in \mathcal{N}(A)$, we get $Bx = 0$. It follows that $b_i^* \in {}^{\perp}\mathcal{N}(A) = \mathcal{R}(A^*)$.
So, there exists $y_i \in R^{n \times 1}$ such that $b_i^* = A^* y_i$, and hence $b_i = y_i^* A$. Therefore

$B = \begin{bmatrix} y_1^* \\ \vdots \\ y_n^* \end{bmatrix} A \in R^{n \times n} A$, showing that $R$ is right FP-injective.

(2)⇔(4) and (3)⇔(5) follow by the fact that ${}^{\perp}\mathcal{R}(A) = \mathcal{N}(A^*)$ for any $A \in R^{m \times n}$.  □

### 1.4.2  *-Regularity

**Definition 1.4.9** Let $R$ be a *-ring. Then $R$ is called *-regular* (resp., *-unit regular*) if $R$ is regular (resp., unit-regular) with * proper.

**Example 1.4.10** The rings in Example 1.3.19 (1), (3) and (5) are *-unit regular.

For the converse of (1.4.1), we can see from the next result that if a right (resp., left) P-injective ring is equipped with a proper involution *, then it is regular (and hence *-regular).

**Theorem 1.4.11 ([51, Theorem 1])** *Let $R$ be a *-proper ring. Then $R$ is regular if and only if $R$ is right (resp., left) P-injective.*

**Proof** It suffices to prove the sufficiency. Suppose that $R$ be right P-injective and * is proper. For any $a, b \in R$ with $a^* ab = 0$, we have $(ab)^* ab = b^* a^* ab = 0$, and hence $ab = 0$ since * is proper. It follows that $\mathbf{r}(a) = \mathbf{r}(a^* a)$. Then by right P-injectivity of $R$, $Ra = \mathbf{l}(\mathbf{r}(a)) = \mathbf{l}(\mathbf{r}(a^* a)) = Ra^* a$. So, we have $a = ca^* a$ for some $c$. Then $ca^* = ca^* ac^*$ is Hermitian. Therefore $a = ca^* a = (ca^*)^* a = ac^* a$ is regular, which means that $R$ is a regular ring.  □

**Proposition 1.4.12 ([27, Lemma 2.5])** *Let $R$ be a *-ring. If $R$ is regular with the GN-property, then $R$ is *-regular.*

**Proof** Let $a \in R$ with $a^* a = 0$. Since $R$ is regular, there exists an idempotent $e$ such that $aR = eR$. So $e = ar$ for some $r \in R$, and $e^* e = r^* a^* ar = 0$. Write $f = 1 - e$. Then $0 = (1 - f)^* (1 - f) = 1 + f^* f - (f + f^*)$. It follows that

$$ff^* = [(1 + f^* f) - (f + f^*)] + ff^* = 1 + (f - f^*)^* (f - f^*)$$

is a unit as $R$ has the GN-property. However, $(1 - f)ff^* = 0$ implies that $e = 1 - f = 0$, and hence $a = 0$ as $aR = eR = 0$.  □

**Proposition 1.4.13 ([28, Proposition 1])** *Let $R$ be a $*$-ring. Then the following statements are equivalent:*

(1) *$R$ is $*$-regular.*
(2) *For any $a \in R$, there exists a projection $p$ such that $aR = pR$.*
(3) *For any $a \in R$, there exists a projection $q$ such that $Ra = Rq$.*
(4) *$Ra = Ra^*a$ for every $a \in R$.*
(5) *$aR = aa^*R$ for every $a \in R$.*
(6) *$R$ is regular and $Re = Re^*e$ for any idempotent $e \in R$.*
(7) *$R$ is regular and $eR = ee^*R$ for any idempotent $e \in R$.*

**Proof** (1)$\Rightarrow$(2). Let $a \in R$. Then there exists $t \in R$ such that $a^*ata^*a = a^*a$ by the regularity of $a^*a$. So, $(ata^*a - a)^*(ata^*a - a) = (ta^*a - 1)^*(a^*ata^*a - a^*a) = 0$. Since $*$ is proper, we have $ata^*a = a$. Hence $a(at)^* = ata^*a(at)^* = [a(at)^*]^*a(at)^*$. It follows that $ata^* = [a(at)^*]^* = a(at)^*$ is Hermitian. Write $p = ata^*$, then clearly $p$ is precisely the projection such that $aR = pR$.

(2)$\Rightarrow$(4). Let $a \in R$. Then $aR = pR$ for a projection $p \in R$. So $a = pa$ and $p = ar$ for some $r \in R$. It follows that $a = p^*a = (ar)^*a = r^*a^*a \in Ra^*a$.

(4)$\Rightarrow$(6). It suffices to show that $R$ is regular. Let $a \in R$. By hypothesis, there exists $r \in R$ such that $a = r^*a^*a$. Then we have $ar = r^*a^*ar = (ar)^*ar$. It follows that $ar = (ar)^*$ and $ara = (ar)^*a = r^*a^*a = a$, as required.

(6)$\Rightarrow$(1). Let $x \in R$ with $x^*x = 0$. Since $R$ is regular, $x = xyx$ for some $y \in R$. Write $e = xy$. Then $e$ is an idempotent and $e^*e = y^*x^*xy = 0$. So $Re = Re^*e = 0$. Thus $e = 0$ and $x = ex = 0$, which implies that the involution of $R$ is proper. Therefore, $R$ is a $*$-regular ring.

Dually, we have (1)$\Leftrightarrow$(3)$\Leftrightarrow$(5)$\Leftrightarrow$(7). $\qquad\square$

**Theorem 1.4.14 ([28, Theorem 4])** *Let $R$ be a $*$-ring and $n \geq 2$. Then the following statements are equivalent:*

(1) *$R^{n \times n}$ is $*$-regular.*
(2) *$R$ is regular with $(n-1)$-GN property.*
(3) *$R$ is regular and $*$ is $n$-proper.*

**Proof** (1)$\Rightarrow$(2). For any $a \in R$, consider the matrix

$$A = \begin{bmatrix} a & 0 & \cdots & 0 \\ 0 & 0 & \cdots & 0 \\ \vdots & \vdots & & \vdots \\ 0 & 0 & \cdots & 0 \end{bmatrix} \in R^{n \times n}.$$

Since $R^{n \times n}$ is regular, there is $B = (b_{ij}) \in R^{n \times n}$ such that $ABA = A$. It follows that $ab_{11}a = a$. Hence $R$ is regular. Take any $x_1, \cdots, x_{n-1} \in R$. Let $E = \begin{bmatrix} 1 & 0 \\ \alpha & 0_{n-1} \end{bmatrix} \in R^{n \times n}$, where $\alpha = [x_1, \cdots, x_{n-1}]^T$. Clearly $E$ is idempotent. In view of Theorem 1.4.13 (4), $R^{n \times n}E^*E = R^{n \times n}E$. Note that $E^*E =$

$$\begin{bmatrix} 1+\sum_{i=1}^{n-1} x_i^* x_i & 0 \\ 0 & 0_{n-1} \end{bmatrix} \text{ and } \begin{bmatrix} 1 & 0 \\ 0 & 0_{n-1} \end{bmatrix} \in R^{n\times n} E. \text{ So there exists } Y = \begin{bmatrix} y_1 & Y_2 \\ Y_3 & Y_4 \end{bmatrix} \in$$

$R^{n\times n}$ such that $YE^*E = \begin{bmatrix} 1 & 0 \\ 0 & 0_{n-1} \end{bmatrix}$, which yields $y_1(1 + \sum_{i=1}^{n-1} x_i^* x_i) = 1$. It

follows that $(1+\sum_{i=1}^{n-1} x_i^* x_i)y_1^* = 1$. So, $1+\sum_{i=1}^{n-1} x_i^* x_i$ is a unit in $R$, and therefore $R$ has the $(n-1)$-GN property.

(2)$\Rightarrow$(3). Let $a_1, \cdots, a_n \in R$ with $\sum_{i=1}^{n} a_i^* a_i = 0$. By Theorem 1.4.13, there is a projection $q_n$ such that $q_n R = a_n R$. Moreover, we have $a_n t a_n = a_n$ for some $t \in R$ by the regularity of $R$. Let $x = t q_n$. Then

$$0 = x^*(\sum_{i=1}^{n} a_i^* a_i)x = \sum_{i=1}^{n-1}(a_i x)^* a_i x + (a_n x)^* a_n x$$

$$= \sum_{i=1}^{n-1}(a_i x)^* a_i x + (q_n)^* q_n = \sum_{i=1}^{n-1}(a_i x)^* a_i x + q_n.$$

Since $R$ has the $(n-1)$-GN property, it follows that $1 - q_n = 1 + \sum_{i=1}^{n-1}(a_i x)^* a_i x$ is a unit. Clearly, $1 - q_n$ is idempotent. Hence $1 - q_n = 1$, and so $q_n = 0$. Then it gives $a_n = 0$ since $a_n R = q_n R$. Similarly, we can get $a_1 = \cdots = a_{n-1} = 0$. Therefore, $*$ is $n$-proper.

(3)$\Rightarrow$(1). It follows by Propositions 1.3.18 and 1.4.4.                    □

**Corollary 1.4.15 ([28, Corollaries 5, 7])**

(1) *Let $R$ be a $*$-ring. Then $R^{2\times 2}$ is $*$-regular if and only if $R$ is regular with GN-property.*
(2) *Let $R$ be a $*$-ring. If $R^{n\times n}$ is $*$-regular, then $m \cdot 1$ is invertible for any positive integer $m \leq n$.*

**Remark 1.4.16 ([28, Remark 8(3)])** Let $R = \mathbb{Z}_3$ and $* = 1_R$. Then $R$ is regular with the GN-property. So, $R^{2\times 2}$ is $*$-regular. But $R^{3\times 3}$ is not $*$-regular since $R$ does not possess the 2-GN property.

**Theorem 1.4.17 ([27, Theorem 3.1])** *Let $R$ be a $*$-ring. Then the following statements are equivalent:*

(1) *$R$ is $*$-unit regular.*
(2) *For every $a \in R$, there exist a unit $u$ and a projection $p$ such that $a = pu$.*
(3) *For every $a \in R$, there exist a unit $v$ and a projection $q$ such that $a = vq$.*
(4) *For every $a \in R$, there exists a unit $u$ such that $a = aua$, $(ua)^* = ua$.*
(5) *For every $a \in R$, there exists a unit $v$ such that $a = ava$, $(av)^* = av$.*

**Proof** (1)$\Rightarrow$(2). Let $a \in R$. Since $R$ is unit regular, there exist an idempotent $e$ and a unit $v$ such that $a = ev$. The $*$-regularity of $R$ implies that there exists a projection $p$ such that $eR = pR$. So $e = pe$ and $p = ep$. It follows that $(1 - e - p)^2 = 1 - e - (1 - e)p - p(1 - e) + p = 1$. Write $1 - e - p = -w$ and $u = wv$. Then

$u$ is a unit and $e = pe = -p(1 - p - e) = pw$. Combining this with $a = ev$, one has $a = (pw)v = p(wv) = pu$.

(2)$\Rightarrow$(3). For any $a \in R$, by the hypothesis, there exist a projection $p$ and a unit $u$ such that $a^* = pu$. Hence $a = u^*p$, as desired.

(3)$\Rightarrow$(4). From $a = vq$, we have $a = aq$ and $q = v^{-1}a$. Put $u = v^{-1}$. Then $a = aua$ and $(ua)^* = a = ua$.

(4)$\Rightarrow$(1). It suffices to show that $*$ is proper. Let $a \in R$ with $aa^* = 0$. Then there exists a unit $u$ such that $a = aua$ and $(ua)^* = ua$. It follows that $a = a(ua)^* = aa^*u^* = 0$. So $*$ is proper, and hence $R$ is $*$-unit regular.

The proof of (2)$\Rightarrow$(5)$\Rightarrow$(1) is similar to that of (3)$\Rightarrow$(4)$\Rightarrow$(1). $\qquad\square$

**Theorem 1.4.18 ([17, Proposition 2.11],[27, Theorem 3.8])** *Let $R$ be a $*$-ring and $n \geq 2$. Then the following statements are equivalent:*

(1) $R^{n \times n}$ *is $*$-unit regular.*
(2) $R$ *is unit-regular with the $(n-1)$-GN property.*
(3) $R$ *is unit-regular and $n$-proper.*

**Proof** It follows by Proposition 1.4.4 and Theorem 1.4.14. $\qquad\square$

**Corollary 1.4.19 ([27, Corollary 3.9])** *Let $R$ be a $*$-ring. Then $R^{2 \times 2}$ is $*$-unit regular if and only if $R$ is unit-regular with the GN-property.*

**Remark 1.4.20 ([27, Example 3.11])**

(1) Let $R = \mathbb{C}^{n \times n}$ and $*$ be the conjugate transpose map of complex matrices. Then $R$ is $*$-unit regular.
(2) Let $R = \mathbb{R}^{n \times n}$ and $*$ be the transpose map of real matrices. Then $R$ is $*$-unit regular.
(3) Let $R = \mathbb{Z}_3$ and $* = 1_R$. Then $R$ is unit-regular with the GN-property. So $R^{2 \times 2}$ is $*$-unit regular. But $R^{3 \times 3}$ is not $*$-unit regular as $R$ does not possess the 2-GN property.

A ring $R$ is $*$-*strongly regular* if $R$ is strongly regular and $*$ is proper.

**Proposition 1.4.21 ([17, Proposition 2.8])** *Let $R$ be a $*$-ring. Then the following statements are equivalent:*

(1) $R$ *is $*$-strongly regular.*
(2) $R$ *is strongly regular and every idempotent in $R$ is a projection.*
(3) *For every $a \in R$, there exist a unit $u$ and a projection $p$ such that $a = up = pu$.*

**Proof** (1)$\Rightarrow$(2). Clearly, $R$ is $*$-regular. For any idempotent $e \in R$. In view of Theorem 1.4.13, there is a projection $p \in R$ such that $eR = pR$. Hence $p = ep, e = pe$. By the strongly regularity, $e$ is central. Hence $e = pe = ep = p$.

(2)$\Rightarrow$(3). For any $a \in R$, by the regularity of $R$, there is $r \in R$ such that $a = ara$. Let $t = rar$. Then $ata = a, tat = t$ and $at, ta$ are idempotent. So, $at, ta$ are central projections by the hypothesis. Thus $at = [a(ta)]t = (ta)at = t[(at)a] = ta$. Write $u = a + 1 - at$ and $p = at = ta$. Then $a = up$ and $u$ is a unit since $u(t + 1 - at) = 1 = (t + 1 - at)u$.

(3)$\Rightarrow$(1). It suffices to show that the involution $*$ is proper. Let $x \in R$ with $x^*x = 0$. Then $x = pu = up$ for a unit $u$ and a projection $p$. So, we have $0 = x^*x = (pu)^*pu = u^*p^*pu = u^*pu$. Hence $p = 0$, and so $a = pu = 0$. This proves that $*$ is proper.                                                                $\square$

## 1.5  Invertibility of the Difference and the Sum of Idempotents

In the last section of this chapter, we collect some research on the invertibility of the difference and the sum of idempotents.

Although the results of this section seems to be not easy to pick up (thus perhaps do not fit the title of this chapter "Algebraic Basic Knowledge"), they have some good guiding significance for the related study in subsequent chapters as idempotents are one of the most fundamental research objects.

### 1.5.1  Invertibility of the Difference of Idempotents

**Lemma 1.5.1 (Kato's Identities, See also [69, Proposition 2.2])** *Let* $f, g \in R$ *be idempotents. Then*

$$(f - g)^2 + (1 - f - g)^2 = 1 \tag{1.5.1}$$

*which amounts to*

$$(f - g)^2 = (f + g)(2 - f - g). \tag{1.5.2}$$

**Theorem 1.5.2 ([69, Theorem 2.8])** *Let* $f, g \in R$ *be idempotents. Then* $f - g \in R^{-1}$ *if and only if* $f + g \in R^{-1}$ *and* $2 - f - g \in R^{-1}$. *In this case,* $f$ *is similar to* $1 - g$.

**Proof** The "if and only if" statement follows by Eq. (1.5.2). Further, we have

$$(f - g)^{-1}(1 - g)(f - g) = (f - g)^{-1}(1 - g)f = (f - g)^{-1}(f - g)f = f.$$

So, $f$ is similar to $1 - g$.                                                                $\square$

Also, it follows from Eq. (1.5.2) that the condition of $f - g \in R^{-1}$ is stronger than that of $f + g \in R^{-1}$.

**Lemma 1.5.3 ([72, Lemma 2.1])** *Let* $f, g \in R$ *be idempotents. Then the following statements are equivalent:*

(1) *There exists an idempotent $h \in R$ such that*

$$hR = fR, \quad (1-h)R = gR, \tag{1.5.3}$$

*that is,*

$$hf = f, \quad fh = h, \quad (1-h)g = g, \quad g(1-h) = 1-h. \tag{1.5.4}$$

(2) $R = fR \oplus gR$.
(3) $R = R(1-f) \oplus R(1-g)$.
(4) $R = {}^\circ f \oplus {}^\circ g$.
(5) $R = (1-f)^\circ \oplus (1-g)^\circ$.

**Proof** (1)$\Rightarrow$(2). It follows from $R = hR \oplus (1-h)R$.

(2)$\Rightarrow$(1). Suppose that $R = fR \oplus gR$. Towards the existence of $h$ we note that the unit element 1 is decomposed as $1 = fu + gv$ for some $u, v \in R$. Let $h = fu$. Then $h = h^2 + fugv = h^2 + gvfu$, and $h - h^2 \in fR \cap gR = \{0\}$. Hence $h$ is idempotent. From $f - hf = f(1 - uf) = gvf \in fR \cap gR$ we conclude that $f - hf = 0$ and $hf = f$. Further, $fh = f^2u = fu = h$. This gives $hR = fR$. A similar argument with $1 - h$ in place of $h$ and $g$ in place of $f$ yields $(1-h)g = g$ and $g(1-h) = 1 - h$, that is, $(1-h)R = gR$.

To prove the uniqueness of $h$, assume that (1.5.3) is satisfied with $h_1, h_2$ in place of $h$. Then $h_1 f = f$ implies $h_1 f h_2 = f h_2$ and $h_1 h_2 = h_2$. On the other hand, $h_1 g = 0$ implies $h_1 g h_2 = 0$, $h_1 (g + h_2 - 1) = 0$ and $h_1 h_2 = h_1$. Hence $h_1 = h_2$.

(1)$\Leftrightarrow$(3). By an argument similar to the one used above we show that condition (3) is equivalent to the existence of an idempotent $k \in R$ satisfying $Rk = R(1 - f)$ and $R(1 - k) = R(1 - g)$, that is,

$$k(1-f) = k, \quad (1-f)k = 1-f,$$

$$(1-k)(1-g) = 1-k, \quad (1-g)(1-k) = 1-g.$$

This is equivalent to (1.5.4) when we set $k = 1 - h$.

The equivalence of (4) with (3) and of (5) with (2) follows from Eq. (1.2.1).  $\square$

**Lemma 1.5.4 ([72, Lemma 3.1])** *Let $f, g \in R$ be idempotents. Then*

$$f - g \in R^{-1} \Rightarrow 1 - fgf \in R^{-1} \Leftrightarrow 1 - fg \in R^{-1}. \tag{1.5.5}$$

**Proof** Suppose that $f - g \in R^{-1}$. Then the element $(f - g)^2$ is invertible and commutes with $f$:

$$f(f-g)^2 = f - fgf = (f-g)^2 f.$$

Writing $1 - fgf = (1 - fgf)f + (1 - fgf)(1 - f)$, we obtain

$$1 - fgf = (f-g)^2 f + 1 - f.$$

From this equation we deduce that $(f - g)^{-2}f + 1 - f$ is the inverse of $1 - fgf$. Moreover, since $1 - fg = 1 - f(fg)$, it follows from Jacobson's lemma that $1 - fg \in R^{-1} \Leftrightarrow 1 - fgf \in R^{-1}$. $\qquad\qquad\qquad\qquad\qquad\qquad\qquad\qquad\qquad\qquad\square$

**Theorem 1.5.5 ([72, Theorem 3.2])** *Let* $f, g \in R$ *be idempotents. Then the following statements are equivalent:*

(1) $f - g \in R^{-1}$.

(2) $R = fR \oplus gR = Rf \oplus Rg$.

(3) *There exist idempotents* $h, k \in R$ *such that* $hR = fR$, $(1 - h)R = gR$, $Rk = Rf$, *and* $R(1 - k) = Rg$; *h and k are unique if they exist.*

(4) $1 - fg \in R^{-1}$, $R = fR + gR$ *and* $f^\circ \cap g^\circ = \{0\}$.

(5) $f + g - fg \in R^{-1}$, $fR \cap gR = \{0\}$ *and* $R = f^\circ + g^\circ$.

(6) $1 - fg \in R^{-1}$ *and* $f + g - fg \in R^{-1}$.

**Proof** (1)$\Rightarrow$(6). Suppose that $f - g \in R^{-1}$. By Lemma 1.5.4, $1 - fg \in R^{-1}$. Since $(1-f)-(1-g) = g-f \in R^{-1}$, we also have $1-(1-f)(1-g) = f+g-fg \in R^{-1}$.

(6)$\Rightarrow$(4). Let $w \in R$ be the inverse of $f + g - fg$. From $1 = (f + g - fg)w = f(1 - g)w + gw$, we obtain $R = fR + gR$, and from $1 = w(f + g - fg) = wf + w(1 - f)g$, we conclude that $f^\circ \cap g^\circ = \{0\}$.

(4)$\Rightarrow$(2). Let $x \in fR \cap gR$. Then $x = fx = gx = fgx$ and $(1 - fg)x = 0$; hence $x = 0$, and $fR \cap gR = \{0\}$. Consequently, $R = fR \oplus gR$. Write $\tilde{f} = 1 - g$ and $\tilde{g} = 1 - g$. Then $\tilde{f} + \tilde{g} - \tilde{f}\tilde{g} = 1 - fg \in R^{-1}$. As in the proof of (6)$\Rightarrow$(4) we show that $R = \tilde{f}R + \tilde{g}R = f^\circ + g^\circ$. Hence $R = f^\circ \oplus g^\circ = Rf \oplus Rg$.

(2)$\Rightarrow$(3). It follows from Lemma 1.5.3.

(3)$\Rightarrow$(1). From the hypothesis, we get

$$hf = f, \quad fh = h, \quad (1 - h)g = g, \quad g(1 - h) = 1 - h, \qquad (1.5.6)$$

$$kf = k, \quad fk = f, \quad (1 - k)g = 1 - k, \quad g(1 - k) = g. \qquad (1.5.7)$$

Using these equations, we show that $h + k - 1$ is the inverse of $f - g$:

$$(h + k - 1)(f - g) = hf + kf - f - hg - kg + g = 1,$$

$$(f - g)(h + k - 1) = fh + fk - f - gh - gk + g = 1.$$

(5)$\Leftrightarrow$(4). It follows by substituting $\tilde{f} = 1 - f$ and $\tilde{g} = 1 - g$ for $f$ and $g$. $\qquad\square$

**Corollary 1.5.6 ([72, Corollary 3.3])** *Let* $f, g \in R$ *be idempotents. If* $f - g \in R^{-1}$, *then*

$$(f - g)^{-1} = h + k - 1,$$

$$h = (f - g)^{-1}(1 - g), \quad k = (1 - g)(f - g)^{-1}$$

*where h and k are idempotents defined in* (3) *of preceding theorem.*

**Theorem 1.5.7 ([72, Theorem 3.5])** *Let* $f, g \in R$ *be idempotents. Then the following statements are equivalent:*

(1) $f - g \in R^{-1}$.
(2) $f - g$ *is regular and*

$$f R \cap g R = \{0\}, \quad f^\circ \cap g^\circ = \{0\}, \quad Rf \cap Rg = \{0\}, \quad {}^\circ f \cap {}^\circ g = \{0\}.$$

*Proof* (1)$\Rightarrow$(2). It is clear from the above discussion.

(2)$\Rightarrow$(1). Suppose that (2) holds. First we show that $(f - g)^\circ = \{0\}$. Let $(f - g)x = 0$. Then $fx = gx \in f R \cap g R = \{0\}$, and $x \in f^\circ \cap g^\circ = \{0\}$. Hence $x = 0$. By the regularity of $f - g$, we obtain $(f - g)t(f - g) = f - g$ for some $t \in R$, and hence $1 - t(f - g) \in (f - g)^\circ = \{0\}$. If follows that $f - g$ is left invertible. By a symmetrical argument we conclude that $f - g$ is right invertible. $\quad\square$

**Corollary 1.5.8 ([72, Corollary 3.6])** *Let* $f, g \in R$ *be idempotents. Then the following statements are equivalent:*

(1) $fg - gf \in R^{-1}$.
(2) $f - g \in R^{-1}$ *and* $1 - f - g \in R^{-1}$.
(3) $R = f R \oplus g R = Rf \oplus Rg = f^\circ \oplus g R = {}^\circ f \oplus Rg$.

*Proof* The equivalence of (1) and (2) follows from

$$fg - gf = (1 - f - g)(f - g) = (f - g)(f + g - 1),$$

and the equivalence of (2) and (3) from Theorem 1.5.5 first applied to the pair $f, g$ and then to the pair $1 - f, g$. $\quad\square$

## 1.5.2 Invertibility of the Sum of Idempotents

**Theorem 1.5.9 ([71, Theorem 3.3])** *Let* $2 \in R^{-1}$ *and* $f, g \in R$ *be idempotents. Then the following statements are equivalent:*

(1) $f + g \in R^{-1}$.
(2) $f + g$ *is regular and*

$$f R \cap g(1 - f)R = \{0\} = f^\circ \cap g^\circ, \tag{1.5.8}$$

$$Rf \cap R(1 - f)g = \{0\} = {}^\circ f \cap {}^\circ g. \tag{1.5.9}$$

*Proof* (1)$\Rightarrow$(2). Suppose that $f + g$ is left invertible in $R$. This implies that $f + g$ is regular. Let $x \in f R \cap g(1 - f)R$. Then $x = fx = gx = g(1 - f)y$ for some $y \in R$, and

$$(f + g)x = 2x = 2g(1 - f)y = (f + g)(1 - f)2y.$$

Hence $x = (1 - f)2y \in (1 - f)R$. Since also $x \in fR$, we conclude that $x = 0$. Therefore, $fR \cap g(1 - f)R = \{0\}$. Next assume that $x \in f^{\circ} \cap g^{\circ}$. Then $fx = gx = 0$, and from $(f + g)x = 0$ we get $x = 0$. Hence $f^{\circ} \cap g^{\circ} = \{0\}$. Dually, by the right invertibility of $f + g$, we can get $Rf \cap R(1 - f)g = \{0\} = {}^{\circ}f \cap {}^{\circ}g$.

(2)$\Rightarrow$(1). By the hypothesis, we have $(f + g)t(f + g) = f + g$ for some $t \in R$. Write $x = 1 - t(f + g)$. Then $(f + g)x = 0$, and so $fx = -gx = -gfx - g(1 - f)x$, $gfx = -gx = fx$. Hence

$$2fx = fx + gfx = -g(1 - f)x,$$

and, since $2 \in R^{-1}$, $fx \in fR \cap g(1 - f)R = \{0\}$. Thus $fx = 0 = gx$, and $x \in f^{\circ} \cap g^{\circ} = \{0\}$. It follows that $1 - t(f + g) = 0$ and hence $f + g$ is left invertible. Similarly, we can show that $f + g$ is right invertible.    $\square$

**Theorem 1.5.10 ([71, Theorem 3.4])** *Let* $f, g \in R$ *be idempotents. If* $f - g \in R^{-1}$, *then so is* $f + g$, *and*

$$(f + g)^{-1} = 1 - h - k + 2kh \tag{1.5.10}$$

*where* $h$ *and* $k$ *are the idempotents satisfying* $hR = fR$, $(1 - h)R = gR$, $Rk = Rf$, $R(1 - k) = Rg$.

**Proof** If $f - g \in R^{-1}$, the existence of $h, k$ satisfying the stated conditions in guaranteed by Theorem 1.5.5. We observe that

$$hf = f, \quad fh = h, \quad (1 - h)g = g, \quad g(1 - h) = 1 - h,$$
$$kf = k, \quad fk = f, \quad (1 - k)g = 1 - k, \quad g(1 - k) = g.$$

Using these equations, we can verify that

$$(f + g)(1 - h - k + 2kh) = 1 = (1 - h - k + 2kh)(f + g).$$

$\square$

**Theorem 1.5.11 ([71, Theorem 3.5])** *Let* $2 \in R^{-1}$ *and* $f, g \in R$ *be idempotents. Then the following statements are equivalent:*

(1) $f - g \in R^{-1}$.
(2) $f + g \in R^{-1}$ *and* $1 - fg \in R^{-1}$.

**Proof** It suffices to show (2)$\Rightarrow$(1). First we prove that $f - g$ is regular. From $(f + g)(1 - g) = f(1 - g)$ we get $1 - g = w(1 - g)$ with $w = (f + g)^{-1}f$. Hence $(f - g)(1 - g) = (f - g)w(f - g)$ and

$$f - g = (f - g)w(f - g) + (f - g)g.$$

We show that $(f - g)g$ is regular and then apply Lemma 1.2.20: Since $(f - g)g = (fg - 1)g$, we have $g = g(fg - 1)^{-1}(f - g)g$, and

$$(f - g)g = (f - g)g(fg - 1)^{-1}(f - g)g.$$

Next, we show that $(f - g)^\circ = \{0\}$. Let $(f - g)x = 0$ for some $x \in R$. Then $fx = gx = gfx$, and

$$(1 - f)x = (f + g)^{-1}(f + g)(1 - f)x = (f + g)^{-1}(gx - gfx) = 0,$$

$$fx = (1 - gf)^{-1}(1 - gf)fx = (1 - gf)^{-1}(fx - gfx) = 0.$$

Hence $x = (1 - f)x + fx = 0$. Symmetrically we show that $^\circ(f - g) = \{0\}$. Then $f - g \in R^{-1}$ by Lemma 1.2.20 again. □

**Theorem 1.5.12 ([71, Theorem 3.6])** *Let* $2 \in R^{-1}$ *and* $f, g \in R$ *be idempotents. Then the following statements are equivalent:*

(1) $fg + gf \in R^{-1}$.
(2) $f + g \in R^{-1}$ *and* $1 - f - g \in R^{-1}$.
(3) $f + g$ *is regular and*

$$R = f^\circ \oplus gR = {}^\circ f \oplus Rg, \tag{1.5.11}$$

$$fR \cap g(1 - f)R = \{0\} = f^\circ \cap g^\circ, \tag{1.5.12}$$

$$Rf \cap R(1 - f)g = \{0\} = {}^\circ f \cap {}^\circ g. \tag{1.5.13}$$

**Proof** The equivalence of (1) and (2) is a consequence of the equation

$$(f + g - 1)(f + g) = fg + gf = (f + g)(f + g - 1).$$

The equivalence of (2) and (3) follows from Theorem 1.5.9 (the invertibility of $f + g$) and parts (1) and (2) of Theorem 1.5.5 (the invertibility of $(1 - f) - g$). □

**Theorem 1.5.13 ([71, Theorem 3.7])** *Let* $R$ *be a* $*$-*ring with* $2 \in R^{-1}$, *and let* $f, g \in R$ *be idempotents. Then the following statements are equivalent:*

(1) $f + g \in R^{-1}$.
(2) $f + g$ *is regular and*

$$fR \cap g(1 - f)R = \{0\} = f^\circ \cap g^\circ, \tag{1.5.14}$$

$$f^*R \cap g^*(1 - f^*)R = \{0\} = (f^*)^\circ \cap (g^*)^\circ. \tag{1.5.15}$$

**Proof** From the proof of Theorem 1.5.9, it follows that the regularity of $f + g$ together with $fR \cap g(1 - f)R = \{0\} = f^\circ \cap g^\circ$ implies that $f + g$ is left invertible.

Since $f^* + g^*$ is regular if $f + g$ is, we conclude by the same argument that the regularity of $f + g$ together with $f^*R \cap g^*(1 - f^*)R = \{0\} = (f^*)^\circ \cap (g^*)^\circ$ implies that $f^* + g^*$ is left invertible. Hence $f + g$ is right invertible, and the result follows.                                                                          □

**Corollary 1.5.14 ([71, Corollary 3.8])** *Let $R$ be a $*$-ring with $2 \in R^{-1}$, and let $f, g \in R$ be projections. Then the following statements are equivalent:*

(1) $f + g \in R^{-1}$,
(2) $f + g$ *is regular and* $fR \cap g(1 - f)R = \{0\} = f^\circ \cap g^\circ$.

# Chapter 2
# Moore-Penrose Inverses

As early as 1920, Eliakim Hastings Moore [92] found that any complex matrix $A$ (square or not) can be associated with a unique matrix $G$ such that $AG$ and $GA$ are the projection operators on the range of $A$ and $G$, respectively. Since such $G$ reduces to the ordinary inverse of $A$ when $A$ is square and nonsingular, it is called by Moore the general reciprocal of $A$. This idea was extended to general operators by Yuan-Yung Tseng [112, 113] in 1949. Unaware of Moore's work, Roger Penrose [102] defined in 1955 a unique generalized inverse $A^\dagger$ for $A$ by using four matrix equations (see Sect. 2.1 below), and showed its applications to solve linear matrix equations. It turns out that Moore's general reciprocal of a matrix $A$ and Penrose's generalized inverse $A^\dagger$ coincide; this unique generalized inverse $A^\dagger$ is now commonly called the Moore-Penrose inverse of $A$.

In [52], the notion of Moore-Penrose inverses was extended by Robert E. Hartwig from complex matrices to matrices over an arbitrary $*$-ring, and in [106] it was further extended by Roland Puystjens and Donald W. Robinson to morphisms of a category with involution.

In this chapter, we first review basic properties and calculation methods for Moore-Penrose inverses of complex matrices, and then turn to Moore-Penrose inverses in the more general setting of rings, semigroups and categories with involution. Properties and characterizations of Moore-Penrose inverses of elements are given. The Moore-Penrose invertibility of products, block matrices, companion matrices, differences (products) of projections and sums of morphisms are described.

Throughout this chapter, unless otherwise stated, $R$ is a $*$-ring with identity and $S$ is a $*$-semigroup.

J. Chen, X. Zhang, *Algebraic Theory of Generalized Inverses*,
https://doi.org/10.1007/978-981-99-8285-1_2

## 2.1 Moore-Penrose Inverses of Complex Matrices

**Definition 2.1.1** Let $A \in \mathbb{C}^{m \times n}$. If $X \in \mathbb{C}^{n \times m}$ satisfies:

(1) $AXA = A$,   (2) $XAX = X$,   (3) $(AX)^* = AX$,   (4) $(XA)^* = XA$,

then $X$ is called the Moore-Penrose inverse of $A$ and denoted by $A^\dagger$.

The next result shows the existence and uniqueness of $A^\dagger$.

**Theorem 2.1.2** Let $A \in \mathbb{C}^{m \times n}$. The Moore-Penrose inverse of $A$ exists and is unique.

**Proof** We first prove the uniqueness of $X$. If both $X_1$ and $X_2$ satisfy conditions (1)–(4) in Definition 2.1.1, then

$$X_1 = X_1 A X_1 = X_1 A X_2 A X_1 = (X_1 A)^* (X_2 A)^* X_1$$

$$= (X_2 A X_1 A)^* X_1 = (X_2 A)^* X_1 = X_2 A X_1$$

$$= X_2 A X_2 A X_1 = X_2 (A X_2)^* (A X_1)^* = X_2 (A X_1 A X_2)^*$$

$$= X_2 (A X_2)^* = X_2 A X_2 = X_2.$$

Next we prove the existence of $A^\dagger$. Here we give two methods.

Method 1:   (using the full rank factorization)
Let $A = GH$ be a full rank factorization of $A$, where $G \in \mathbb{C}^{m \times r}$, $H \in \mathbb{C}^{r \times n}$ with $\mathrm{rank}(G) = \mathrm{rank}(H) = \mathrm{rank}(A) = r$. Then $HH^*$ and $G^*G$ are nonsingular. Let

$$X = H^* (HH^*)^{-1} (G^*G)^{-1} G^*.$$

It is easy to verify that $X$ satisfies the conditions (1)–(4) in Definition 2.1.1.
Method 2:   (using the singular value decomposition)
Let $A = U \begin{bmatrix} \Sigma & 0 \\ 0 & 0 \end{bmatrix} V^*$ be the singular value decomposition of $A$, and let $X = V \begin{bmatrix} \Sigma^{-1} & 0 \\ 0 & 0 \end{bmatrix} U^*$. Then it is straightforward to check that $X$ satisfies the conditions (1)–(4) in Definition 2.1.1.

$\square$

We list some properties of Moore-Penrose inverse here.

**Theorem 2.1.3** Let $A \in \mathbb{C}^{m \times n}$. Then:

(1) $(A^\dagger)^\dagger = A$.
(2) $(A^*)^\dagger = (A^\dagger)^*$.
(3) $A^* = A^\dagger A A^* = A^* A A^\dagger$.

(4) $A^\dagger = (A^*A)^\dagger A^* = A^*(AA^*)^\dagger$.

(5) $\mathcal{R}(A^\dagger) = \mathcal{R}(A^*)$, $\mathcal{N}(A^\dagger) = \mathcal{N}(A^*)$.

(6) $\text{rank}(A) = \text{rank}(A^\dagger) = \text{rank}(AA^\dagger) = \text{rank}(A^\dagger A)$.

(7) $(A^*A)^\dagger = A^\dagger(A^*)^\dagger$, $(AA^*)^\dagger = (A^*)^\dagger A^\dagger$.

(8) $(UAV)^\dagger = V^*A^\dagger U^*$, where $U$ and $V$ are unitary matrices.

(9) If $\lambda \in \mathbb{C}$, then $(\lambda A)^\dagger = \lambda^\dagger A^\dagger$, where $\lambda^\dagger = \begin{cases} \lambda^{-1}, \ \lambda \neq 0, \\ 0, \quad \lambda = 0. \end{cases}$

(10) If $\text{rank}(A) = n$, then $A^\dagger = (A^*A)^{-1}A^*$; If $\text{rank}(A) = m$, then $A^\dagger = A^*(AA^*)^{-1}$.

Consider the question of finding a $x$ minimizing $||Ax - b||$ when $b \notin \mathcal{R}(A)$.

**Definition 2.1.4** Let $A \in \mathbb{C}^{m \times n}$ and $b \in \mathbb{C}^m$. A vector $u \in \mathbb{C}^n$ is called a least-squares solution of $Ax = b$ if $||Au - b|| \leq ||Av - b||$ for all $v \in \mathbb{C}^n$.

**Theorem 2.1.5** Let $A \in \mathbb{C}^{m \times n}$ and $b \in \mathbb{C}^m$. Then the following conditions are equivalent:

(1) $\eta$ is a least-squares solution of $Ax = b$.

(2) $A^*A\eta = A^*b$.

(3) $A^\dagger A\eta = A^\dagger b$.

**Proof** We first prove $\mathcal{R}(A)^\perp = \mathcal{N}(A^*)$. Since $A^*x = 0$ implies that $\langle Ay, x \rangle = (Ay)^*x = y^*A^*x = 0$ for any $y \in \mathbb{C}^n$, we have $\mathcal{N}(A^*) \subseteq \mathcal{R}(A)^\perp$. Conversely, if $x \in \mathcal{R}(A)^\perp$, then $y^*A^*x = \langle Ay, x \rangle = 0$ for any $y \in \mathbb{C}^n$. Taking $y = A^*x$, we have $y^*y = 0$. It follows that $A^*x = y = 0$. Thus, $\mathcal{R}(A)^\perp \subseteq \mathcal{N}(A^*)$.

(1) $\Rightarrow$ (2). Since $\mathbb{C}^m = \mathcal{R}(A) \oplus \mathcal{R}(A)^\perp$, there exist $u \in \mathbb{C}^n$ and $\xi \in \mathcal{R}(A)^\perp$ such that $b = Au + \xi$. So $Au - b = -\xi \in \mathcal{R}(A)^\perp$. Then we have

$$||A\eta - b||^2 = ||A\eta - Au + Au - b||^2 = ||A\eta - Au||^2 + ||Au - b||^2 \geq ||Au - b||^2.$$

By the choice of $\eta$, we get $||A\eta - b||^2 = ||Au - b||^2$, and thus $A\eta = Au$. It follows that $A\eta - b = Au - b \in \mathcal{R}(A)^\perp = \mathcal{N}(A^*)$. So, $A^*A\eta = A^*b$.

(2) $\Rightarrow$ (1). If $A^*A\eta = A^*b$, then $A\eta - b \in \mathcal{N}(A^*) = \mathcal{R}(A)^\perp$. It follows that $\langle A\eta - b, Ay \rangle = 0$ for any $y \in \mathbb{C}^n$. For any $x \in \mathbb{C}^n$, we have that $\langle A\eta - b, A(x - \eta) \rangle = 0$, which means that $A\eta - b \perp A(x - \eta)$. So

$$||Ax - b||^2 = ||Ax - A\eta + A\eta - b||^2 = ||Ax - A\eta||^2 + ||A\eta - b||^2 \geq ||A\eta - b||^2.$$

This proves that $\eta$ is a least-squares solution of $Ax = b$.

(2) $\Leftrightarrow$ (3). It suffices to prove that $\mathcal{N}(A^*) = \mathcal{N}(A^\dagger)$, which means that $A^*x = 0$ if and only if $A^\dagger x = 0$. Since $A^* = (AA^\dagger A)^* = A^*(AA^\dagger)^* = A^*AA^\dagger$, we have that $\mathcal{N}(A^\dagger) \subseteq \mathcal{N}(A^*)$. And $A^\dagger = A^\dagger AA^\dagger = A^\dagger(AA^\dagger)^* = A^\dagger(A^\dagger)^*A^*$ yields that $\mathcal{N}(A^*) \subseteq \mathcal{N}(A^\dagger)$.                    □

A linear system $Ax = b$ may have many least-squares solutions. We are interested in the solution with minimum norm.

**Definition 2.1.6** Let $A \in \mathbb{C}^{m \times n}$ and $b \in \mathbb{C}^m$. A vector $u \in \mathbb{C}^n$ is called the minimum-norm least-squares solution of $Ax = b$ if $u$ is a least-squares solution of $Ax = b$ and $||u|| \leq ||w||$ for any least-squares solution $w$.

The next result reveals the link between $A^\dagger$ and the minimum-norm least-squares solution of $Ax = b$.

**Theorem 2.1.7** *Let $A \in \mathbb{C}^{m \times n}$ and $b \in \mathbb{C}^m$. Then $A^\dagger b$ is the minimum-norm least-squares solution of $Ax = b$.*

**Proof** Since $A^\dagger A A^\dagger b = A^\dagger b$, $A^\dagger b$ is a least-squares solution of $Ax = b$ by Theorem 2.1.5.

It is easy to check that $\mathcal{N}(A) = \mathcal{N}(A^\dagger A) = \{(I - A^\dagger A)h : h \in \mathbb{C}^n\}$, so the general solution of the consistent system $Ax = AA^\dagger b$ is given by

$$x = A^\dagger b + (I - A^\dagger A)h, \quad h \in \mathbb{C}^n.$$

When $(I - A^\dagger A)h \neq 0$, since

$$
\begin{aligned}
\langle A^\dagger b, (I - A^\dagger A)h \rangle &= b^*(A^\dagger)^*(I - A^\dagger A)h \\
&= b^*(A^\dagger)^*(I - A^\dagger A)^*h \\
&= b^*((I - A^\dagger A)A^\dagger)^*h \\
&= 0,
\end{aligned}
$$

we get $A^\dagger b \perp (I - A^\dagger A)h$, and then

$$
\begin{aligned}
||A^\dagger b||^2 &< ||A^\dagger b||^2 + ||(I - A^\dagger A)h||^2 \\
&= ||A^\dagger b + (I - A^\dagger A)h||^2.
\end{aligned}
$$

Therefore, $x = A^\dagger b$ is the minimum-norm least-squares solution of $Ax = b$. □

## 2.2  Characterizations of Moore-Penrose Inverses of Elements in Semigroups or Rings

In this section, we consider Moore-Penrose inverses of elements in a $*$-semigroup, and of matrices over a $*$-ring.

### 2.2.1 Moore-Penrose Inverses of Elements in a Semigroup

**Definition 2.2.1** An element $a \in S$ is Moore-Penrose invertible if there exists $x \in S$ satisfying the following equations:

$$(1)\ axa = a, \quad (2)\ xax = x, \quad (3)\ (ax)^* = ax, \quad (4)\ (xa)^* = xa.$$

If such an $x$ exists, then it is unique and called the Moore-Penrose inverse of $a$ (denoted by $a^\dagger$).

If an element $x$ satisfies the equations $(i)$, $(j)$, $\cdots$, $(k)$ from above equations (1)–(4), then $x$ is called an $\{i, j, \cdots, k\}$-inverse of $a$ and denoted by $a^{(i,j,\cdots,k)}$. Sometimes a $\{1\}$-inverse of $a$ is also known as an inner inverse and denoted by $a^-$, a $\{2\}$-inverse of $a$ is also known as an outer inverse, and a $\{1, 2\}$-inverse of $a$ is also known as a reflexive inverse (denoted by $a^+$).

The set of all $\{i, j, \cdots, k\}$-inverses of $a$ is denoted by $a\{i, j, \cdots, k\}$. The sets of all invertible elements, Moore-Penrose invertible elements and $\{i, j, \cdots, k\}$-invertible elements in $S$ are denoted by $S^{-1}$, $S^\dagger$ and $S^{\{i,j,\cdots,k\}}$, respectively.

The following lemma was first given by Penrose [102] in complex matrices, it indeed holds in a $*$-semigroup.

**Lemma 2.2.2** Let $a \in S$. Then $a$ is Moore-Penrose invertible if and only if $a \in S^{\{1,3\}} \cap S^{\{1,4\}}$. In this case, $a^\dagger = a^{(1,4)}aa^{(1,3)}$, for any $a^{(1,3)} \in a\{1, 3\}$ and $a^{(1,4)} \in a\{1, 4\}$.

**Proof** If $a$ is Moore-Penrose invertible, then obviously $a \in S^{\{1,3\}} \cap S^{\{1,4\}}$. Conversely, if $a \in S^{\{1,3\}} \cap S^{\{1,4\}}$ with $a^{(1,3)} \in a\{1, 3\}$ and $a^{(1,4)} \in a\{1, 4\}$, let $x = a^{(1,4)}aa^{(1,3)}$. It is easy to verify that $x$ satisfies the equations (1)–(4) in Definition 2.2.1. $\square$

The following result gives a characterization of $\{1, 3\}$ (resp., $\{1, 4\}$)-inverse, which will be frequently used in sequel discussions.

**Lemma 2.2.3 ([52])** Let $a, x \in S$. Then:

(1) $x$ is a $\{1, 3\}$-inverse of $a$ if and only if $x^*a^*a = a$.
(2) $x$ is a $\{1, 4\}$-inverse of $a$ if and only if $aa^*x^* = a$.

**Proof** (1) If $x$ is a $\{1, 3\}$-inverse of $a$, then $a = axa = (ax)^*a = x^*a^*a$. Conversely, if $x^*a^*a = a$, then $ax = x^*a^*ax$. It follows that $ax$ is Hermitian, then we have that $a = x^*a^*a = (ax)^*a = axa$. So $x$ is a $\{1, 3\}$-inverse of $a$.

Similarly, we can prove (2). $\square$

From Lemma 2.2.3, we immediately have the following result.

**Lemma 2.2.4** Let $a \in S$. Then:

(1) $a$ is $\{1, 3\}$-invertible if and only if $Sa^*a = Sa$.
(2) $a$ is $\{1, 4\}$-invertible if and only if $aa^*S = aS$.

(3) *a is Moore-Penrose invertible if and only if $Sa^*a = Sa$ and $aa^*S = aS$.*

We can also characterize a $\{1, 3\}$ (resp., $\{1, 4\}$)-invertible element by a projection.

**Lemma 2.2.5** *Let $a \in S$. Then:*

(1) *$a$ is $\{1, 3\}$-invertible if and only if there exists a unique projection $p$ such that $aS = pS$.*
(2) *$a$ is $\{1, 4\}$-invertible if and only if there exists a unique projection $q$ such that $Sa = Sq$.*

**Proof** (1) If $a$ is $\{1, 3\}$-invertible, suppose that $a^{(1,3)}$ is a $\{1, 3\}$-inverse of $a$. It is easy to verify that $p = aa^{(1,3)}$ is a projection satisfying $aS = pS$. If $p_1S = p_2S = aS$, then $p_1 = p_2$ by Lemma 1.2.16.

Conversely, if there exists a unique projection $p$ such that $aS = pS$, then $a = ps$ and $p = aa'$ for some $s, a' \in S$. Let $x = a'$. Then $ax$ is Hermitian and

$$axa = aa'a = pa = pps = ps = a.$$

Similarly, we can prove (2). □

In 2016, Zhu et al. [135] gave an interesting characterization of Moore-Penrose invertible elements.

**Theorem 2.2.6 ([135, Theorem 2.16])** *Let $a \in S$. Then the following conditions are equivalent:*

(1) *$a$ is Moore-Penrose invertible.*
(2) *$aS = aa^*aS$.*
(3) *$Sa = Saa^*a$.*

*In this case,*

$$a^\dagger = (ax)^* = (ya)^*,$$

*where $aa^*ax = a = yaa^*a$.*

**Proof** (1) $\Rightarrow$ (2). Let $a^\dagger$ be the Moore-Penrose inverse of $a$. Then $a = a(a^\dagger a)^* = aa^*(a^\dagger aa^\dagger)^* = aa^*aa^\dagger(a^\dagger)^*$ and hence $aS = aa^*aS$.

(2) $\Rightarrow$ (1). Suppose that $a = aa^*ax$ for some $x \in S$. Next we show that $w = (ax)^*$ is the Moore-Penrose inverse of $a$. Noting that

$$wa = (ax)^*a = (ax)^*aa^*ax,$$

so $wa$ is Hermitian. This implies that $(ax)^*a = ((ax)^*a)^* = a^*ax$. Then we have that

$$awa = a(ax)^*a = aa^*ax = a.$$

Noting that $a^* = (ax)^*aa^*$, we have

$$aw = a(ax)^* = ax^*a^* = ax^*(ax)^*aa^* = ax^*a^*axa^*.$$

So $aw$ is Hermitian. Noting that $a(ax)^* = (a(ax)^*)^* = axa^*$, then we get

$$waw = (ax)^*a(ax)^* = x^*a^*a(ax)^* = x^*a^*axa^* = x^*(ax)^*aa^* = (ax)^* = w.$$

Similarly, we can prove (1) $\Leftrightarrow$ (3).                                    □

Recall that $a \in S$ is left $*$-cancellable if $a^*ab = a^*ac \Rightarrow ab = ac$ for each $b, c \in S$, right $*$-cancellable if $baa^* = caa^* \Rightarrow ba = ca$ for each $b, c \in S$, and $*$-cancellable if it is both left and right $*$-cancellable. Moore-Penrose invertibility can be characterized as follows.

**Theorem 2.2.7** *For $a \in S$, the following conditions are equivalent:*

(1) *$a$ is Moore-Penrose invertible.*
(2) *$a$ is $*$-cancellable and $a^*a \in S^\dagger$.*
(3) *$a$ is $*$-cancellable and $aa^* \in S^\dagger$.*
(4) *$a$ is $*$-cancellable and $a^*aa^*$ is regular.*
(5) *$a$ is $*$-cancellable and $aa^*$, $a^*a$ are regular.*

*In this case, $a^\dagger = (a^*a)^\dagger a^* = a^*(aa^*)^\dagger = a^*(a^*aa^*)^- a^* = a^*(aa^*)^- a(a^*a)^- a^*$.*

**Proof** The equivalence of (2) and (3) is clear.

(1)$\Rightarrow$(2). Suppose that $a^*ab = a^*ac$. Then we have

$$ab = aa^\dagger ab = (aa^\dagger)^*ab = (a^\dagger)^*a^*ab = (a^\dagger)^*a^*ac = ac,$$

which means that $a$ is left $*$-cancellable. If $baa^* = caa^*$, then

$$ba = baa^\dagger a = ba(a^\dagger a)^* = baa^*(a^\dagger)^* = caa^*(a^\dagger)^* = ca.$$

So $a$ is $*$-cancellable. Since $a \in S^\dagger$, it is easy to check that $(a^*a)^\dagger = a^\dagger(a^*)^\dagger$.

(2)$\Rightarrow$(1). Since $a$ is $*$-cancellable and $a^*a \in S^\dagger$, let $x = (a^*a)^\dagger a^*$, then we have

$$xax = (a^*a)^\dagger a^*a(a^*a)^\dagger a^* = (a^*a)^\dagger a^* = x,$$

$$(ax)^* = [a(a^*a)^\dagger a^*]^* = a[(a^*a)^\dagger a^*]^* = a[(a^*a)^*]^\dagger a^* = a(a^*a)^\dagger a^* = ax$$

and

$$(xa)^* = [(a^*a)^\dagger a^*a]^* = (a^*a)^\dagger a^*a = xa.$$

Noting that $a^*a = a^*a(a^*a)^\dagger a^*a$, it follows that $a^*a(1 - (a^*a)^\dagger a^*a) = 0$. Since $a$ is $*$-cancellable, we have that $a(1 - (a^*a)^\dagger a^*a) = 0$. So $axa = a(a^*a)^\dagger a^*a = a$.

(1)$\Rightarrow$(4). It suffices to prove that $a^*aa^*$ is regular. This follows from

$$a^*aa^*((a^\dagger)^*a^\dagger(a^\dagger)^*)a^*aa^* = a^*aa^\dagger aa^* = a^*aa^*.$$

(4)$\Rightarrow$(5). If there exists $c \in S$ such that $a^*aa^* = a^*aa^*ca^*aa^*$, then $aa^* = aa^*ca^*aa^*$ and $a^*a = a^*aa^*ca^*a$ by $a$ being $*$-cancellable. It follows that $aa^*$ and $a^*a$ both are regular.

(5)$\rightarrow$(1) If there exist $x, y \in S$ such that $a^*a = a^*axa^*a$ and $aa^* = aa^*yaa^*$, then $a = axa^*a$ and $a = aa^*ya$ follow from $a$ being $*$-cancellable. So $a \subset aa^*S \cap Sa^*a$, which implies that $a \in S^\dagger$ by Lemma 2.2.4.

The equality $a^\dagger = a^*(a^*aa^*)^- a^* = a^*(aa^*)^- a(a^*a)^- a^*$ follows by Lemmas 2.2.2 and 2.2.3.                                                                          $\square$

## 2.2.2  Moore-Penrose Inverses of Elements in a Ring

In a ring $R$, we have the following important theorem of $\{1\}$-inverses, which is due to Penrose.

**Theorem 2.2.8 ([102, Theorem 2])** *Let* $a, b \in R^{\{1\}}, d \in R$. *Then the equation*

$$axb = d \tag{2.2.1}$$

*is consistent if and only if, for some* $a^{(1)} \in a\{1\}, b^{(1)} \in b\{1\},$

$$aa^{(1)}db^{(1)}b = d, \tag{2.2.2}$$

*in which case the general solution is*

$$x = a^{(1)}db^{(1)} + y - a^{(1)}aybb^{(1)} \tag{2.2.3}$$

*for arbitrary* $y \in R$.

**Proof** If equality (2.2.2) holds, then $x = a^{(1)}db^{(1)}$ is a solution of equation (2.2.1). Conversely, if $x$ is any solution of equation (2.2.1), then

$$aa^{(1)}db^{(1)}b = aa^{(1)}axbb^{(1)}b = axb = d.$$

Moreover, it follows from (2.2.2) that every element $x$ of the form (2.2.3) satisfies (2.2.1). On the other hand, let $x$ be any solution of (2.2.1). Then,

$$x = x + a^{(1)}db^{(1)} - a^{(1)}db^{(1)} = x + a^{(1)}db^{(1)} - a^{(1)}axbb^{(1)},$$

which is of the form (2.2.3).                                                            $\square$

**Corollary 2.2.9** *Let $a \in R^{\{1\}}$ and $a^{(1)} \in a\{1\}$. Then*

$$a\{1\} = \{a^{(1)} + z - a^{(1)}azaa^{(1)} : z \in R\}.$$

*__Proof__* The above set is obtained by writing $y = a^{(1)} + z$ in the set of solution of $axa = a$ as given by Theorem 2.2.8. □

With Lemma 2.2.5 and Theorem 2.2.8, we have the following expression of $\{1, 3\}$-inverses and $\{1, 4\}$-inverses.

**Corollary 2.2.10** *Let $a \in R$.*

(1) *If $a \in R^{\{1,3\}}$ with $a^{(1,3)} \in a\{1, 3\}$, then*

$$a\{1, 3\} = \{a^{(1,3)} + (1 - a^{(1,3)}a)z : z \in R\}.$$

(2) *If $a \in R^{\{1,4\}}$ with $a^{(1,4)} \in a\{1, 4\}$, then*

$$a\{1, 4\} = \{a^{(1,4)} + y(1 - aa^{(1,4)}) : y \in R\}.$$

*__Proof__*

(1) From the proof of Lemma 2.2.5, we know that $x$ is a $\{1, 3\}$-inverse of $a$ if and only if $ax = aa^{(1,3)}$. By Theorem 2.2.8, we have the general solution of $ax = aa^{(1,3)}$ is

$$x = a^{(1,3)}aa^{(1,3)} + s - a^{(1,3)}as$$

for arbitrary $s \in R$. Substituting $z + a^{(1,3)}$ for $s$ in above equality, we obtain the expression of $a\{1, 3\}$.
(2) It can be proved in a similar method.

□

Also, we can characterize a $\{1, 3\}$-(resp., $\{1, 4\}$-) invertible element $a$ by right (resp., left) ideal generated $a$ and the right (resp., left) annihilator of $a^*$.

**Theorem 2.2.11 ([52, Proposition 5])** *Let $a \in R$. Then:*

(1) *$a$ is $\{1, 3\}$-invertible if and only if $R = aR \oplus (a^*)^\circ$. In this case*

$$a\{1, 3\} = \{r + (1 - ra)w : w \in R\},$$

*where $1 = ar + u$ for some $r \in R$ and $u \in (a^*)^\circ$.*
(2) *$a$ is $\{1, 4\}$-invertible if and only if $R = Ra \oplus {}^\circ(a^*)$. In this case*

$$a\{1, 4\} = \{s + w(1 - as) : w \in R\},$$

*where $1 = sa + v$ for some $s \in R$ and $v \in {}^\circ(a^*)$.*

(3)  *a is Moore-Penrose invertible if and only if*

$$R = aR \oplus (a^*)^\circ \ and \ R = Ra \oplus {}^\circ(a^*).$$

## Proof

(1)  If $a$ is $\{1, 3\}$-invertible, then there exists a unique projection $p$ such that $aR = pR$ by Lemma 2.2.5. So $pa = a$ and $p = aa'$ for some $a' \in R$. And for any idempotent $p$, we have $R = pR \oplus (1 - p)R$. It follows that $a^* = a^* p$, hence $a^*(1 - p)R = a^* p(1 - p)R = 0$, which means that $(1 - p)R \subseteq (a^*)^\circ$. Meanwhile, $a^* x = 0$ implies that $px = p^* x = (a')^* a^* x = 0$. So $x = x - px \in (1 - p)R$. This proves that $R = aR \oplus (a^*)^\circ$.

Conversely, if $R = aR \oplus (a^*)^\circ$, then we have that $a^* R = a^* aR$. It follows that $Ra = Ra^* a$, which means that $a$ is $\{1, 3\}$-invertible by Lemma 2.2.4.

Suppose that $1 = ar + u$ for some $r \in R$ and $u \in (a^*)^\circ$. Then $a^* = a^* ar$, it follows that $r \in a\{1, 3\}$ by Lemma 2.2.4. By Corollary 2.2.10, we obtain the expression of $a\{1, 3\}$.

(2)  The proof is similar to (1).

(3)  It follows by (1) and (2).

□

The next result gives a characterization of Moore-Penrose inverses by annihilators and one-sided principal ideals.

**Theorem 2.2.12 ([108, Theorem 2.8])** *Let $a, x \in R$. Then the following statements are equivalent:*

(1)  $x = a^\dagger$.

(2)  $axa = a, \ xR = a^* R \ and \ Rx = Ra^*$.

(3)  $axa = a, \ {}^\circ x = {}^\circ(a^*) \ and \ x^\circ = (a^*)^\circ$.

(4)  $axa = a, \ xR \subseteq a^* R \ and \ Rx \subseteq Ra^*$.

(5)  $axa = a, \ {}^\circ(a^*) \subseteq {}^\circ x \ and \ (a^*)^\circ \subseteq x^\circ$.

**Proof** $(1) \Rightarrow (2)$. Since $x = a^\dagger$, we easily obtain $a^* = xaa^* = a^* ax$ and $x = a^* x^* x = xx^* a^*$. So $xR = a^* R$ and $Rx = Ra^*$.

$(2) \Rightarrow (3) \Rightarrow (4) \Rightarrow (5)$ follows by Lemma 1.2.21.

$(5) \Rightarrow (1)$. Since $a^* x^* a^* = a^*$, we see that $(1 - x^* a^*) \in (a^*)^\circ \subseteq x^\circ$ and $(1 - a^* x^*) \in {}^\circ(a^*) \subseteq {}^\circ x$. Therefore, $x = xx^* a^*$ and $x = a^* x^* x$. This yields $ax = ax(ax)^*$ and $xa = (xa)^* xa$, hence $ax$ and $xa$ are Hermitian. Finally, $xax = x(ax)^* = xx^* a^* = x$. It follows that $x = a^\dagger$.

□

If $p, q \in R$ are idempotents, then arbitrary $a \in R$ can be written as $a = paq + pa(1 - q) + (1 - p)aq + (1 - p)a(1 - q)$ or in the matrix form $a = \begin{bmatrix} a_{11} & a_{12} \\ a_{21} & a_{22} \end{bmatrix}_{p \times q}$, where $a_{11} = paq, a_{12} = pa(1 - q), a_{21} = (1 - p)aq$ and $a_{22} = (1 - p)a(1 - q)$.

**Theorem 2.2.13 ([108, Theorem 2.12])** *Let $a \in R$. Then the following statements are equivalent:*

(1) *$a$ is Moore-Penrose invertible.*
(2) *There exist projections $p, q \in R$ such that $pR = aR$ and $Rq = Ra$.*
(3) *$a \in R^{\{1\}}$ and there exist projections $p, q \in R$ such that $^{\circ}a = {}^{\circ}p$ and $a^{\circ} = q^{\circ}$.*

*If the previous statements are valid then the statements (2) and (3) deal with the same pair of unique projections $p$ and $q$. Moreover, $qa^{(1)}p$ is invariant under the choice of $a^{(1)} \in a\{1\}$ and*

$$a = \begin{bmatrix} a & 0 \\ 0 & 0 \end{bmatrix}_{p \times q}, \quad a^{\dagger} = \begin{bmatrix} qa^{(1)}p & 0 \\ 0 & 0 \end{bmatrix}_{q \times p}. \tag{2.2.4}$$

*Proof* (1)$\Rightarrow$(2). By Lemmas 2.2.2 and 2.2.5.

(2)$\Rightarrow$(3). By Lemma 1.2.21.

(3)$\Rightarrow$(1). From the proof of Lemma 2.2.5, we know that $a = pa = aq$, $p = aa^{(1)}p$ and $q = qa^{(1)}a$. Set $x = qa^{(1)}p$. We have $x = a^{\dagger}$ because

$$\begin{aligned} ax &= aqa^{(1)}p = aa^{(1)}p = p = p^{*}, \\ xa &= qa^{(1)}pa = qa^{(1)}a = q = q^{*}, \\ axa &= pa = a, \\ xax &= qx = x. \end{aligned}$$

Now the invariance of $qa^{(1)}p$ under the choice of $a^{(1)} \in a\{1\}$ follows because it is known that Moore-Penrose inverse is unique when it exists. Note that we have also proved representations (2.2.4) since $a = paq$ and $a^{\dagger} = x = qa^{(1)}p$. The uniqueness of $p$ and $q$ follows by Lemma 2.2.5.                                                    □

The following theorem characterizes a Moore-Penrose invertible element by invertible elements.

**Theorem 2.2.14 ([93, Theorem 1])** *Let $a$ be a regular element in a ring $R$ with inner inverse $a^{-}$. Then the following statements are equivalent:*

(1) *$a^{\dagger}$ exists.*
(2) *$u = a^{*}a + 1 - a^{-}a \in R^{-1}$.*
(3) *$v = aa^{*} + 1 - aa^{-} \in R^{-1}$.*
(4) *$s = a^{-}aa^{*}a + 1 - a^{-}a \in R^{-1}$.*
(5) *$t = aa^{*}aa^{-} + 1 - aa^{-} \in R^{-1}$.*

*In this case,*

$$a^{\dagger} = (au^{-1})^{*} = (v^{-1}a)^{*} = a^{*}a(s^{*}s)^{-1}a^{*} = a^{*}(tt^{*})^{-1}aa^{*}.$$

**Proof** $(1)\Rightarrow(2)$. Let $m = a^-(a^\dagger)^* + 1 - a^-a$ and $n = a^-aa^*a + 1 - a^-a$. We have that

$$mn = (a^-(a^\dagger)^* + 1 - a^-a)(a^-aa^*a + 1 - a^-a)$$
$$= (a^-(a^\dagger aa^\dagger)^* + 1 - a^-a)(a^-aa^*a + 1 - a^-a)$$
$$= (a^-(a^\dagger)^*a^\dagger a + 1 - a^-a)(a^-aa^*a + 1 - a^-a)$$
$$= a^-(a^\dagger)^*a^\dagger aa^-aa^*a + 1 - a^-a$$
$$= a^-a + 1 - a^-a = 1.$$

Let $x = a^-a$ and $y = a^*a - 1$. Then $n = 1 + xy$. By Jacobson's lemma, $1 + yx = a^*a + 1 - a^-a = u$ is left invertible.

Similarly, we can check that

$$(aa^*aa^- + 1 - aa^-)((a^\dagger)^*a^- + 1 - aa^-) = 1.$$

So $v = aa^* + 1 - aa^-$ is right invertible. By Jacobson's lemma, $u = a^*a + 1 - a^-a$ is also right invertible. Thus, $u = a^*a + 1 - a^-a \in R^{-1}$.

$(2)\Rightarrow(1)$. Since $u = a^*a + 1 - a^-a \in R^{-1}$, we have that $au = aa^*a$. Then $a = aa^*au^{-1}$. It follows that $a^\dagger$ exists and $a^\dagger = (au^{-1})^*$ by Theorem 2.2.6.

$(2)\Leftrightarrow(3)\Leftrightarrow(4)\Leftrightarrow(5)$ follows by Jacobson's lemma.                    □

The following result is the one-sided version of Theorem 2.2.14.

**Theorem 2.2.15 ([135, Corollary 3.3])** *Let $a$ be a regular element in a ring $R$ with inner inverse $a^-$. Then the following statements are equivalent:*

(1) *$a^\dagger$ exists.*
(2) *$u = a^*a + 1 - a^-a$ is left invertible.*
(3) *$v = aa^* + 1 - aa^-$ is left invertible.*
(4) *$u = a^*a + 1 - a^-a$ is right invertible.*
(5) *$v = aa^* + 1 - aa^-$ is right invertible.*

**Proof** $(1)\Rightarrow(2)$, (3), (4), (5). It follows by Theorem 2.2.14.

Using Jacobson's lemma, we also have $(2)\Rightarrow(3)$ and $(4)\Rightarrow(5)$.

$(3)\Rightarrow(1)$. As $v = aa^* + 1 - aa^-$ is left invertible (say with a left inverse $v^-$), we have $va = aa^*a$, and then $a = v^-va = v^-aa^*a$. Hence $a$ is Moore-Penrose invertible by Theorem 2.2.6.

$(5)\Rightarrow(1)$. It is similar to $(3)\Rightarrow(1)$.                    □

**Theorem 2.2.16 ([107, Corollary 2.1])** *Let $a \in R$. Then the following conditions are equivalent:*

(1) *$a \in R^\dagger$.*
(2) *There exists a projection $p$ such that $pa = 0$ and $u = aa^* + p \in R^{-1}$.*
(3) *There exists a projection $q$ such that $aq = 0$ and $v = a^*a + q \in R^{-1}$.*

*In this case, $a^\dagger = a^*u^{-1} = v^{-1}a^*$.*

**Proof** (1) $\Rightarrow$ (2). Let $p = 1 - aa^\dagger$. Then $u = aa^* + (1 - aa^\dagger)$, obviously, $p$ is a projection and $pa = 0$. Now, we prove the invertibility of $u$. It is easy to verify that

$$(aa^* + (1 - aa^\dagger))((a^\dagger)^* a^\dagger + (1 - aa^\dagger)) = 1.$$

Noting that $aa^* + (1 - aa^\dagger)$ is Hermitian, so $u = aa^* + p = aa^* + (1 - aa^\dagger) \in R^{-1}$.

(2)$\Rightarrow$(1). Since $u = aa^* + p$ and $pa = 0$, we have $ua = aa^*a + pa = aa^*a$. It follows that $a = u^{-1}aa^*a \in Raa^*a$. So $a \in R^\dagger$ with $a^\dagger = (u^{-1}a)^* = a^* u^{-1}$ by Theorem 2.2.6.

(1)$\Leftrightarrow$(3). Let $q = 1 - a^\dagger a$. Then the equivalence of conditions (1) and (3) can be proved by a similar way. $\square$

**Definition 2.2.17** Let $f \in R$ be an idempotent. An element $p \in R$ is called a range projection of $f$ if $p$ is a projection satisfying

$$pf = f \text{ and } fp = p.$$

The range projection of an idempotent $f$ will be denoted by $f^\perp$.

In view of Corollary 1.2.18 and Lemma 2.2.5, $f$ has such a range projection if and only if $f \in R^{(1,3)}$, in which case $p$ is a range projection of $f$ if and only if $fR = pR$.

**Theorem 2.2.18 ([73, Theorem 2.1])** *Let $f \in R$ be an idempotent. Then the following conditions are equivalent:*

(1) $f + f^* - 1 \in R^{-1}$.
(2) $f^\perp$ and $(f^*)^\perp$ exist.

*The range projections are unique, given by the formulae*

$$f^\perp = f(f + f^* - 1)^{-1}, \quad (f^*)^\perp = (f + f^* - 1)^{-1} f. \tag{2.2.5}$$

**Proof** (1) $\Rightarrow$ (2). Let $g = f + f^* - 1$ be invertible in $R$, and let $p = fg^{-1} = g^{-1} f^*$. We need to show that $p$ is a projection, and $fp = p$, $pf = f$.

$$\begin{aligned}
p^2 &= (fg^{-1})(fg^{-1}) = (g^{-1} f^* f)g^{-1} = fg^{-1} = p, \\
p^* &= (fg^{-1})^* = g^{-1} f^* = fg^{-1} = p, \\
fp &= f^2 g^{-1} = fg^{-1} = p, \\
pf &= fg^{-1} f = ff^* g^{-1} = fgg^{-1} = f.
\end{aligned}$$

This proves that $f^\perp = p$. Interchanging $f$ and $f^*$ in the preceding argument, we get $(f^*)^\perp = f^*(f + f^* - 1)^{-1} = (f + f^* - 1)^{-1} f$ since $(f^*)^\perp$ is Hermitian.

$(2) \Rightarrow (1)$. Let $p = f^{\perp}$ and $q = (f^*)^{\perp}$. Then

$$(p + q - 1)(f + f^* - 1) = pf + (fp)^* - p + (fq)^* + qf^* - q - f - f^* + 1$$
$$= f + p - p + q + f^* - q - f - f^* + 1 = 1.$$

Similarly, $(f + f^* - 1)(p + q - 1) = 1$, which proves that $f + f^* - 1$ is invertible with

$$(f + f^* - 1)^{-1} = f^{\perp} + (f^*)^{\perp} - 1.$$

To prove the uniqueness of $f^{\perp}$, assume that there exists a projection $u$ such that $fu = u$ and $uf = f$. Then

$$u(f + f^* - 1) = uf + (fu)^* - u = f + u - u = f;$$

since $f + f^* - 1 \in R^{-1}$, we have $u = f(f + f^* - 1)^{-1}$. The existence of $(f^*)^{\perp}$ follows by symmetry.                                                                                          □

**Corollary 2.2.19** *Let $R$ be with the GN-property. Then every idempotent has a unique range projection given by (2.2.5).*

**Proof** If $R$ has the GN-property, then for any idempotent $f$,

$$(f + f^* - 1)^2 = 1 + (f^* - f)^*(f^* - f) \in R^{-1},$$

which implies that $f + f^* - 1$ is invertible in $R$.                                                       □

**Theorem 2.2.20 ([73, Theorem 3.1])** *Let $f \in R$ be an idempotent. Then $f$ is Moore-Penrose invertible if and only if $f + f^* - 1 \in R^{-1}$. In this case $f^{\dagger} = (f^*)^{\perp} f^{\perp}$, and $ff^{\dagger} = f^{\perp}$, $f^{\dagger} f = (f^*)^{\perp}$.*

**Proof** Suppose that $f + f^* - 1$ is invertible. By Theorem 2.2.18 the range projections $f^{\perp}$ and $(f^*)^{\perp}$ exist. Let $u = (f^*)^{\perp} f^{\perp}$:

$$fuf = f(f^*)^{\perp} f^{\perp} f = ((f^*)^{\perp} f^*)^* f = f^2 = f,$$
$$ufu = (f^*)^{\perp} f^{\perp} f(f^*)^{\perp} f^{\perp} = (f^*)^{\perp} f^{\perp} ff^{\perp} = (f^*)^{\perp} f^{\perp} = u,$$
$$fu = f(f^*)^{\perp} f^{\perp} = ff^{\perp} = f^{\perp},$$
$$uf = (f^*)^{\perp} f^{\perp} f = (f^*)^{\perp} f = (f^*)^{\perp}.$$

Hence $u = f^{\dagger}$.

Conversely assume that $f$ is Moore-Penrose invertible. A direct verification of the properties of range projections shows that

$$ff^{\dagger} = f^{\perp}, \quad f^{\dagger} f = (f^*)^{\perp}.$$

                                                                                                    □

The following theorem shows when the Moore-Penrose invertibility of an element is equivalent to its regularity.

**Theorem 2.2.21 ([74, Theorem 2])** *Let $R$ be with the GN-property. Then $a \in R$ is Moore-Penrose invertible if and only if $a$ is regular.*

**Proof** Any Moore-Penrose invertible element $a$ is regular as $a = aa^{\dagger}a$.

Suppose that $a$ is regular. Then there exists $b \in R$ such that $aba = a$ and $bab = b$. Set $f = ba$ and $g = ab$. Then $f, g$ are idempotents, $af = a$, $ga = a$, and the Moore-Penrose inverses $f^{\dagger} = (ba)^{\dagger}$ and $g^{\dagger} = (ab)^{\dagger}$ exist by the preceding theorem. Set $r = f^{\dagger}f$, $s = gg^{\dagger}$ and $x = rbs$. We verify that $x = a^{\dagger}$. First note that $ar = (af)r = aff^{\dagger}f = af = a$. Similarly, $sa = a$. Then

$$axa = (ar)b(sa) = aba = a,$$
$$xax = (rbs)a(rbs) = rb(sar)bs = rbabs = rbs = x,$$
$$xa = rbsa = rba = f^{\dagger}f = (f^*)^{\perp}f^{\perp}f = (f^*)^{\perp}f = (f^*)^{\perp},$$
$$ax = arbs = abs = gg^{\dagger} = g(g^*)^{\perp}g^{\perp} = gg^{\perp} = g^{\perp}.$$

This proves the Moore-Penrose invertibility of $a$.                                  □

### 2.2.3  Moore-Penrose Inverses of Matrices over a Ring

We give some characterizations of Moore-Penrose inverses of matrices over a ring $R$.

**Theorem 2.2.22 ([51, Theorem 1])** *Let $R$ be a $*$-ring. Then the following statements are equivalent:*

(1)  $R$ *is regular, and $*$ is $n$-proper for any positive integer $n$.*
(2)  *Every $A \in R^{m \times n}$ has a $\{1, 3\}$-inverse for each positive integers $m, n$.*
(3)  *Every $A \in R^{m \times n}$ has a $\{1, 4\}$-inverse for each positive integers $m, n$.*
(4)  *Every $A \in R^{m \times n}$ has a Moore-Penrose inverse for each positive integers $m, n$.*

**Proof** (1)$\Rightarrow$(4). Since $R^{n \times n}$ is a regular ring (Proposition 1.4.4), every $A$ in $R^{m \times n}$ has a $\{1\}$-inverse. This follows simply by embedding each $A$ in a square matrix by adding zeros and then using block multiplication, so (4) follows by Theorem 2.2.7.

(4)$\Rightarrow$(2),(3). It is trivial.

(2)$\Rightarrow$(1). By assumption, for every $a \in R$ has a $\{1\}$-inverse, so $R$ is a regular ring. Suppose $A \in R^{m \times n}$ with $A^*A = 0$. Let $X$ be a $\{1,3\}$-inverse of $A$. Then

$$A = AXA = (AX)^*A = X^*A^*A = 0,$$

hence $*$ is $n$-proper for any positive integer $n$.

(3)$\Rightarrow$(1). It is similar to (2)$\Rightarrow$(1).                          □

**Theorem 2.2.23 ([51, Theorem 2])** *The following statements are equivalent for* $A \in R^{m \times n}$:

(1) $A\{1, 3\} \neq \emptyset$ *or, equivalently,* $A\{1, 2, 3\} \neq \emptyset$.
(2) *There exists a unique projection* $P \in R^{m \times m}$ *such that* $\mathcal{R}(A) = \mathcal{R}(P)$.
(3) $R^m = \mathcal{R}(A) \oplus \mathcal{N}(A^*)$.
(4) $R^m = \mathcal{R}(A) + \mathcal{N}(A^*)$.
(5) $\mathcal{R}(A) = \mathcal{R}(A^*A)$.

*In this case,* $(A^*A)^- A^* \in A\{1, 2, 3\}$.

**Proof** $(1) \Rightarrow (2)$. Assume $X \in A\{1, 3\}$. Let $P = AX$. We observe first that $P^2 = P = P^*$ such that $\mathcal{R}(A) = \mathcal{R}(P)$. If $Q^2 = Q = Q^*$ with $\mathcal{R}(A) = \mathcal{R}(Q)$. Let $I_m = [e_1, \cdots, e_m]$ and $I_n = [f_1, \cdots, f_n]$. Then there exist $y_i \in R^n$ and $z_j \in R^m$ such that $Qe_i = Ay_i$, $Af_j = Qz_j$. It is clear that $Q = AY$, $A = QZ$ in which $Y = [y_1, \cdots, y_m]$, $Z = [z_1, \cdots, z_n]$, hence $QA = Q^2 Z = QZ = A$,

$$Q = Q^* = (AY)^* = Y^*(AXA)^* = (AY)^*(AX)^* = Q^* AX = QAX = AX = P.$$

$(2) \Rightarrow (3)$. Let $R^m = \mathcal{R}(P) \oplus \mathcal{R}(1 - P)$. Now, Since $R(A) = R(P)$, we have $A = PA$ and $P = AY$ for some $Y \in R^{n \times m}$, hence

$$A^*(I - P) = (PA)^*(I - P) = A^* P(I - P) = 0$$

(i.e., $\mathcal{R}(I - P) \subseteq \mathcal{N}(A^*)$). Also $A^* y = 0$ implies

$$Py = P^* y = (AY)^* y = Y^* A^* y = 0.$$

So $y = (I - P)y \in R(I - P)$, which implies that $\mathcal{N}(A^*) \subseteq \mathcal{R}(I - P)$. Thus $\mathcal{N}(A^*) = \mathcal{R}(I - P)$, hence (3) follows.

$(3) \Rightarrow (4)$. Obviously.

$(4) \Rightarrow (5)$. By hypothesis, there exist $x_i \in R^n$ and $y_i \in \mathcal{N}(A^*)$ such that $e_i = Ax_i + y_i$, hence $I = AX + Y$, where $X = [x_1, \cdots, x_m]$, and $Y = [y_1, \cdots, y_m]$. So $A^* = A^* AX + A^* Y = A^* AX$, and thus $\mathcal{R}(A^*) = \mathcal{R}(A^*A)$.

$(5) \Rightarrow (1)$. Since $\mathcal{R}(A^*) = \mathcal{R}(A^*A)$, there exists $X$ with $A^* = A^* AX$, hence $XAX \in A\{1, 2, 3\}$, and (1) follows.

Finally we prove that $(A^*A)^- A^* \in A\{1, 2, 3\}$. Assume $X \in A\{1, 3\}$, then $A = X^* A^* A$, $A^* = A^* AX$. Notice that $XX^* \in A^* A\{1\}$, i.e., $(A^*A)\{1\} \neq \emptyset$, since

$$A^* A = A^* A(A^*A)^- A^* A.$$

Multiplying on the left by $X^*$ and on the right by $X$, respectively. We obtain

$$A = A(A^*A)^- A^* A, \quad A^* = A^* A(A^*A)^- A^*.$$

Suppose $Z = (A^*A)^- A^*$. Then $Z \in A\{1\}$, and

$$(AZ)^* = (A(A^*A)^- A^*)^* = A(A(A^*A)^-)^* = A(A^*A)^- A^* A(A(A^*A)^-)^*$$

$$= A(A^*A)^- (A(A^*A)^- A^*A)^* = A(A^*A)^- A^* = AZ,$$

$$ZAZ = (A^*A)^- A^* A(A^*A)^- A^* = (A^*A)^- A^* = Z,$$

hence $Z \in A\{1, 2, 3\}$. Thus we complete the proof.                    □

Dually, we have the following theorem.

**Theorem 2.2.24 ([51, Theorem 3])** *The following statements are equivalent for* $A \in R^{m \times n}$:

(1) $A\{1, 4\} \neq \emptyset$ *or, equivalently,* $A\{1, 2, 4\} \neq \emptyset$.
(2) *There exists a unique projection* $Q \in R^{n \times n}$ *such that* $\mathcal{R}(A^*) = \mathcal{R}(Q)$.
(3) $R^n = \mathcal{R}(A^*) \oplus \mathcal{N}(A)$.
(4) $R^m = \mathcal{R}(A^*) + \mathcal{N}(A)$.
(5) $\mathcal{R}(A) = \mathcal{R}(AA^*)$.

*In this case,* $A^*(AA^*)^- \in A\{1, 2, 4\}$.

**Theorem 2.2.25 ([51, Theorem 4])** *The following statements are equivalent for* $A \in R^{m \times n}$:

(1) $A^\dagger$ *exists.*
(2) $R^m = \mathcal{R}(A) \oplus \mathcal{N}(A^*)$ *and* $R^n = \mathcal{R}(A^*) \oplus \mathcal{N}(A)$.
(3) $R^m = \mathcal{R}(A) + \mathcal{N}(A^*)$ *and* $R^m = \mathcal{R}(A^*) + \mathcal{N}(A)$.
(4) $\mathcal{R}(A^*) = \mathcal{R}(A^*A)$ *and* $\mathcal{R}(A) = \mathcal{R}(AA^*)$.

*In this case,*

$$A^\dagger = (A^*A)^{(1,4)} A^* = A^*(AA^*)^{(1,3)} = A^*(A^*A)^- A(A^*A)^- A^*$$

$$= A^*(A^*AA^*)^- A^*.$$

***Proof*** We prove only that $A^\dagger = A^*(A^*AA^*)^- A^*$. Now, let $Y = A^*(A^*AA^*)^- A^*$, and $X = A^\dagger$. Then

$$A = X^*A^*A = AA^*X^*, \quad A^* = A^*AX = XAA^*$$

and it follows that

$$AYA = AA^*(A^*AA^*)^- A^*A = X^*A^*AA^*(A^*AA^*)^- A^*AA^*X^*$$

$$= X^*A^*AA^*X^* = AA^*X^* = A,$$

$$YAY = YAA^*(A^*AA^*)^- A^* = YAA^*(A^*AA^*)^- A^*AX = YAX$$

$$= A^*(A^*AA^*)^- A^*AX = Y,$$

$$(AY)^* = (AA^*(A^*AA^*)^- A^*)^* = AYA(AA^*(A^*AA^*)^-)^*$$

$$= AA^*(A^*AA^*)^- A^*A(AA^*(A^*AA^*)^-)^*$$

$$= AA^*(A^*AA^*)^- (AA^*(A^*AA^*)^- A^*A)^*$$

$$= AA^*(A^*AA^*)^- A^* = AY.$$

Analogously, it can be shown that $(YA)^* = YA$, hence $A^\dagger = A^*(A^*AA^*)^- A^*$.  □

**Theorem 2.2.26 ([51, Theorem 5])** *Let $R$ be a right FP-injective ring, $A \in R^{m \times n}$. Then:*

(1) $A\{1, 3\} \neq \emptyset$ *if and only if* $\mathcal{N}(A) = \mathcal{N}(A^*A)$.
(2) $A\{1, 4\} \neq \emptyset$ *if and only if* $\mathcal{N}(A^*) = \mathcal{N}(AA^*)$.
(3) *A has a Moore-Penrose inverse if and only if* $\mathcal{N}(A) = \mathcal{N}(A^*A)$ *and* $\mathcal{N}(A^*) = \mathcal{N}(AA^*)$.

*Proof*

(1)  Assume $X \in A\{1, 3\}$, then

$$A = AXA = (AX)^*A = X^*A^*A,$$

hence $\mathcal{N}(A) = \mathcal{N}(A^*A)$.
    Conversely, assume $\mathcal{N}(A) = \mathcal{N}(A^*A)$.

(a)  If $m \geq n$, we can verify that

$$[A, 0] \in l(\mathbf{r}\begin{bmatrix} A^*A & 0 \\ 0 & 0 \end{bmatrix}) = R^{m \times m}\begin{bmatrix} A^*A & 0 \\ 0 & 0 \end{bmatrix}$$

by right FP-injectivity. So, there exists $[X, Z] \in R^{m \times m}$ such that

$$[A, 0] = [X, Z]\begin{bmatrix} A^*A & 0 \\ 0 & 0 \end{bmatrix} = [XA^*A, 0],$$

hence $A = XA^*A$ and $A^* = A^*AX^*$. Consequently, $\mathcal{R}(A^*) = \mathcal{R}(A^*A)$.
(b)  If $m \leq n$. It is easy to verify that

$$\begin{bmatrix} A \\ 0 \end{bmatrix} \in l(\mathbf{r}(A^*A)) = R^{n \times n}A^*A$$

by right FP-injectivity. So, there exists $\begin{bmatrix} X \\ Z \end{bmatrix} \in R^{n \times n}$ such that

$$\begin{bmatrix} A \\ 0 \end{bmatrix} = \begin{bmatrix} X \\ Z \end{bmatrix} A^* A = \begin{bmatrix} X A^* A \\ Z A^* A \end{bmatrix},$$

hence $A = X A^* A$ and $A^* = A^* A X^*$. Consequently, $\mathcal{R}(A^*) = \mathcal{R}(A^* A)$.

From (a) and (b), we obtain $A\{1, 3\} \neq \emptyset$ by Lemma 2.2.4.

(2)  Notice that $A\{1, 4\} \neq \emptyset$ if and only if $A^*\{1, 3\} \neq \emptyset$.

(3)  Obviously.

□

**Theorem 2.2.27 ([51, Theorem 6])**  *Let $A \in R^{m \times n}$. Then:*

(1)  *If $A\{1, 3\} \neq \emptyset$, then $A^* A$ is invertible if and only if $\mathcal{N}(A) = 0$.*

(2)  *If $A\{1, 4\} \neq \emptyset$, then $AA^*$ is invertible if and only if $\mathcal{N}(A^*) = 0$.*

(3)  *If $A$ has a Moore-Penrose inverse, then $A^* A + I - A^\dagger A$ and $AA^* + I - AA^\dagger$ are invertible.*

*Moreover,*

$$A^\dagger = A^*(AA^* + I - AA^\dagger)^{-1} = (A^* A + I - A^\dagger A)^{-1} A^*.$$

*Proof*

(1)  Assume that $A^* A$ is invertible, it is clear that $\mathcal{N}(A) = 0$. Conversely, if $\mathcal{N}(A) = 0$, $X \in A\{1, 3\}$. From the equation $A(XA - I) = 0$, we obtain

$$I = XA = XAXA = X(AX)^* A = XX^* X^* A,$$

$$I = I^* = (XX^* A^* A)^* = A^* A XX^*,$$

hence $A^* A$ is invertible.

(2)  It is similar to (1).

(3)  Let $X = AA^* + I - AA^\dagger$ and $Y = (A^\dagger)^* A^\dagger + I - AA^\dagger$. Then $X^* = X$, $Y^* = Y$ and

$$XY = (AA^* + I - AA^\dagger)((A^\dagger)^* A^\dagger + I - AA^\dagger)$$

$$= AA^*(A^\dagger)^* A^\dagger + AA^*(I - AA^\dagger)^* + (I - AA^\dagger)^*(A^\dagger)^* A^\dagger + I - AA^\dagger$$

$$= A(A^\dagger A)^* A^\dagger + A((I - AA^\dagger)A)^* + (A^\dagger(I - AA^\dagger))^* A^\dagger + I - AA^\dagger$$

$$= AA^\dagger AA^\dagger + I - AA^\dagger = I,$$

$$YX = Y^* X^* = (XY)^* = I^* = I.$$

Consequently, $X$ is invertible. Since

$$A^\dagger(AA^* + I - AA^\dagger) = (A^\dagger A)^* A^* + A^\dagger(I - AA^\dagger) = (AA^\dagger A)^* = A^*,$$

it follows that

$$A^\dagger = A^*(AA^* + I - AA^\dagger)^{-1}.$$

Similarly, $A^*A + I - A^\dagger A$ is invertible and

$$A^\dagger = (A^*A + I - A^\dagger A)^{-1} A^*.$$

$\square$

Finally, let $*$ be an involution in ordinary sense. Then we have

**Theorem 2.2.28 ([51, Theorem 7])** *The following statements are equivalent for $A \in R^{m \times n}$:*

(1) $I + A^*A$ *is invertible.*

(2) $\begin{bmatrix} I \\ A \end{bmatrix}$ *has a $\{1, 3\}$-inverse (Moore-Penrose inverse).*

(3) $I + AA^*$ *is invertible.*

(4) $[I, \ A]$ *has a $\{1, 4\}$-inverse (Moore-Penrose inverse).*

**Proof** $(1) \Rightarrow (2)$. Suppose $M = \left[(I + A^*A)^{-1}, (I + A^*A)^{-1} A^*\right]$. It is easy to verify that $M$ is the Moore-Penrose inverse of $\begin{bmatrix} I \\ A \end{bmatrix}$.

$(2) \Rightarrow (1)$. Suppose $N = \begin{bmatrix} I \\ A \end{bmatrix}$, then

$$N = \begin{bmatrix} I & 0 \\ A & I \end{bmatrix}\begin{bmatrix} I \\ 0 \end{bmatrix} = P\begin{bmatrix} I \\ 0 \end{bmatrix}.$$

We may assume that $K = (I, X)P^{-1} \in N\{1, 3\}$, then

$$NK = P\begin{bmatrix} I \\ 0 \end{bmatrix}[I, X]P^{-1} = P_1(p_1 + Xp_2),$$

where $P = [P_1, P_2]$ and $P^{-1} = \begin{bmatrix} p_1 \\ p_2 \end{bmatrix}$, hence

$$P_1(p_1 + Xp_2) = (P_1(p_1 + Xp_2))^* = (p_1 + Xp_2)^* P_1^*.$$

Notice that $p_2 P_1 = 0$, $p_1 P_1 = I$, we obtain

$$I = p_1 P_1 = p_1 P_1 (p_1 + X p_2) P_1 = p_1 (p_1 + X p_2)^* P_1^* P_1,$$
$$I = I^* = P_1^* P_1 (p_1 + X p_2) p_1^*,$$

hence $P_1^* P_1$ is invertible. Since

$$I + A^* A = \begin{bmatrix} I^*, A^* \end{bmatrix} \begin{bmatrix} I \\ A \end{bmatrix} = P_1^* P_1,$$

we have (1).

(3)$\Leftrightarrow$(4). Using the same way as (1)$\Leftrightarrow$(2).

The equivalence of (1) and (3) depends on the following two formulae

$$(I + A^* A)^{-1} = I - A^* (I + A A^*)^{-1} A,$$

$$(I + A A^*)^{-1} = I - A (I + A^* A)^{-1} A^*.$$

$\square$

**Corollary 2.2.29** *The following statements are equivalent for any ring R:*

(1) *R has the GN-property.*

(2) $\begin{bmatrix} 1 \\ a \end{bmatrix}$ *has the Moore-Penrose inverse for each $a \in R$.*

(3) $\begin{bmatrix} 1, a \end{bmatrix}$ *has the Moore-Penrose inverse for each $a \in R$.*

## 2.3 The Moore-Penrose Inverse of a Product

### 2.3.1 The Moore-Penrose Inverse of a Product paq

The existence criterion and representation for $\{1, 3\}$-inverses, $\{1, 4\}$-inverses and the Moore-Penrose inverse of a product *paq* are discussed in this section.

First, we consider the $\{1, 4\}$-invertibility of *paq* when $a \in R^{\{1,4\}}$.

**Theorem 2.3.1 ([111, Theorem 3.8])** *Let $p, a, q \in R$ with $p' p a = a = a q q'$ for some $p', q' \in R$. If $a \in R^{\{1,4\}}$, then the following conditions are equivalent:*

(1) $paq \in R^{\{1,4\}}$.

(2) $v = a^{(1,4)} a q q^* + 1 - a^{(1,4)} a \in R^{-1}$.

*In this case, $q^* v^{-1} a^{(1,4)} p' \in (paq)\{1, 4\}$.*

**Proof** $(1) \Rightarrow (2)$. Suppose that $x$ is a $\{1, 4\}$-inverse of $paq$, then $paq(paq)^*x^* = paq$ by Lemma 2.2.3. We have

$$
(a^{(1,4)}aqq^*a^{(1,4)}a + 1 - a^{(1,4)}a)(a^*p^*x^*q' + 1 - a^{(1,4)}a)
$$
$$
= a^{(1,4)}aqq^*a^{(1,4)}aa^*p^*x^*q' + 1 - a^{(1,4)}a
$$
$$
= a^{(1,4)}aqq^*a^*p^*x^*q' + 1 - a^{(1,4)}a
$$
$$
= a^{(1,4)}p'paqq^*a^*p'x'q' + 1 - a^{(1,4)}a
$$
$$
= a^{(1,4)}p'paqq' + 1 - a^{(1,4)}a
$$
$$
= a^{(1,4)}aqq' + 1 - a^{(1,4)}a
$$
$$
= 1.
$$

Noting that $a^{(1,4)}aqq^*a^{(1,4)}a + 1 - a^{(1,4)}a$ is Hermitian, it is invertible. Let $x = a^{(1,4)}aqq^* - 1$ and $y = a^{(1,4)}a$. Then $1 + xy = a^{(1,4)}aqq^*a^{(1,4)}a + 1 - a^{(1,4)}a$. As $1 + xy \in R^{-1}$, we have $1 + yx = a^{(1,4)}aqq^* + 1 - a^{(1,4)}a = v \in R^{-1}$ by Theorem 1.2.19.

$(2) \Rightarrow (1)$. Since $v \in R^{-1}$, it follows that $v^* = qq^*a^{(1,4)}a + 1 - a^{(1,4)}a \in R^{-1}$. Multiplying by $a$ on the left of $v^* = qq^*a^{(1,4)}a + 1 - a^{(1,4)}a$ yields that $av^* = aqq^*a^{(1,4)}a$, and consequently $a = aqq^*a^{(1,4)}a(v^{-1})^*$. Then we have

$$
paq = paqq^*a^{(1,4)}a(v^{-1})^*q
$$
$$
= paqq^*a^*(a^{(1,4)})^*(v^{-1})^*q
$$
$$
= paqq^*a^*p^*(p')^*(a^{(1,4)})^*(v^{-1})^*q \in paq(paq)^*R.
$$

Thus, by Lemma 2.2.3, $paq \in R^{\{1,4\}}$ with $q^*v^{-1}a^{(1,4)}p' \in (paq)\{1, 4\}$.                    □

**Corollary 2.3.2**  Let $a, q \in R$ with $a = aqq'$ for some $q' \in R$. If $a \in R^{\{1,4\}}$, then the following conditions are equivalent:

(1) $aq \in R^{\{1,4\}}$.
(2) $v = a^{(1,4)}aqq^* + 1 - a^{(1,4)}a \in R^{-1}$.

In this case, $q^*v^{-1}a^{(1,4)} \in (aq)\{1, 4\}$.

Dual to Theorem 2.3.1, we have the following result.

**Theorem 2.3.3 ([111, Theorem 4.3])**  Let $p, a, q \in R$ with $p'pa = a = aqq'$ for some $p', q' \in R$. If $a \in R^{\{1,3\}}$, then the following conditions are equivalent:

(1) $paq \in R^{\{1,3\}}$.
(2) $u = p^*paa^{(1,3)} + 1 - aa^{(1,3)} \in R^{-1}$.

In this case, $q'a^{(1,3)}u^{-1}p^* \in (paq)\{1, 3\}$.

**Proof** The proof of this theorem is dual to Theorem 2.3.1 and so is omitted. $\qquad\square$

**Corollary 2.3.4** *Let* $p, a \in R$ *with* $p'pa = a$ *for some* $p' \in R$. *If* $a \in R^{\{1,3\}}$, *then the following conditions are equivalent:*

(1) $pa \in R^{\{1,3\}}$.
(2) $u = p^* paa^{(1,3)} + 1 - aa^{(1,3)} \in R^{-1}$.

*In this case,* $a^{(1,3)}u^{-1}p^* \in (pa)\{1, 3\}$.

**Theorem 2.3.5** *Let* $p, a, q \in R$ *with* $p'pa = a = aqq'$ *for some* $p', q' \in R$. *If* $a \in R^\dagger$, *then the following statements are equivalent:*

(1) $paq \in R^\dagger$.
(2) $u = p^* paa^\dagger + 1 - aa^\dagger \in R^{-1}$ *and* $v = a^\dagger aqq^* + 1 - a^\dagger a \in R^{-1}$.

*In this case,* $(paq)^\dagger = q^* v^{-1}a^\dagger u^{-1}p^*$.

**Proof** It follows by the above discussion. $\qquad\square$

Next, we discuss the $\{1, 4\}$-invertibility of $paq$ when $a \in R^{\{1,3\}}$.

**Theorem 2.3.6 ([16, Theorem 3.2])** *Let* $p, a, q \in R$ *with* $p'pa = a = aqq'$ *for some* $p', q' \in R$. *If* $a \in R^{\{1,3\}}$, *then the following conditions are equivalent:*

(1) $paq \in R^{\{1,4\}}$.
(2) $s = aq(aq)^* + 1 - aa^{(1,3)} \in R^{-1}$.

*In this case,*

$$(aq)^* s^{-1}p' \in (paq)\{1, 4\}.$$

**Proof** $(1)\Rightarrow(2)$. Suppose that $x$ is a $\{1, 4\}$-inverse of $paq$, then $xpaq(paq)^* = (paq)^*$ by Lemma 2.2.3. We have

$$((a^{(1,3)})^*(q')^* xpaa^{(1,3)} + 1 - aa^{(1,3)})(aq(aq)^* + 1 - aa^{(1,3)})$$
$$= (a^{(1,3)})^*(q')^* xpaa^{(1,3)} aq(aq)^* + 1 - aa^{(1,3)}$$
$$= (a^{(1,3)})^*(q')^* xpaq(paq)^*(p')^* + 1 - aa^{(1,3)}$$
$$= (a^{(1,3)})^*(q')^*(paq)^*(p')^* + 1 - aa^{(1,3)}$$
$$= (a^{(1,3)})^*(q')^*(aq)^* + 1 - aa^{(1,3)}$$
$$= (aa^{(1,3)})^* + 1 - aa^{(1,3)}$$
$$= 1.$$

Noting that $s = aq(aq)^* + 1 - aa^{(1,3)}$ is Hermitian, it is invertible.

$(2) \Rightarrow (1)$. Multiplying by $aa^{(1,3)}$ on the left of $s = aq(aq)^* + 1 - aa^{(1,3)}$ yields that $aa^{(1,3)}s = aq(aq)^*$, and consequently $aa^{(1,3)} = aq(aq)^*s^{-1}$. Then we have

$$paq = paa^{(1,3)}aq$$
$$= paq(aq)^*s^{-1}aq$$
$$= paq(paq)^*(p')^*s^{-1}aq \in paq(paq)^*R.$$

Thus, by Lemma 2.2.3, $paq \in R^{\{1,4\}}$ with $(aq)^*s^{-1}p' \in (paq)\{1,4\}$.                 ⊔

**Corollary 2.3.7** *Let* $a, q \in R$ *with* $a = aqq'$ *for some* $q' \in R$. *If* $a \in R^{\{1,3\}}$, *then the following conditions are equivalent:*

(1) $aq \in R^{\{1,4\}}$.
(2) $s = aq(aq)^* + 1 - aa^{(1,3)} \in R^{-1}$.

*In this case,*

$$(aq)^*s^{-1} \in (aq)\{1, 4\}.$$

Dual to Theorem 2.3.6, the following result is obtained.

**Theorem 2.3.8** ([16, Theorem 3.2]) *Let* $p, a, q \in R$ *with* $p'pa = a = aqq'$ *for some* $p', q' \in R$. *If* $a \in R^{\{1,4\}}$, *then the following conditions are equivalent:*

(1) $paq \in R^{\{1,3\}}$.
(2) $t = (pa)^*pa + 1 - a^{(1,4)}a \in R^{-1}$.

*In this case,*

$$q't^{-1}(pa)^* \in (paq)\{1, 3\}. \tag{2.3.1}$$

**Proof** The proof of this theorem is dual to the one of Theorem 2.3.6 and so is omitted.                 □

**Corollary 2.3.9** *Let* $p, a \in R$ *with* $p'pa = a$ *for some* $p' \in R$. *If* $a \in R^{\{1,4\}}$, *then the following conditions are equivalent:*

(1) $pa \in R^{\{1,3\}}$.
(2) $t = (pa)^*pa + 1 - a^{(1,4)}a \in R^{-1}$.

*In this case,*

$$t^{-1}(pa)^* \in (pa)\{1, 3\}.$$

With above results, we can give equivalent conditions of the Moore-Penrose invertibility of $paq$ as follows.

**Theorem 2.3.10 ([16, Corollary 3.4])** *Let* $p, a, q \in R$. *If* $a \in R^\dagger$, *then the following conditions are equivalent:*

(1) $paq \in R^\dagger$ *and there exist* $p', q' \in R$ *such that* $p'pa = a = aqq'$.
(2) $u = p^*paa^\dagger + 1 - aa^\dagger \in R^{-1}$ *and* $v = a^\dagger aqq^* + 1 - a^\dagger a \in R^{-1}$.
(3) $u = p^*paa^\dagger + 1 - aa^\dagger \in R^{-1}$ *and* $s = aq(aq)^* + 1 - aa^\dagger \in R^{-1}$.
(4) $t = (pa)^*pa + 1 - a^\dagger a \in R^{-1}$ *and* $v = a^\dagger aqq^* + 1 - a^\dagger a \in R^{-1}$.
(5) $t = (pa)^*pa + 1 - a^\dagger a \in R^{-1}$ *and* $s = aq(aq)^* + 1 - aa^\dagger \in R^{-1}$.

*In this case,*

$$(paq)^\dagger = q^*v^{-1}a^\dagger u^{-1}p^* = (aq)^*s^{-1}aa^\dagger u^{-1}p^*$$
$$= q^*v^{-1}a^\dagger at^{-1}(pa)^* = (aq)^*s^{-1}at^{-1}(pa)^*.$$

***Proof*** (1)$\Rightarrow$(2), (3), (4), (5). By Theorems 2.3.1, 2.3.3, 2.3.6, 2.3.8 and Lemma 2.2.2.

(2)$\Rightarrow$(1). Let $p' = u^{-1}p^*$ and $q' = q^*v^{-1}$. Then

$$p'pa = u^{-1}p^*pa = u^{-1}(p^*paa^\dagger + 1 - aa^\dagger)a = u^{-1}ua = a$$

and

$$aqq' = aqq^*v^{-1} = a(a^\dagger aqq^* + 1 - a^\dagger a)v^{-1} = avv^{-1} = a.$$

Thus, $paq \in R^\dagger$ by Theorems 2.3.1, 2.3.3 and Lemma 2.2.2.

(5)$\Rightarrow$(1). Let $p' = at^{-1}(pa)^*$ and $q' = (aq)^*s^{-1}a$. Then

$$p'pa = at^{-1}(pa)^*pa = at^{-1}((pa)^*pa + 1 - a^\dagger a)a^\dagger a = at^{-1}ta^\dagger a = a$$

and

$$aqq' = aq(aq)^*s^{-1}a = aa^\dagger(aq(aq)^* + 1 - aa^\dagger)s^{-1}a = aa^\dagger ss^{-1}a = a.$$

Thus, $paq \in R^\dagger$ by Theorems 2.3.6, 2.3.8 and Lemma 2.2.2.

(3)$\Rightarrow$(1) and (4)$\Rightarrow$(1) can be proved similarly. □

We consider the Moore-Penrose invertibility of $paq$ when $a$ is regular.

**Theorem 2.3.11** *Let* $p, a, q \in R$. *If* $a$ *is regular with* $a^- \in a\{1\}$. *Then the following conditions are equivalent:*

(1) $u = aq(paq)^*paa^- + 1 - aa^- \in R^{-1}$.
(2) $v = a^-aq(paq)^*pa + 1 - a^-a \in R^{-1}$.
(3) $paq \in R^\dagger$ *and there exist* $p', q' \in R$ *such that* $p'pa = a = aqq'$.

*Moreover,*

$$(paq)^\dagger = (pu^{-1}aq)^* = (pav^{-1}q)^*.$$

**Proof** (1) ⇔ (2). By Jacobson's lemma.

(3) ⇒ (1). Since $paq \in R^\dagger$, there exists $x \in R$ such that $paq = paq(paq)^*paqx$ by Theorem 2.2.6.

Note that

$$\begin{aligned}
&\left(aq(paq)^*paa^- + 1 - aa^-\right)\left(aqxq'a^- + 1 - aa^-\right) \\
&= aq(paq)^*paqxq'a^- + 1 - aa^- \\
&= p'paq(paq)^*paqxq'a^- + 1 - aa^- \\
&= p'paqq'a^- + 1 - aa^- \\
&= aqq'a^- + 1 - aa^- \\
&= aa^- + 1 - aa^- \\
&= 1,
\end{aligned}$$

which means that $u$ is right invertible.

Note also that $paq \in R^\dagger$ implies $(paq)^* = ypaq(paq)^*paq$ for some $y \in R$ from Theorem 2.2.6. Then $a^-p'ypa + 1 - a^-a$ is a left inverse of $v = a^-aq(paq)^*pa + 1 - a^-a$. Jacobson's lemma guarantees that $u$ is left invertible.

(1) ⇒ (3). Since $u \in R^{-1}$ implies $v \in R^{-1}$, let $p' = av^{-1}a^-ap(paq)^*$ and $q' = (paq)^*qaa^-u^{-1}a$. Then

$$\begin{aligned}
p'pa &= av^{-1}a^-ap(paq)^*pa = av^{-1}(a^-ap(paq)^*pa + 1 - a^-a)a^-a \\
&= av^{-1}va^-a = a
\end{aligned}$$

and

$$\begin{aligned}
aqq' &= aq(paq)^*qaa^-u^{-1}a = aa^-(aq(paq)^*qaa^- + 1 - aa^-)u^{-1}a \\
&= aa^-uu^{-1}a = a.
\end{aligned}$$

Since $u$ is invertible, we get $ua = aq(paq)^*pa$, and consequently $a = u^{-1}aq(paq)^*pa$. Multiplying by $p$ on the left and by $q$ on the right yield

$$paq = pu^{-1}aq(paq)^*paq = pu^{-1}p'paq(paq)^*paq.$$

Thus, by Theorem 2.2.6, $paq \in R^\dagger$ and $(paq)^\dagger = (pu^{-1}aq)^*$. Similarly, we can prove that $(paq)^\dagger = (pav^{-1}q)^*$.                                                                 □

**Theorem 2.3.12** *Let $p, a, q \in R$. If there exist $p', q' \in R$ such that $p'pa = a = aqq'$, then $paq \in R^\dagger$ if and only if $pa \in R^{\{1,3\}}$ and $aq \in R^{\{1,4\}}$. In this case,*

$$(paq)^\dagger = (aq)^{(1,4)}a(pa)^{(1,3)}.$$

**Proof** If $pa \in R^{\{1,3\}}$ and $aq \in R^{\{1,4\}}$, then

$$aq = aq(aq)^*[(aq)^{(1,4)}]^*$$

and

$$pa = [(pa)^{(1,3)}]^*(pa)^* pa$$

by Lemma 2.2.4. Substituting $a$ respectively by $p'pa$ and $aqq'$ in above equalities, we have that

$$paq = paq(paq)^*(p')^*[(aq)^{(1,4)}]^*$$

and

$$paq = [(pa)^{(1,3)}]^*(q')^*(paq)^* paq.$$

It follows that $paq \in R^{\{1,4\}}$ with $(aq)^{(1,4)} p' \in (paq)\{1, 4\}$ and $paq \in R^{\{1,3\}}$ with $q'(pa)^{(1,3)} \in (paq)\{1, 3\}$. By Lemma 2.2.2, we have $paq \in R^\dagger$ with

$$(paq)^\dagger = (paq)^{(1,4)} paq(paq)^{(1,3)} = (aq)^{(1,4)} a(pa)^{(1,3)}.$$

Conversely, if $paq \in R^\dagger$, then $paq = paq(paq)^*x$ for some $x$. Then $aq = p'paq = aq(aq)^* p^*x$, it follows that $aq \in R^{\{1,4\}}$. Similarly, we can prove that $pa \in R^{\{1,3\}}$.                                                                                   □

### 2.3.2   The Moore-Penrose Inverse of a Matrix Product

Let $P$, $A$ and $Q$ be matrices over a ring $R$. We explore the existence of $(PA)^{(1,3)}$, $(AQ)^{(1,4)}$ and the Moore-Penrose invertibility of $PAQ$ from the view-point of invertible matrices.

It will be convenient to introduce the following sets. For any idempotent $e \in R$, we consider

$$eRe + 1 - e = \{exe + 1 - e : x \in R\},$$

which is a submonoid of $R$ under multiplication, and the group $U_e$ of e-units in the subring $eRe$ (corner ring) given by

$$U_e = \{exe : exeR = eR, Rexe = Re\}.$$

The next known result links invertible elements in $eRe + 1 - e$ and elements of $U_e$.

**Lemma 2.3.13 ([15, Lemma 2.1])** *Let $a \in R$ and $e = e^2 \in R$. Then the following conditions are equivalent:*

(1) $e \in eaeR \cap Reae$.
(2) $eae + 1 - e$ *is invertible.*
(3) $ae + 1 - e$ *is invertible.*
(4) $eae \in U_e$.

*In this case, the $e$-inverse of $eae$ in $U_e$ is given by*

$$(eae)^{-1}_{eRe} = e(eae + 1 - e)^{-1}e.$$

We can now present our main result of this section. We begin with a characterization for the matrix product $PA$ to have a $\{1, 3\}$-inverse when $A\{1, 3\} \neq \emptyset$.

**Theorem 2.3.14 ([15, Theorem 3.1])** *Let $A \in R^{m \times n}$ be $\{1, 3\}$-invertible, $E = AA^{(1,3)}$, and let $P \in R^{m \times m}$ be invertible. If $P(I - E) = I - E$ then the following conditions are equivalent:*

(1) $PA$ *has a $\{1, 3\}$-inverse.*
(2) $E \in R^{m \times m} Z \cap Z R^{m \times m}$ *where* $Z = EP^*PE$.
(3) $U = P^*PE + I - E$ *is invertible.*
(4) $I + S^*S$ *is invertible with* $S = (I - E)(I - P^{-1})$.

*In this case, there exists a $\{1, 3\}$-inverse of $PA$ of the form*

$$(PA)^{(1,3)} = A^{(1,3)}U^{-1}P^* = A^{(1,3)}P^{-1}(I + S^*S)^{-1}(I + S^*). \qquad (2.3.2)$$

**Proof** Observe that if $X$ is an arbitrary element of $A\{1, 3\}$, then $X = A^{(1,3)} + (I - A^{(1,3)}A)Z$ with $Z \in R^{n \times m}$ and, thus, $AX = AA^{(1,3)}$. Therefore, if $P(I - E) = I - E$ we also have $P(I - AX) = I - AX$.

(1) $\Rightarrow$ (2). If $PA$ has a $\{1, 3\}$-inverse $Y$ then $Y^*A^*P^* = PAY$ and $PA = PAYPA$. Hence, $E = AA^{(1,3)} = AYPE = P^{-1}(PAY)PE = SEP^*PE$, where $S = P^{-1}Y^*A^*$. Since $E^* = E$, it also follows that $E = EP^*PES^* \in EP^*PER^{m \times m}$. Dually, we can prove $E \in R^{m \times m}EP^*PE$.

(2) $\Leftrightarrow$ (3). This equivalence follows from Lemma 2.3.13.

(3) $\Leftrightarrow$ (4). Let $S = (I - E)(I - P^{-1})$. Using $S^2 = 0$, we can write

$$I + S^*S = I + S^* - S^*P^{-1} = (I + S^*)(I - S^*P^{-1})$$

$$= (I + S^*)(PE + (P^{-1})^*(I - E))P^{-1} = (I + S^*)(P^{-1})^* UP^{-1},$$

where $U = P^*PE + I - E$. Since $I + S^*$ is invertible, $I + S^*S$ is invertible if and only if $U$ is invertible. Moreover,

$$(I + S^*S)^{-1} = PU^{-1}P^*(I - S^*).$$

From this, we also obtain that $U^{-1}P^* = P^{-1}(I + S^*S)^{-1}(I + S^*)$ whenever (3) holds and, thus, the second equality of (2.3.2) holds.

(3) $\Rightarrow$ (1). Define $Y = A^{(1,3)}U^{-1}P^*$. We will prove that $Y$ is a $\{1, 3\}$-inverse of $PA$. Firstly, we see that

$$YPA = A^{(1,3)}U^{-1}P^*PA = A^{(1,3)}U^{-1}\left(P^*PE + I - E\right)A = A^{(1,3)}A.$$

Then $PAYPA = PA$, and thus, $Y$ is a $\{1\}$-inverse of $PA$. Since $U^*E = EU$ it follows that $(PAY)^* = P\left(U^{-1}\right)^*EP^* = PEU^{-1}P^* = PAY$ and so $Y \in (PA)\{1, 3\}$.                                    □

Now we state an analogue of the above theorem concerning the $\{1, 4\}$-inverse of the matrix product $AQ$.

**Theorem 2.3.15 ([15, Theorem 3.4])** *Let $A \in R^{m \times n}$ be $\{1, 4\}$-invertible, $F = A^{(1,4)}A$, and let $Q \in R^{n \times n}$ be invertible. If $(I - F)Q = I - F$, then the following conditions are equivalent:*

(1) *$AQ$ has a $\{1, 4\}$-inverse.*
(2) *$F \in R^{n \times n}W \cap WR^{n \times n}$ where $W = FQQ^*F$.*
(3) *$V = FQQ^* + I - F$ is invertible.*
(4) *$I + TT^*$ is invertible with $T = (I - Q^{-1})(I - F)$.*

*In this case, there exists a $\{1, 4\}$-inverse of $AQ$ of the form*

$$(AQ)^{(1,4)} = Q^*V^{-1}A^{(1,4)} = \left(I + T^*\right)\left(I + TT^*\right)^{-1}Q^{-1}A^{(1,4)}. \qquad (2.3.3)$$

***Proof*** We first note that $\left(A^{(1,4)}\right)^*$ is a $\{1, 3\}$-inverse of $A^*$ and that

$$Q^*\left(I - A^*\left(A^{(1,4)}\right)^*\right) = I - A^*\left(A^{(1,4)}\right)^*.$$

An application of Theorem 2.3.14 to the product $Q^*A^*$ shows that the following conditions are equivalent:

(1') *$Q^*A^*$ has a $\{1, 3\}$-inverse.*
(2') *$F \in R^{n \times n}FQQ^*F \cap FQQ^*FR^{n \times n}$.*
(3') *$U = QQ^*F + I - F$ is invertible.*
(4') *$I + S^*S$ is invertible with $S = (I - F)\left(I - (Q^*)^{-1}\right)$.*

From these relations, we conclude that (1), (2), (3), and (4) in this theorem are equivalent. Finally, by (2.3.2) we have $Y = \left(A^{(1,4)}\right)^*U^{-1}Q = \left(A^{(1,4)}\right)^*(Q^*)^{-1}(I + S^*S)^{-1}(I + S^*)$ is a $\{1, 3\}$-inverse of $(AQ)^*$. Hence, $Y^*$ is a $\{1, 4\}$-inverse of $AQ$, and thus (2.3.3) holds.                                    □

Based on previous results, we derive a characterization of existence of the Moore-Penrose inverse of a matrix product $PAQ$ in the case that $A^\dagger$ exists.

**Theorem 2.3.16 ([15, Theorem 3.6])** *Let $A \in R^{m \times n}$ be such that $A^{\dagger}$ exists, let $E = AA^{\dagger}$, $F = A^{\dagger}A$ and let $P \in R^{m \times m}$ and $Q \in R^{n \times n}$ be invertible matrices. If $P(I - E) = I - E$ and $(I - F)Q = I - F$, then the following conditions are equivalent:*

(1) $(PAQ)^{\dagger}$ *exists.*
(2) $E \in R^{m \times m} Z \cap ZR^{m \times m}$ *and* $F \in R^{n \times n} W \cap WR^{n \times n}$, *where* $Z = EP^*PE$ *and* $W = FQQ^*F$.
(3) $U = P^*PE + I - E$ *and* $V = FQQ^* + I - F$ *are invertible.*
(4) $I + S^*S$ *and* $I + TT^*$ *are invertible with* $S = (I - E)(I - P^{-1})$ *and* $T = (I - Q^{-1})(I - F)$.

*In this case,*

$$(PAQ)^{\dagger} = Q^*V^{-1}A^{\dagger}U^{-1}P^*$$

$$= \left(I + T^*\right)\left(I + TT^*\right)^{-1} Q^{-1}A^{\dagger}P^{-1}\left(I + S^*S\right)^{-1}\left(I + S^*\right).$$

$$(2.3.4)$$

**Proof** We know that the Moore-Penrose inverse of $PAQ$ exists if and only if $PA$ has a $\{1, 3\}$-inverse and $AQ$ has a $\{1, 4\}$-inverse, in which case

$$(PAQ)^{\dagger} = (AQ)^{(1,4)}A(PA)^{(1,3)}.$$

Now, the proof of the theorem is a consequence of Theorems 2.3.14 and 2.3.15.    □

## 2.4   Moore-Penrose Inverses of Differences and Products of Projections

Moore-Penrose inverses of sums, differences, and products of projections in various settings attract wide interest.

Let us remind the reader that, in what follows, $p$ and $q$ are always two projections; $\bar{p} = 1 - p$ and $\bar{q} = 1 - q$. We also fix the notations $a = pqp$, $b = pq\bar{p}$, $d = \bar{p}q\bar{p}$.

The following lemma will be used in the sequel.

**Lemma 2.4.1**

(1) $bb^* = (p - a) - (p - a)^2$.
(2) $b^*b = d - d^2$.
(3) $db^* = b^*(p - a)$.
(4) $b = ab + bd$.

**Proof** By a direct verification.                                                                                    □

## Lemma 2.4.2 ([129, Lemma 3])

(1) If $p\overline{q} \in R^\dagger$, then $p - a = p\overline{q}p \in R^\dagger$ and $(p - a)(p - a)^\dagger b = b$.
(2) If $\overline{p}q \in R^\dagger$, then $d = \overline{p}q\overline{p} \in R^\dagger$ and $bdd^\dagger = b$.
(3) If $p\overline{q}, \overline{p}q \in R^\dagger$, then $bd^\dagger = (p - a)^\dagger b$ and $d^\dagger b^* = b^*(p - a)^\dagger$.
(4) If $p\overline{q}, \overline{p}q \in R^\dagger$, then $p - q \in R^\dagger$ and $(p - q)^\dagger = \overline{q}(p\overline{q}p)^\dagger - q(\overline{p}q\overline{p})^\dagger$.

## Proof

(1) Since $p\overline{q} \in R^\dagger$, we have $p - a = p\overline{q}p = (p\overline{q})(p\overline{q})^* \in R^\dagger$. Moreover,

$$
\begin{aligned}
(p - a)(p - a)^\dagger b &= (p\overline{q})(p\overline{q})^*[(p\overline{q})(p\overline{q})^*]^\dagger(pq\overline{p}) \\
&= (p\overline{q})(p\overline{q})^*[(p\overline{q})^\dagger]^*[(p\overline{q})]^\dagger(pq\overline{p}) \\
&= -p\overline{q}(p\overline{q})^\dagger p\overline{q}\,\overline{p} = -p\overline{q}\,\overline{p} \\
&= pq\overline{p} = b.
\end{aligned}
$$

(2) Since $\overline{p}q \in R^\dagger$, it follows that $d = \overline{p}q\overline{p} = (\overline{p}q)(\overline{p}q)^* \in R^\dagger$. Replacing $p$ and $q$ by $\overline{p}$ and $\overline{q}$ respectively in (1), one can see that $dd^\dagger \overline{p}\,\overline{q}p = (\overline{p}q\overline{p})(\overline{p}q\overline{p})^\dagger\overline{p}\,\overline{q}p \overset{(1)}{=} \overline{p}\,\overline{q}p$. Hence $dd^\dagger b^* = -dd^\dagger\overline{p}\,\overline{q}p = -\overline{p}\,\overline{q}p = b^*$. Therefore, $bdd^\dagger = b$.

(3) By Lemma 2.4.1(4), one can see that $bd^\dagger = (ab + bd)d^\dagger = abd^\dagger + bdd^\dagger \overset{(2)}{=} abd^\dagger + b$ and hence $b = bd^\dagger - abd^\dagger = (p - a)bd^\dagger$. Consequently, we have $(p - a)^\dagger b = (p - a)^\dagger(p - a)bd^\dagger \overset{(1)}{=} bd^\dagger$, which implies that $d^\dagger b^* = b^*(p - a)^\dagger$ since $(d^\dagger)^* = d^\dagger$ and $[(p - a)^\dagger]^* = (p - a)^\dagger$.

(4) Let $z = (p - a)(p - a)^\dagger - bd^\dagger - d^\dagger b^* - dd^\dagger$. In view of Lemma 2.4.1(1), (2) and Lemma 2.4.2(1)–(3), it is straightforward to check $p - q \in R^\dagger$ with $(p - q)^\dagger = z$. Now, we have

$$
\begin{aligned}
(p - q)^\dagger &= z = (p - a)(p - a)^\dagger - bd^\dagger - d^\dagger b^* - dd^\dagger \\
&\overset{(3)}{=} (p - a)(p - a)^\dagger - bd^\dagger - b^*(p - a)^\dagger - dd^\dagger \\
&= [(p - a) - b^*](p - a)^\dagger - (b + d)d^\dagger \\
&= [p(1 - q)p + (1 - p)(1 - q)p](p\overline{q}p)^\dagger - q(1 - p)(\overline{p}q\overline{p})^\dagger \\
&= \overline{q}(p\overline{q}p)^\dagger - q(\overline{p}q\overline{p})^\dagger. \qquad\qquad (2.4.1)
\end{aligned}
$$

This completes the proof.

□

## Theorem 2.4.3 ([129, Theorem 4]) *The following statements are equivalent for any two projections $p, q \in R$:*

   (1) $1 - pq \in R^\dagger$.   (2) $1 - pqp \in R^\dagger$.   (3) $p - pqp \in R^\dagger$.
   (4) $1 - qp \in R^\dagger$.   (5) $1 - qpq \in R^\dagger$.   (6) $q - qpq \in R^\dagger$.

**Proof** (1)⇔(4) is clear by $1 - qp = (1 - pq)^*$. We need only prove that (1)–(3) are equivalent.

(1)⇒(3). Let $x = (1 - pq)^\dagger$. Then we have

$$(1 - p)x(1 - pq)p = (1 - p)(1 - pq)x(1 - pq)p$$
$$= (1 - p)(1 - pq)p = 0. \tag{2.4.2}$$

Consequently,

$$px(1 - pq)(1 - p) = [(1 - p)x(1 - pq)p]^* = 0. \tag{2.4.3}$$

On the other hand, by a direct verification, we have

$$p(1 - pq)xp = p(1 - pq)[p + (1 - p)]xp$$
$$= p(1 - pq)pxp + p(1 - pq)(1 - p)xp$$
$$= (p - pqp)pxp - pq(1 - p)xp. \tag{2.4.4}$$

Now, we verify $p - pqp \in R^\dagger$ and $(p - pqp)^\dagger = pxp$.

*Step 1.* $(p - pqp)pxp(p - pqp) = p - pqp$. Indeed,

$$(p - pqp)pxp(p - pqp)$$
$$\overset{(2.4.2)}{=\!=\!=} (p - pqp)pxp(1 - pq)p - pq[(1 - p)x(1 - pq)p]$$
$$= [(p - pqp)pxp - pq(1 - p)xp](1 - pq)p$$
$$\overset{(2.4.4)}{=\!=\!=} p(1 - pq)xp(1 - pq)p$$
$$= p(1 - pq)x(1 - pq)p$$
$$= p(1 - pq)p = p - pqp. \tag{2.4.5}$$

*Step 2.* $[(p - pqp)pxp]^* = (p - pqp)pxp$.
Actually, (2.4.2) implies

$$(1 - p)x^*p - (1 - p)qpx^*p = (1 - p)(1 - qp)x^*p = (1 - p)[x(1 - pq)]^*p$$
$$= (1 - p)x(1 - pq)p = 0.$$

Hence

$$(1 - p)x^*p = (1 - p)qpx^*p. \tag{2.4.6}$$

Meanwhile,

$$(p - pqp)px^*p(p - pqp) = [(p - pqp)pxp(p - pqp)]^*$$

$$\overset{(2.4.5)}{=\!=\!=} [p - pqp]^* = p - pqp,$$

that is,

$$(p - a)px^*p(p - a) = p - a. \tag{2.4.7}$$

In view of (2.4.6), (2.4.7) and Lemma 2.4.1, one can get

$$
\begin{aligned}
pq(1 - p)xp &= pq(1 - p)(1 - pq)xp = pq(1 - p)x^*(1 - qp)p \\
&= pq(1 - p)x^*p(1 - qp)p + pq(1 - p)x^*(1 - p)(1 - qp)p \\
&\overset{(2.4.6)}{=\!=\!=} pq(1 - p)qpx^*p(1 - q)p - pq(1 - p)x^*(1 - p)qp \\
&= bb^*x^*(p - a) - pq(1 - p)x^*(1 - qp)(1 - p)qp \\
&= [(p - a) - (p - a)^2]x^*(p - a) - b(1 - pq)xb^* \\
&\overset{(2.4.7)}{=\!=\!=} (p - a) - (p - a)^2 - b(1 - pq)xb^*,
\end{aligned}
$$

where $(p - a)^* = p - a$ and $(b(1 - pq)xb^*)^* = b(1 - pq)xb^*$. This guarantees

$$[pq(1 - p)xp]^* = pq(1 - p)xp. \tag{2.4.8}$$

So we have

$$
\begin{aligned}
[(p - pqp)pxp]^* &\overset{(2.4.4)}{=\!=\!=} [p(1 - pq)xp + pq(1 - p)xp]^* \\
&\overset{(2.4.8)}{=\!=\!=} p(1 - pq)xp + pq(1 - p)xp \\
&\overset{(2.4.4)}{=\!=\!=} (p - pqp)pxp.
\end{aligned}
$$

*Step 3.* $[pxp(p - pqp)]^* = [px(1 - pq)p]^* = px(1 - pq)p = pxp(p - pqp)$.
*Step 4.* $pxp(p - pqp)pxp = pxp$. Indeed,

$$
\begin{aligned}
pxp(p - pqp)pxp &\overset{(2.4.3)}{=\!=\!=} px(1 - pq)pxp + [px(1 - pq)(1 - p)]xp \\
&= px(1 - pq)[p + (1 - p)]xp \\
&= px(1 - pq)xp = pxp.
\end{aligned}
$$

$(3) \Rightarrow (1)$. Suppose $p - a = p - pqp \in R^\dagger$. Let

$$x = [1 + b^*(p - a)](p - a)^\dagger(1 + b) - b^* - b^*b + 1 - p.$$

We will show that $(1-pq)^\dagger = x$ in accordance with the definition of Moore-Penrose inverse.

Firstly, note that $(p-a)^\dagger \in pRp$ and

$$(1-pq)x$$
$$= (1-pq)p(p-a)^\dagger(1+b) + (1-pq)b^*(p-a)(p-a)^\dagger(1+b)$$
$$\quad (1 \quad pq)b^* - (1-pq)b^*b + (1-pq)(1-p)$$
$$= (p-a)(p-a)^\dagger(1+b) + b^*(p-a)(p-a)^\dagger(1+b)$$
$$-bb^*(p-a)(p-a)^\dagger(1+b) - b^* + bb^* - b^*b + bb^*b + 1 - p - b,$$

where $bb^* = (p-a) - (p-a)^2$ (see Lemma 2.4.1). Hence

$$(1-pq)x = (p-a)(p-a)^\dagger + (p-a)(p-a)^\dagger b + b^*(p-a)(p-a)^\dagger$$
$$+b^*(p-a)(p-a)^\dagger b - b^*b - b^* - b + 1 - p. \tag{2.4.9}$$

Now, it is straightforward to check

$$[(1-pq)x]^* = (1-pq)x. \tag{2.4.10}$$

Similarly, we have

$$x(1-pq)$$
$$= (p-a)^\dagger(1-pq+b) + b^*(p-a)(p-a)^\dagger(1-pq+b)$$
$$-b^*(1-pq+b) + 1 - p$$
$$= (p-a)^\dagger(p-a) + b^*(p-a) - b^*(1-pqp) + 1 - p$$
$$= (p-a)^\dagger(p-a) + 1 - p, \tag{2.4.11}$$

from which it is easy to see that

$$[x(1-pq)]^* = x(1-pq). \tag{2.4.12}$$

Secondly, it follows from (2.4.9) that

$$(1-pq)x(1-pq)$$
$$= (p-a)(p-a)^\dagger(1-pq+b) + b^*(p-a)(p-a)^\dagger(1-pq+b)$$
$$-b^*(1-pq+b) - b + 1 - p$$
$$= (p-a) + b^*(p-a) - b^*(p-a) - b + 1 - p$$
$$= p - a - b + 1 - p = 1 - pq. \tag{2.4.13}$$

Finally, we have

$$x(1-pq)x$$

$$\overset{(2.4.11)}{=\!=\!=} [(p-a)^\dagger(p-a)+1-p]([1+b^*(p-a)](p-a)^\dagger(1+b)$$

$$-b^*-b^*b+1-p)$$

$$= [1+b^*(p-a)](p-a)^\dagger(1+b)-b^*-b^*b+1-p=x. \quad (2.4.14)$$

Combining (2.4.10), (2.4.12)–(2.4.14), one can see that

$$(1-pq)^\dagger = x = [1+b^*(p-a)](p-a)^\dagger(1+b)-b^*-b^*b+1-p.$$
$$(2.4.15)$$

(2)$\Rightarrow$(3). Since $(1-pqp)^* = 1-pqp \in R^\dagger$ and $p(1-pqp) = (1-pqp)p$, we have $p(1-pqp)^\dagger = (1-pqp)^\dagger p$. One can check $(p-pqp)^\dagger = p(1-pqp)^\dagger$. (3)$\Rightarrow$(2). It is trivial to verify that $(1-pqp)^\dagger = (p-pqp)^\dagger + 1 - p$.  □

**Corollary 2.4.4** *The following conditions are equivalent for any two projections $p$ and $q$ in a $*$-proper ring $R$:*
  (1) $1-pq \in R^\dagger$.  (2) $1-pqp \in R^\dagger$.  (3) $p-pqp \in R^\dagger$.
  (4) $p-pq \in R^\dagger$.  (5) $p-qp \in R^\dagger$.   (6) $1-qp \in R^\dagger$.
  (7) $1-qpq \in R^\dagger$.  (8) $q-qpq \in R^\dagger$.  (9) $q-qp \in R^\dagger$.
  (10) $q-pq \in R^\dagger$.
*Moreover, $(p-pqp)^\dagger = (1-pq)^\dagger p$ when any one of these conditions is satisfied.*

**Proof** (1)$\Leftrightarrow$(2)$\Leftrightarrow$(3)$\Leftrightarrow$(6) has been proved in Theorem 2.4.3. We will prove (3)$\Leftrightarrow$(4)$\Leftrightarrow$(5).
  (4)$\Leftrightarrow$(5) is obvious since $p-pq = (p-qp)^*$.
  (3)$\Leftrightarrow$(4). Since $R$ is $*$-proper and $p-pqp = p(1-q)p = p\overline{q}(p\overline{q})^*$, it is easy to see that $p(1-q) \in R^\dagger$ if and only if $p-pqp \in R^\dagger$.
  Moreover, if any one of the above conditions is satisfied, then we have

$$(1-pq)^\dagger p \overset{(2.4.15)}{=\!=\!=} \{[1+b^*(p-a)](p-a)^\dagger(1+b)-b^*-b^*b+1-p\}p$$

$$= (p-a)^\dagger + b^*(p-a)(p-a)^\dagger - b^*$$

$$= (p-a)^\dagger = (p-pqp)^\dagger. \quad \text{(see Lemma 2.4.2)}$$

This completes the proof.  □

One can see that (4)$\Rightarrow$(3) and (5)$\Rightarrow$(3) in Corollary 2.4.4 are valid even if $R$ is not $*$-proper. However, the following example shows that (3) does not imply (4) or (5) in general if $R$ is not $*$-proper.

**Example 2.4.5** Let $R$ be the ring in Example 1.3.11 and let $p = x, q = 1 - y$. Then $p, q$ are projections. In addition, $p(1 - q)p = xyx = 0 \in R^{\dagger}$. But $p(1 - q) = xy \notin p(1 - q)[p(1 - q)]^{*}R = \{0\}$. Therefore, $p(1 - q) \notin R^{\dagger}$.

Replacing $p$ and $q$ by $\bar{p} = 1 - p$ and $\bar{q} = 1 - q$ respectively in Corollary 2.4.4, we obtain the following corollary.

**Corollary 2.4.6** *The following statements are equivalent for any two projections $p$ and $q$ in a $*$-proper ring $R$:*

  (1) $p + q - pq \in R^{\dagger}$. (2) $p + \bar{p}q\bar{p} \subset R^{\dagger}$. (3) $\bar{p}q\bar{p} \in R^{\dagger}$.
  (4) $q - pq \in R^{\dagger}$.     (5) $q - qp \in R^{\dagger}$.    (6) $p + q - qp \in R^{\dagger}$.
  (7) $q + \bar{q}p\bar{q} \in R^{\dagger}$. (8) $\bar{q}p\bar{q} \in R^{\dagger}$. (9) $p - qp \in R^{\dagger}$.
  (10) $p - pq \in R^{\dagger}$.

**Theorem 2.4.7 ([129, Theorem 8])** *The following statements are equivalent for any two projections $p$ and $q$ in $R$:*

(1)  $p(1 - q) \in R^{\dagger}$ *and* $(1 - p)q \in R^{\dagger}$.
(2)  $p - q \in R^{\dagger}$.

*Proof* (1)$\Rightarrow$(2). By Lemma 2.4.2(4).

(2)$\Rightarrow$(1). It follows that $(p - q)^{\dagger}(p - q) = (p - q)(p - q)^{\dagger}$ since $p - q \in R^{\dagger}$ and $(p - q)^{*} = p - q$. Now, it is easy to check that $[(p - q)^{2}]^{\dagger} = [(p - q)^{\dagger}]^{2}$. Also note that $[(p - q)^{2}]^{*} = (p - q)^{2}$ and $p(p - q)^{2} = (p - q)^{2}p = p - pqp$. We have $p[(p - q)^{2}]^{\dagger} = [(p - q)^{2}]^{\dagger}p$, i.e., $p[(p - q)^{\dagger}]^{2} = [(p - q)^{\dagger}]^{2}p$. Let $x = (p - q)^{\dagger}p$. We will prove that $(p\bar{q})^{\dagger} = x$.

First, $(p\bar{q})x = p(p - q)(p - q)^{\dagger}p$ implies $[(p\bar{q})x]^{*} = (p\bar{q})x$.
Moreover, we have

$$(p\bar{q})x(p\bar{q}) = p(p - q)(p - q)^{\dagger}pp(p - q)$$
$$= p(p - q)(p - q)(p - q)^{\dagger}(p - q)^{\dagger}pp(p - q)$$
$$= p(p - q)(p - q)(p - q)^{\dagger}(p - q)^{\dagger}(p - q)$$
$$= p(p - q) = p\bar{q}.$$

Furthermore, $x(p\bar{q}) = (p - q)^{\dagger}p(p - q)$ implies

$$[x(p\bar{q})]^{*} = [(p - q)^{\dagger}p(p - q)]^{*} = (p - q)p(p - q)^{\dagger}$$
$$= (p - q)p(p - q)^{\dagger}(p - q)^{\dagger}(p - q)$$
$$= (p - q)(p - q)^{\dagger}(p - q)^{\dagger}p(p - q)$$
$$= (p - q)^{\dagger}p(p - q) = x(p\bar{q}),$$

and hence

$$x(p\overline{q})x = [x(p\overline{q})]^*x = [(p-q)^\dagger p(p-q)]^*(p-q)^\dagger p$$
$$= (p-q)p(p-q)^\dagger(p-q)^\dagger p$$
$$= (p-q)(p-q)^\dagger(p-q)^\dagger p$$
$$= (p-q)^\dagger p = x.$$

This proves $(p\overline{q})^\dagger = (p-q)^\dagger p$.  (2.4.16)

Thus, $\overline{p} - \overline{q} = -(p-q) \in R^\dagger$ implies $\overline{p}q \in R^\dagger$.  □

In case $R$ is $*$-proper, $p(1-q) \in R^\dagger$ if and only if $(1-p)q \in R^\dagger$ by Corollary 2.4.4(3)$\Leftrightarrow$(9). Whence we have the following corollary.

**Corollary 2.4.8** *The following conditions are equivalent for any two projections $p$ and $q$ in a $*$-proper ring $R$:*

(1) $p(1-q) \in R^\dagger$.
(2) $p-q \in R^\dagger$.
(3) $(1-p)q \in R^\dagger$.

Let $p$ and $q$ be projections in a $*$-proper ring $R$. It is clear that all the conditions in Corollaries 2.4.4, 2.4.6 and 2.4.8 are mutually equivalent. The following corollary shows the relations among those Moore-Penrose inverses when they exist.

**Corollary 2.4.9** *Let $p$ and $q$ be projections in a $*$-proper ring $R$. If any one of the conditions in Corollaries 2.4.4, 2.4.6 and 2.4.8 is satisfied, then*

(1) $(1-pq)^\dagger p\overline{q}p = p\overline{q}(p\overline{q}p)^\dagger = p\overline{q}p(1-qp)^\dagger = p(p\overline{q})^\dagger = (\overline{q}p)^\dagger p = p(p-q)^\dagger p$.
(2) $(p+\overline{p}q)^\dagger \overline{p}q\overline{p} = (q+p\overline{q})^\dagger \overline{p}q\overline{p} = \overline{p}q(\overline{p}q\overline{p})^\dagger = \overline{p}q\overline{p}(p+q\overline{p})^\dagger = \overline{p}(\overline{p}q)^\dagger = (q\overline{p})^\dagger\overline{p} = \overline{p}(q-p)^\dagger\overline{p}$.

*Proof*

(1) First of all, we have $(p\overline{q}p)^\dagger = (p-pqp)^\dagger = (1-pq)^\dagger p$ by Corollary 2.4.4. Hence

$$(1-pq)^\dagger p\overline{q}p = (p\overline{q}p)^\dagger p\overline{q}p = p\overline{q}p(p\overline{q}p)^\dagger = p\overline{q}(p\overline{q}p)^\dagger.$$

Consequently,

$$p\overline{q}(p\overline{q}p)^\dagger = p\overline{q}p(p\overline{q}p)^\dagger = [p\overline{q}p(p\overline{q}p)^\dagger]^* = [(1-pq)^\dagger p\overline{q}p]^* = p\overline{q}p(1-qp)^\dagger.$$

Next, it follows by (2.4.1) that

$$p\overline{q}(p\overline{q}p)^\dagger = p\overline{q}(p\overline{q}p)^\dagger p = p[\overline{q}(p\overline{q}p)^\dagger - q(\overline{p}q\overline{p})^\dagger]p \stackrel{(2.4.1)}{=\!=\!=} p(p-q)^\dagger p.$$

Finally, from (2.4.16) we can see that $p(p\overline{q})^{\dagger} = p(p-q)^{\dagger}p = (\overline{q}p)^{\dagger}p$.

(2)  Replace $p$ and $q$, respectively, by $\overline{p}$ and $\overline{q}$ in (1).

$\square$

## 2.5   Jacobson's Lemma for Moore-Penrose Inverses

It is well known as Jacobson's lemma that, if $1 - ab$ is invertible, then so is $1 - ba$, for any $a, b \in R$. Moreover, $(1 - ba)^{-1} = 1 + b(1 - ab)^{-1}a$. In the next example we show that the existence of the Moore-Penrose inverse of $1 - ab$ does not imply the existence of the Moore-Penrose inverse of $1 - ba$.

**Example 2.5.1 ([13, Example 3.10])** Let $R = \mathbb{C}^{2 \times 2}$ and $*$ be the transpose of matrices. Let $A = \begin{bmatrix} 0 & -i \\ 1 & 0 \end{bmatrix}$ and $B = \begin{bmatrix} 1 & 0 \\ 0 & i \end{bmatrix}$. Then $I - AB = \begin{bmatrix} 1 & -1 \\ -1 & 1 \end{bmatrix}$ is Moore-Penrose invertible with $(I - AB)^{\dagger} = \begin{bmatrix} 1/4 & -1/4 \\ -1/4 & 1/4 \end{bmatrix}$. But $I - BA = \begin{bmatrix} 1 & i \\ -i & 1 \end{bmatrix}$ is not Moore-Penrose invertible since $(I - BA)^*(I - BA) = 0$.

### 2.5.1   Jacobson's Lemma for Moore-Penrose Inverses in a (Generalized) GN Ring

We shall call $R$ a GN ring if it has the GN-property with respect to the involution $*$.

**Proposition 2.5.2 ([130, Proposition 1])** *Let $R$ be a $*$-ring and define the involution on $R^{2 \times 2}$ by $(a_{ij})^* = (a^*_{ji})$. Then $R$ is a GN ring if Jacobson's lemma for Moore-Penrose inverses holds in $R^{2 \times 2}$.*

**Proof** For any $a \in R$, let $A = \begin{bmatrix} 0 & 1 \\ 1 & a \end{bmatrix}$ and $B = \begin{bmatrix} 0 & 1 \\ 0 & 0 \end{bmatrix} \in R^{2 \times 2}$. Then $AB = \begin{bmatrix} 0 & 0 \\ 0 & 1 \end{bmatrix}$ and $(I - AB)^{\dagger} = \begin{bmatrix} 1 & 0 \\ 0 & 0 \end{bmatrix}$. By hypothesis, $(I - BA)^{\dagger}$ also exists. Let $C = I - BA = \begin{bmatrix} 0 & -a \\ 0 & 1 \end{bmatrix}$ and $X = (C^{\dagger})^* = \begin{bmatrix} x_1 & x_2 \\ x_3 & x_4 \end{bmatrix}$. Then

$$\begin{bmatrix} 0 & -a \\ 0 & 1 \end{bmatrix} = C = XC^*C = \begin{bmatrix} 0 & x_2(1 + a^*a) \\ 0 & x_4(1 + a^*a) \end{bmatrix}.$$

Hence $x_4(1 + a^*a) = 1$ and $(1 + a^*a)x_4^* = [x_4(1 + a^*a)]^* = 1$. This shows $1 + a^*a \in R^{-1}$. Therefore, $R$ is a GN ring.

$\square$

Next, we extend the class of GN rings to a larger one and consider the relationship between them.

**Definition 2.5.3 ([130, Definition 1])** Let $R$ be a $*$-ring. If $1 - (u^* - u)^2 \in R^{-1}$ for all $u \in R^{-1}$ then $R$ is called a *generalized GN ring*.

Now, we consider Jacobson's lemma for Moore-Penrose inverses.

**Lemma 2.5.4** *Let $a, b \in R$. If $\alpha = 1 - ab$ is regular with an inner inverse $\alpha^-$, then $\beta = 1 - ba$ is regular with an inner inverse given by $\beta^- = 1 + b\alpha^- a$.*

*Proof* It can be easily verified.                                                    □

**Proposition 2.5.5** *Let $R$ be a GN ring and $a, b \in R$. If $1 - ab$ is Moore–Penrose invertible then so is $1 - ba$.*

*Proof* It follows by Theorem 2.2.21 and Lemma 2.5.4.                                 □

According to Example 2.5.1, there exists a ring $R$ in which an element $1 - ab$ is Moore-Penrose invertible while $1 - ba$ is not. Actually, it can be seen from the next example that even if every element is Moore-Penrose invertible in a ring, the formula $(1 - ba)^\dagger = 1 + b(1 - ab)^\dagger a$ does not hold in general.

**Example 2.5.6** Consider the involution on $\mathbb{C}^{2\times 2}$ induced from the conjugate complex involution on $\mathbb{C}$. It is well known that every matrix in $\mathbb{C}^{2\times 2}$ is Moore-Penrose invertible. Let $A = \begin{bmatrix} 1 & 0 \\ 0 & 0 \end{bmatrix}$ and $B = \begin{bmatrix} 1 & 0 \\ 1 & 0 \end{bmatrix}$. Then $I - AB = \begin{bmatrix} 0 & 0 \\ 0 & 1 \end{bmatrix}$ and $I - BA = \begin{bmatrix} 0 & 0 \\ -1 & 1 \end{bmatrix}$. By computation, we have $(I - AB)^\dagger = \begin{bmatrix} 0 & 0 \\ 0 & 1 \end{bmatrix}$ and $(I - BA)^\dagger = \begin{bmatrix} 0 & -1/2 \\ 0 & 1/2 \end{bmatrix}$. This shows $(I - BA)^\dagger \neq I + B(I - AB)^\dagger A$. In fact, one can check that $(I - BA)^\dagger \neq I + BCA$ for any $C \in \mathbb{C}^{2\times 2}$.

To provide some conditions under which the Moore-Penrose invertibility of $1 - ba$ and the formula for $(1 - ba)^\dagger$ can be interpreted in terms of $(1 - ab)^\dagger$, we need the following lemmas, which are of interest in their own right.

**Lemma 2.5.7 ([130, Lemma 4])** *Let $e \in R$ be an idempotent. If $u = 1 + (e - e^*)(e - e^*)^* \in R^{-1}$ then $e$ has Moore–Penrose inverse and $e^\dagger = u^{-1}e^* = e^*u^{-1}$.*

*Proof* Since $e^2 = e$ and $u = 1 + (e - e^*)(e - e^*)^*$ it follows that $u = u^*$, $eu = ee^*e = ue$ and $e^*u = ue^*$. Moreover, since $u \in R^{-1}$ one can verify that $(u^{-1})^* = u^{-1}$, $eu^{-1} = u^{-1}e$, $u^{-1}e^* = e^*u^{-1}$ and $e = ee^*(eu^{-1}) = (u^{-1}e)e^*e$. Now it is straight forward to check that $e^\dagger = u^{-1}e^* = e^*u^{-1}$.                                                    □

**Lemma 2.5.8 ([130, Lemma 5])** *Let $a, b \in R$ and $b$ be a reflexive inverse of $a$. If both $ab$ and $ba$ are Moore-Penrose invertible then $a$ is also Moore-Penrose invertible and $a^\dagger = (ba)^\dagger b(ab)^\dagger$.*

**Proof** Let $x = (ba)^\dagger b(ab)^\dagger$. Then we have

$$ax = a[(ba)^\dagger b(ab)^\dagger] = (aba)[(ba)^\dagger (bab)(ab)^\dagger] = (ab)(ab)^\dagger.$$

Hence $(ax)^* = ax$ and $axa = [(ab)(ab)^\dagger](aba) = a$. Similarly, we have $xa = (ba)^\dagger(ba)$, $(xa)^* = xa$ and $xax = [(ba)^\dagger(ba)][(ba)^\dagger b(ab)^\dagger] = (ba)^\dagger b(ab)^\dagger = x$. Thus, $a^\dagger = x = (ba)^\dagger b(ab)^\dagger$.                                                                $\square$

**Theorem 2.5.9 ([130, Theorem 6])** *Let $R$ be a generalized GN ring and $a, b \in R$. Suppose that $\alpha = 1 - ab$ has Moore-Penrose inverse. Then the following statements are equivalent:*

(1) $\beta = 1 - ba$ *has Moore–Penrose inverse.*
(2) *Both $u = 1 - [(bpa)^* - bpa]^2$ and $v = 1 - [(bqa)^* - bqa]^2$ are invertible, where $p = 1 - \alpha^\dagger \alpha$ and $q = 1 - \alpha \alpha^\dagger$.*

*In this case, $\beta^\dagger = u^{-1} e^* [1 + b(\alpha^\dagger - pq)a] f^* v^{-1}$, where $e = 1 - bpa$ and $f = 1 - bqa$.*

**Proof** Let $\beta^+ = 1 + b(\alpha^\dagger - pq)a$. Then it is direct to verify that $\beta^+$ is a reflexive inverse of $\beta$. Moreover, $\beta^+ \beta = \beta + b(\alpha^\dagger - pq)a\beta = \beta + b(\alpha^\dagger - pq)\alpha a = \beta + b\alpha^\dagger \alpha a = 1 - bpa$. Similarly, $\beta\beta^+ = 1 - bqa$.

(1)$\Rightarrow$(2). Let $w_1 = \beta^* \beta + bpa$ and $w_2 = \beta\beta^* + bqa$. Then $w_1 = \beta^* \beta + 1 - \beta^+ \beta$ and $w_2 = \beta\beta^* + 1 - \beta\beta^+$. Since $\beta$ has Moore-Penrose inverse, it follows that $w_1$ and $w_2$ are units by Theorem 2.2.14. Consequently, $u = 1 - [(bpa)^* - bpa]^2 = 1 - (w_1^* - w_1)^2$ and $v = 1 - [(bqa)^* - bqa]^2 = 1 - (w_2^* - w_2)^2$ are units as $R$ is a generalized GN ring.

(2)$\Rightarrow$(1). Since $e = 1 - bpa = \beta^+ \beta$ is an idempotent and $1 + (e - e^*)(e - e^*)^* = 1 - [(bpa)^* - bpa]^2 = u$ is a unit, it follows that $e$ has Moore-Penrose inverse and $e^\dagger = u^{-1} e^*$ by Lemma 2.5.7. Similarly, $f = \beta\beta^+$ has Moore–Penrose inverse and $f^\dagger = f^* v^{-1}$. In view of Lemma 2.5.8, we have $\beta$ is Moore-Penrose invertible and $\beta^\dagger = e^\dagger \beta^+ f^\dagger = u^{-1} e^* [1 + b(\alpha^\dagger - pq)a] f^* v^{-1}$.                                                                $\square$

### 2.5.2  Jacobson's Lemma for Moore-Penrose Inverses in a Ring

Next we give some equivalent conditions under which $1 - ba$ are Moore-Penrose invertible when $1 - ab$ is Moore-Penrose invertible in a ring.

**Lemma 2.5.10 ([111, Lemma 5.1])** *If $a$ is regular with an inner inverse $a^-$, then the following conditions are equivalent:*

(1) $a \in R^{\{1,3\}}$.
(2) $aa^- \in R^{\{1,3\}}$.

*In this case, $aa^{(1,3)} \in (aa^-)\{1,3\}$ and*

$$a^-(aa^-)^{(1,3)} \in a\{1,3\},$$

*for any $(aa^-)^{(1,3)} \in (aa^-)\{1,3\}$.*

**Proof** (1) $\Rightarrow$ (2). In order to prove $aa^- \in R^{\{1,3\}}$, it suffices to show that $aa^{(1,3)} \in (aa^-)\{1,3\}$. Indeed, we have $aa^-aa^{(1,3)}aa^- = aa^{(1,3)}aa^- = aa^-$ and $aa^-aa^{(1,3)} = aa^{(1,3)}$.

(2) $\Rightarrow$ (1). For any $(aa^-)^{(1,3)} \in (aa^-)\{1,3\}$, let $x = a^-(aa^-)^{(1,3)}$. First noting that $ax = aa^-(aa^-)^{(1,3)}$, we have

$$axa = (ax)a = aa^-(aa^-)^{(1,3)}a = aa^-(aa^-)^{(1,3)}aa^-a = aa^-a = a,$$

which shows that $x$ is a $\{1,3\}$-inverse of $a$.                                   □

Dually, we have the following result.

**Lemma 2.5.11 ([111, Lemma 5.2])** *If $a$ is regular with an inner inverse $a^-$, then the following conditions are equivalent:*

(1) $a \in R^{\{1,4\}}$.
(2) $a^-a \in R^{\{1,4\}}$.

*In this case, $a^{(1,4)}a \in (a^-a)\{1,4\}$ and*

$$(a^-a)^{(1,4)}a^- \in a\{1,4\},$$

*for any $(a^-a)^{(1,4)} \in (a^-a)\{1,4\}$.*

Combining preceding two lemmas with Lemma 2.2.2, we can easily derive the following corollary, which generalizes Lemma 2.5.8.

**Corollary 2.5.12** *If $a$ is regular with an inner inverse $a^-$, then the following conditions are equivalent:*

(1) $a \in R^\dagger$.
(2) $aa^- \in R^{\{1,3\}}$ *and* $a^-a \in R^{\{1,4\}}$.

*In this case, $aa^\dagger \in (aa^-)\{1,3\}$, $a^\dagger a \in (a^-a)\{1,4\}$ and*

$$a^\dagger = (a^-a)^{(1,4)}a^-aa^-(aa^-)^{(1,3)}$$

*for any $(aa^-)^{(1,3)} \in (aa^-)\{1,3\}$ and $(a^-a)^{(1,4)} \in (a^-a)\{1,4\}$.*

**Example 2.5.13** Let $\mathbb{C}^{2\times 2}$ be the ring of all $2 \times 2$ matrices over the complex field $\mathbb{C}$, with transpose as the involution. Suppose that $a = \begin{bmatrix} 1 & -i \\ i & -1 \end{bmatrix}$ and $b = \begin{bmatrix} 0 & 0 \\ 0 & -1 \end{bmatrix}$, then we have $1 - ab = \begin{bmatrix} 1 & -i \\ 0 & 0 \end{bmatrix}$ and $1 - ba = \begin{bmatrix} 1 & 0 \\ i & 0 \end{bmatrix}$. It is easy to verify that $1 - ab$

has $\begin{bmatrix} 1 & 0 \\ 0 & 0 \end{bmatrix}$ as a $\{1, 3\}$-inverse while $1 - ba$ is not $\{1, 3\}$-invertible.

From Example 2.5.13, we can see that $1 - ba$ may not be $\{1, 4\}$-invertible when $1 - ab$ is. Dually, $1 - ab \in R^{\{1,3\}}$ does not imply $1 - ba \in R^{\{1,3\}}$, either.

**Lemma 2.5.14 ([111, Lemma 3.2])** *Let $e \in R$ be an idempotent. Then $e \in R^{\{1,3\}}$ if and only if $1 - e \in R^{\{1,4\}}$. In this case, $1 - ee^{(1,3)} \in (1 - e)\{1, 4\}$ and $1 - (1 - e)^{(1,4)}(1 - e) \in e\{1, 3\}$.*

**Proof** If $e \in R^{\{1,3\}}$, in order to prove that $1 - e \in R^{\{1,4\}}$, we only need to check that $1 - ee^{(1,3)} \in (1 - e)\{1, 4\}$. Indeed, we have

$$(1 - ee^{(1,3)})(1 - e) = 1 - ee^{(1,3)} - e + ee^{(1,3)}e = 1 - ee^{(1,3)}$$

and

$$(1-e)(1-ee^{(1,3)})(1-e) = (1-e)(1-ee^{(1,3)}) = 1-e-ee^{(1,3)}+eee^{(1,3)} = 1-e.$$

The converse statement can be proved similarly by verifying that $1-(1-e)^{(1,4)}(1-e) \in e\{1, 3\}$. The details of verification will be omitted. $\square$

**Theorem 2.5.15 ([111, Theorem 5.6])** *Let $a, b \in R$. If $\alpha = 1 - ab \in R^{\{1,3\}}$ with a $\{1, 3\}$-inverse $\alpha^{(1,3)}$, then the following conditions are equivalent:*

(1) $\beta = 1 - ba \in R^{\{1,3\}}$.
(2) $1 - ba_r^{\pi} a \in R^{\{1,3\}}$.
(3) $ba_r^{\pi} a \in R^{\{1,4\}}$.
(4) $v = \alpha_r^{\pi} aa^* + 1 - \alpha_r^{\pi} \in R^{-1}$.

*In this case, $(1 + b\alpha^{(1,3)}a)(1 - a^*v^{-1}\alpha_r^{\pi} a) \in \beta\{1, 3\}$, where $\alpha_r^{\pi} = 1 - \alpha\alpha^{(1,3)}$.*

**Proof** As $\alpha = 1 - ab \in R^{\{1,3\}}$ with a $\{1, 3\}$-inverse $\alpha^{(1,3)}$, it is clear that $\alpha\alpha^{(1,3)} \in R^{\{1,3\}}$ with $\alpha\alpha^{(1,3)} \in (\alpha\alpha^{(1,3)})\{1, 3\}$. It follows that $\alpha_r^{\pi} \in R^{\{1,4\}}$ with $1 - \alpha\alpha^{(1,3)}\alpha\alpha^{(1,3)} = \alpha_r^{\pi} \in \alpha_r^{\pi}\{1, 4\}$ by Lemma 2.5.14.

(1) $\Leftrightarrow$ (2). Noting that $\beta$ is regular with an inner inverse $\beta^- = 1 + b\alpha^{(1,3)}a$ by Lemma 2.5.4, we have

$$\beta\beta^- = (1 - ba)(1 + b\alpha^{(1,3)}a)$$
$$= 1 - ba + b\alpha^{(1,3)}a - bab\alpha^{(1,3)}a$$
$$= 1 - b(1 - \alpha^{(1,3)} + ab\alpha^{(1,3)})a$$
$$= 1 - b(1 - \alpha\alpha^{(1,3)})a$$
$$= 1 - ba_r^{\pi} a.$$

Then, by Lemma 2.5.10, $\beta \in R^{\{1,3\}}$ if and only if $1 - b\alpha_r^\pi a \in R^{\{1,3\}}$.

(2) $\Leftrightarrow$ (3). By Lemma 2.5.14.

(3) $\Leftrightarrow$ (4). First we have

$$\alpha_r^\pi ab\alpha_r^\pi = (1 - \alpha\alpha^{(1,3)})ab(1 - \alpha\alpha^{(1,3)}) = 1 - \alpha\alpha^{(1,3)} = \alpha_r^\pi$$

and

$$\alpha_r^\pi ab = (1 - \alpha\alpha^{(1,3)})ab = 1 - \alpha\alpha^{(1,3)} = \alpha_r^\pi.$$

Since $\alpha_r^\pi \in R^{\{1,4\}}$, it follows that $b\alpha_r^\pi a \in R^{\{1,4\}}$ if and only if $v = (\alpha_r^\pi)^{(1,4)}\alpha_r^\pi aa^* + 1 - (\alpha_r^\pi)^{(1,4)}\alpha_r^\pi \in R^{-1}$ by Theorem 2.3.1. Noting that $\alpha_r^\pi \in \alpha_r^\pi\{1,4\}$, we have $v = \alpha_r^\pi aa^* + 1 - \alpha_r^\pi$.

At last, we give a $\{1,3\}$-inverse of $\beta$. If (3) holds, first we know that $\beta$ is regular with an inner inverse $\beta^- = 1 + b\alpha^{(1,3)}a$. Then, for any $(b\alpha_r^\pi a)^{(1,4)} \in (b\alpha_r^\pi a)\{1,4\}$, we have $1 - (b\alpha_r^\pi a)^{(1,4)}b\alpha_r^\pi a \in (1 - b\alpha_r^\pi a)\{1,3\} = (\beta\beta^-)\{1,3\}$ by Lemma 2.5.14. Combining these with $\beta^-(\beta\beta^-)^{(1,3)} \in \beta\{1,3\}$ which follows from Lemma 2.5.10, we get

$$(1 + b\alpha^{(1,3)}a)[1 - (b\alpha_r^\pi a)^{(1,4)}b\alpha_r^\pi a] \in \beta\{1,3\}.$$

If (4) holds, for any $(\alpha_r^\pi)^{(1,4)} \in \alpha_r^\pi\{1,4\}$, we have that $a^*v^{-1}(\alpha_r^\pi)^{(1,4)}\alpha_r^\pi \in (b\alpha_r^\pi a)\{1,4\}$ by Theorem 2.3.1. Combining this with $\alpha_r^\pi \in (\alpha_r^\pi)\{1,4\}$, we get $(b\alpha_r^\pi a)^{(1,4)}b\alpha_r^\pi a = a^*v^{-1}\alpha_r^\pi a$. It follows that

$$(1 + b\alpha^{(1,3)}a)[1 - (b\alpha_r^\pi a)^{(1,4)}b\alpha_r^\pi a]$$
$$= (1 + b\alpha^{(1,3)}a)(1 - a^*v^{-1}\alpha_r^\pi a) \in \beta\{1,3\}.$$

$\square$

Dually, we can prove the following result by a similar way.

**Theorem 2.5.16 ([111, Theorem 5.7])** *Let* $a, b \in R$. *If* $\alpha = 1 - ab \in R^{\{1,4\}}$ *with* $a \{1,4\}$-*inverse* $\alpha^{(1,4)}$, *then the following conditions are equivalent:*

(1) $\beta = 1 - ba \in R^{\{1,4\}}$.

(2) $1 - b\alpha_l^\pi a \in R^{\{1,4\}}$.

(3) $b\alpha_l^\pi a \in R^{\{1,3\}}$.

(4) $u = b^*b\alpha_l^\pi + 1 - \alpha_l^\pi \in R^{-1}$.

*In this case,* $(1 - b\alpha_l^\pi u^{-1}b^*)(1 + b\alpha^{(1,4)}a) \in \beta\{1,4\}$, *where* $\alpha_l^\pi = 1 - \alpha^{(1,4)}\alpha$.

Naturally, we will consider under what conditions $1 - ab \in R^\dagger$ implies that $1 - ba \in R^\dagger$.

If $\alpha \in R^\dagger$, we have $1 - \alpha\alpha^\dagger = 1 - \alpha\alpha^{(1,3)}$ and $1 - \alpha^\dagger\alpha = 1 - \alpha^{(1,4)}\alpha$, for any $\alpha^{(1,3)} \in \alpha\{1,3\}$ and $\alpha^{(1,4)} \in \alpha\{1,4\}$. So we can still denote $\alpha_r^\pi = 1 - \alpha\alpha^\dagger$ and $\alpha_l^\pi = 1 - \alpha^\dagger\alpha$ without causing any ambiguity.

Combining Theorems 2.5.15, 2.5.16 and Lemma 2.2.2, we obtain the following theorem.

**Theorem 2.5.17 ([111, Theorem 5.8])** *Let $a, b \in R$. If $\alpha = 1 - ab \in R^\dagger$, then the following conditions are equivalent:*

(1) $\beta = 1 - ba \in R^\dagger$.
(2) $1 - ba_r^\pi a \in R^{\{1,3\}}$ and $1 - ba_l^\pi a \in R^{\{1,4\}}$.
(3) $ba_r^\pi a \in R^{\{1,4\}}$ and $ba_l^\pi a \in R^{\{1,3\}}$.
(4) $v = \alpha_r^\pi aa^* + 1 - \alpha_r^\pi \in R^{-1}$ and $u = b^*b\alpha_l^\pi + 1 - \alpha_l^\pi \in R^{-1}$.

*In this case,*

$$\beta^\dagger = (1 - b\alpha_l^\pi u^{-1}b^*)(1 + ba^\dagger a)(1 - a^*v^{-1}\alpha_r^\pi a).$$

*Proof* The equivalence of conditions (1)–(4) clearly follows from Theorems 2.5.15, 2.5.16 and Lemma 2.2.2. Next we give a formula for $\beta^\dagger$.

It is clear that $\beta$ is regular with an inner inverse $\beta^- = 1 + ba^\dagger a$. If (4) holds, according to the proof of Theorem 2.5.15, we have

$$\beta\beta^-(\beta\beta^-)^{(1,3)} = (1 - b\alpha_r^\pi a)[1 - (b\alpha_r^\pi a)^{(1,4)}b\alpha_r^\pi a]$$
$$= 1 - (b\alpha_r^\pi a)^{(1,4)}b\alpha_r^\pi a$$
$$= 1 - a^*v^{-1}\alpha_r^\pi a.$$

Dually, we get $(\beta^-\beta)^{(1,4)}\beta^-\beta = 1 - b\alpha_l^\pi u^{-1}b^*$. Combining these with Corollary 2.5.12, we have

$$\beta^\dagger = (\beta^-\beta)^{(1,4)}\beta^-\beta\beta^-(\beta\beta^-)^{(1,3)}$$
$$= [(\beta^-\beta)^{(1,4)}\beta^-\beta]\beta^-[\beta\beta^-(\beta\beta^-)^{(1,3)}]$$
$$= (1 - b\alpha_l^\pi u^{-1}b^*)(1 + ba^\dagger a)(1 - a^*v^{-1}\alpha_r^\pi a).$$

$\square$

## 2.6  The Moore-Penrose Inverse of a 2 × 2 Block Matrix

In this section, some applications of Theorems 2.3.14, 2.3.15 and 2.3.16 are indicated.

First, we characterize the existence of a $\{1, 3\}$-inverse of a $2 \times 2$ block matrix $M$

$$M = \begin{bmatrix} a & c \\ b & d \end{bmatrix} \tag{2.6.1}$$

over a ring $R$, where $a \in R^{m \times m}$ is invertible, $b, c$ and $d$ are matrices over $R$ of orders $k \times m$, $m \times l$ and $k \times l$, respectively. We denote by $1_m$ the identity matrix in $R^{m \times m}$.

Consider the factorization

$$M = \begin{bmatrix} a & c \\ b & d \end{bmatrix} = \begin{bmatrix} 1_m & 0 \\ ba^{-1} & 1_l \end{bmatrix} \begin{bmatrix} a & 0 \\ 0 & s \end{bmatrix} \begin{bmatrix} 1_m & a^{-1}c \\ 0 & 1_l \end{bmatrix} = PAQ. \tag{2.6.2}$$

**Theorem 2.6.1 ([15, Theorem 4.1])** *Let $M$ be as in (2.6.1) and let $s = d - ba^{-1}c$. Assume that $s\{1, 3\} \neq \emptyset$ and let $s^{(1,3)} \in s\{1, 3\}$ and $e = 1_k - ss^{(1,3)}$. Then $M\{1, 3\} \neq \emptyset$ if and only if $u = 1_m + (ba^{-1})^*eba^{-1}$ is invertible. In this case, a $\{1, 3\}$-inverse of $M$ is given by*

$$M^{(1,3)} = \begin{bmatrix} \alpha u^{-1} & \alpha u^{-1}(ba^{-1})^*e - a^{-1}cs^{(1,3)} \\ -s^{(1,3)}ba^{-1}u^{-1} & s^{(1,3)}(1_k - ba^{-1}u^{-1}(ba^{-1})^*e) \end{bmatrix}, \tag{2.6.3}$$

*where $\alpha = (1_m + a^{-1}cs^{(1,3)}b)a^{-1}$.*

**Proof** Let $P$, $A$ and $Q$ be as in (2.6.2). It is easy to check that a $\{1, 3\}$-inverse of $A$ is of the form $A^{(1,3)} = \begin{bmatrix} a^{-1} & 0 \\ 0 & s^{(1,3)} \end{bmatrix}$. Then

$$I - AA^{(1,3)} = \begin{bmatrix} 0 & 0 \\ 0 & e \end{bmatrix},$$

where $e = 1_k - ss^{(1,3)}$. Hence, $P(I - AA^{(1,3)}) = I - AA^{(1,3)}$ holds. Set

$$S = (I - AA^{(1,3)})(I - P^{-1}) = \begin{bmatrix} 0 & 0 \\ eba^{-1} & 0 \end{bmatrix}, \quad I + S^*S = \begin{bmatrix} u & 0 \\ 0 & 1_k \end{bmatrix},$$

where $u = 1_m + (ba^{-1})^*eba^{-1}$. By Theorem 2.3.14 there exists a $\{1, 3\}$-inverse of $PA$ if and only if $u$ is invertible in the ring $R^{m \times m}$. By (2.6.1), a $\{1, 3\}$-inverse of $PA$ is of the form $(PA)^{(1,3)} = A^{(1,3)}P^{-1}(I + S^*S)^{-1}(I + S^*)$. Therefore,

$$(PA)^{(1,3)} = \begin{bmatrix} a^{-1} & 0 \\ 0 & s^{(1,3)} \end{bmatrix} \begin{bmatrix} 1_m & 0 \\ -ba^{-1} & 1_l \end{bmatrix} \begin{bmatrix} u^{-1} & 0 \\ 0 & 1_k \end{bmatrix} \begin{bmatrix} 1_m & (ba^{-1})^*e \\ 0 & 1_k \end{bmatrix}$$

$$= \begin{bmatrix} a^{-1}u^{-1} & a^{-1}u^{-1}(ba^{-1})^*e \\ -s^{(1,3)}ba^{-1}u^{-1} & s^{(1,3)}\left(1_k - ba^{-1}u^{-1}(ba^{-1})^*e\right) \end{bmatrix}. \tag{2.6.4}$$

Now, since $Q$ is invertible, it follows that $PAQ\{1,3\} \neq \emptyset$ if and only if $PA\{1,3\} \neq \emptyset$ in which case $(PAQ)^{(1,3)} = Q^{-1}(PA)^{(1,3)}$. Pre-multiplying (2.6.4) by $Q^{-1}$, (2.6.3) is proved.                               □

We can state the analogue of previous theorem for the characterization of the existence of a $\{1,4\}$-inverse of matrix $M$.

**Theorem 2.6.2 ([15, Theorem 4.2])** *Let $M$ be as in (2.6.1) and let $s = d - ba^{-1}c$. Assume that $s\{1,4\} \neq \emptyset$ and let $s^{(1,4)} \in s\{1,4\}$ and $f = 1_l - s^{(1,4)}s$. Then $M\{1,4\} \neq \emptyset$ if and only if $v = 1_m + a^{-1}cf(a^{-1}c)^*$ is invertible. In this case, a $\{1,4\}$-inverse of $M$ is given by*

$$M^{(1,4)} = \begin{bmatrix} v^{-1}\beta & -v^{-1}a^{-1}cs^{(1,4)} \\ f(a^{-1}c)^*v^{-1}\beta - s^{(1,4)}ba^{-1} & (1_l - f(a^{-1}c)^*v^{-1}a^{-1}c)s^{(1,4)} \end{bmatrix},$$
(2.6.5)

*where $\beta = a^{-1}(1_m + cs^{(1,4)}ba^{-1})$.*

**Proof** We use the factorization (2.6.2). Choose $A^{(1,4)} = \begin{bmatrix} a^{-1} & 0 \\ 0 & s^{(1,4)} \end{bmatrix}$, which gives

$$I - A^{(1,4)}A = \begin{bmatrix} 0 & 0 \\ 0 & f \end{bmatrix}, \text{ where } f = 1_l - s^{(1,4)}s. \text{ Now, set}$$

$$T = (I - Q^{-1})(I - A^{(1,4)}A) = \begin{bmatrix} 0 & a^{-1}cf \\ 0 & 0 \end{bmatrix}, \quad I + TT^* = \begin{bmatrix} v & 0 \\ 0 & 1_l \end{bmatrix},$$

where $v = 1_m + a^{-1}cf(a^{-1}c)^*$. With an application of Theorem 2.3.15 we obtain $AQ\{1,4\} \neq \emptyset$ if and only if $v$ is invertible and

$$(AQ)^{(1,4)} = \begin{bmatrix} v^{-1}a^{-1} & -v^{-1}a^{-1}cs^{(1,4)} \\ f(a^{-1}c)^*v^{-1}a^{-1} & (1_l - f(a^{-1}c)^*v^{-1}a^{-1}c)s^{(1,4)} \end{bmatrix}. \quad (2.6.6)$$

Since $(PAQ)^{(1,4)} = (AQ)^{(1,4)}P^{-1}$. Pre-multiplying (2.6.6) by $P^{-1}$, (2.6.5) is proved.                               □

If $s = d - ba^{-1}c$ is regular, by Theorem 2.2.14 we have that $s^\dagger$ exists if and only if $s^*s + 1_l - s^{(1)}s$ is invertible. In this case, using previous results we can characterize the existence of the Moore-Penrose inverse of matrix $M$.

**Theorem 2.6.3 ([15, Theorem 4.3])** *Let $M$ be as in (2.6.1) and let $s = d - ba^{-1}c$. Assume that $s^\dagger$ exists and set $e = 1_k - ss^\dagger$ and $f = 1_l - s^\dagger s$. Then $M^\dagger$ exists if and only if both $u = 1_m + (ba^{-1})^*eba^{-1}$ and $v = 1_m + a^{-1}cf(a^{-1}c)^*$ are invertible. In this case,*

$$M^\dagger = \begin{bmatrix} \gamma & \gamma(ba^{-1})^*e - v^{-1}a^{-1}cs^\dagger \\ f(a^{-1}c)^*\gamma - s^\dagger ba^{-1}u^{-1} & \delta \end{bmatrix}, \quad (2.6.7)$$

*where*

$$\gamma = v^{-1}\left(a^{-1} + a^{-1}cs^{\dagger}ba^{-1}\right)u^{-1},$$
$$\delta = s^{\dagger} + f\left(a^{-1}c\right)^{*}\gamma\left(ba^{-1}\right)^{*}e - f\left(a^{-1}c\right)^{*}v^{-1}a^{-1}cs^{\dagger} - s^{\dagger}ba^{-1}u^{-1}\left(ba^{-1}\right)^{*}e.$$

**Proof** In view of the factorization (2.6.2), we have that $M$ is Moore-Penrose invertible if and only if $PA$ has a $\{1, 3\}$-inverse and $AQ$ has a $\{1, 4\}$-inverse. Now, since $s^{\dagger}$ exists we can consider $s^{(1,3)} = s^{\dagger}$ in Theorem 2.6.1 and $s^{(1,4)} = s^{\dagger}$ in Theorem 2.6.2 to conclude that $M$ is Moore-Penrose invertible if and only if both $u$ and $v$ defined as in the statement of this theorem are invertible. Using expressions (2.6.4) and (2.6.6), we compute $M^{\dagger} = (AQ)^{(1,4)}A(PA)^{(1,3)}$ and we obtain the expression (2.6.7).                                                           □

In the sequel, let $T$ be a matrix over $R$ of the form

$$T = \begin{bmatrix} a & 0 \\ b & d \end{bmatrix}, \tag{2.6.8}$$

where $a, b$ and $d$ are matrices over $R$ of orders $m \times n$, $k \times n$, and $k \times l$, respectively.

We will characterize the existence of a $\{1, 3\}$-inverse of $T$ when $a\{1, 2, 3\} \neq \emptyset$ and $d\{1, 3\} \neq \emptyset$. Set

$$e = 1_k - dd^{(1,3)}, \quad f = 1_n - a^{(1,2,3)}a, \quad c = ebf. \tag{2.6.9}$$

Consider the factorization

$$T = \begin{bmatrix} a & 0 \\ b & d \end{bmatrix} = \begin{bmatrix} 1_m & 0 \\ eba^{(1,2,3)} & 1_k \end{bmatrix}\begin{bmatrix} a & 0 \\ c & d \end{bmatrix}\begin{bmatrix} 1_n & 0 \\ d^{(1,3)}b & 1_l \end{bmatrix} = PAQ. \tag{2.6.10}$$

**Theorem 2.6.4 ([15, Theorem 4.4])** *Let $e, f,$ and $c$ be as in (2.6.9) and $T$ be as in (2.6.10). Assume that $c\{1, 3\} \neq \emptyset$ and let $c^{(1,3)} \in c\{1, 3\}$. Then $T\{1, 3\} \neq \emptyset$ if and only if $u = 1_m + (ba^{(1,2,3)})^{*}egba^{(1,2,3)}$ is invertible, where $g = 1_k - cc^{(1,3)}$. In this case,*

$$T^{(1,3)} = \begin{bmatrix} 1_n & 0 \\ -d^{(1,3)}b & 1_l \end{bmatrix}\begin{bmatrix} \sigma & \sigma(ba^{(1,2,3)})^{*}eg + fc^{(1,3)} \\ -\eta & d^{(1,3)} - \eta(ba^{(1,2,3)})^{*}eg \end{bmatrix}, \tag{2.6.11}$$

*where $\sigma = (1 - fc^{(1,3)}eb)a^{(1,2,3)}u^{-1}$ and $\eta = d^{(1,3)}eba^{(1,2,3)}u^{-1}$.*

**Proof** Let $P, A$ and $Q$ be as in (2.6.10). We observe that $cc^{(1,3)} = cc^{(1,3)}e$. Using this, it is easy to check that a $\{1, 3\}$-inverse of $A$ is given by

$$A^{(1,3)} = \begin{bmatrix} a^{(1,2,3)} & fc^{(1,3)} \\ 0 & d^{(1,3)} \end{bmatrix}.$$

Then

$$I - AA^{(1,3)} = \begin{bmatrix} 1_m - aa^{(1,2,3)} & 0 \\ 0 & eg \end{bmatrix},$$

where $g = 1_k - cc^{(1,3)}$. We can see that $P(I - AA^{(1,3)}) = I - AA^{(1,3)}$. Then we can apply Theorem 2.3.14 to the product $PA$. With the notation $S = (I - AA^{(1,3)}) (I \quad P^{-1})$, using that $ege = ge$, we have

$$S = \begin{bmatrix} 0 & 0 \\ geba^{(1,2,3)} & 0 \end{bmatrix}, \quad S^* = \begin{bmatrix} 0 & \left(ba^{(1,2,3)}\right)^* eg \\ 0 & 0 \end{bmatrix}, \quad I + S^*S = \begin{bmatrix} u & 0 \\ 0 & 1_k \end{bmatrix},$$

where $u = 1_m + (ba^{(1,2,3)})^* egba^{(1,2,3)}$. By Theorem 2.3.14, $PA$ has a $\{1, 3\}$-inverse if and only if $I + S^*S$ is invertible, or equivalently, $u$ is invertible. In this case, a $\{1, 3\}$-inverse of $PA$ has the form

$$(PA)^{(1,3)} = A^{(1,3)} P^{-1} \left(I + S^*S\right)^{-1} \left(I + S^*\right).$$

Substituting into this expression the matrix products

$$A^{(1,3)} P^{-1} = \begin{bmatrix} a^{(1,2,3)} & fc^{(1,3)} \\ 0 & d^{(1,3)} \end{bmatrix} \begin{bmatrix} 1_m & 0 \\ -eba^{(1,2,3)} & 1_k \end{bmatrix}$$

$$= \begin{bmatrix} (1 - fc^{(1,3)}eb)a^{(1,2,3)} & fc^{(1,3)} \\ -d^{(1,3)}eba^{(1,2,3)} & d^{(1,3)} \end{bmatrix}$$

and

$$(I + S^*S)^{-1}(I + S^*) = \begin{bmatrix} u^{-1} & u^{-1}(ba^{(1,2,3)})^* eg \\ 0 & 1 \end{bmatrix},$$

we obtain, after an easy computation,

$$(PA)^{(1,3)} = \begin{bmatrix} \sigma & \sigma(ba^{(1,2,3)})^* eg + fc^{(1,3)} \\ -\eta & d^{(1,3)} - \eta(ba^{(1,2,3)})^* eg \end{bmatrix},$$

where $\sigma = (1 - fc^{(1,3)}eb)a^{(1,2,3)}u^{-1}$ and $\eta = d^{(1,3)}eba^{(1,2,3)}u^{-1}$. Now, since $Q$ is invertible, $PAQ\{1, 3\} \neq \emptyset$ if and only if $PA\{1, 3\} \neq \emptyset$ in which case $(PAQ)^{(1,3)} = Q^{-1}(PA)^{(1,3)}$. The latter establishes the formula (2.6.11).                      □

In order to characterize the existence of a $\{1,4\}$-inverse of $T$ when $a\{1, 4\} \neq \emptyset$ and $d\{1, 2, 4\} \neq \emptyset$, set

$$e = 1_k - dd^{(1,2,4)}, \quad f = 1_n - a^{(1,4)}a, \quad c = ebf \tag{2.6.12}$$

and consider the factorization

$$T = \begin{bmatrix} a & 0 \\ b & d \end{bmatrix} = \begin{bmatrix} 1_m & 0 \\ ba^{(1,4)} & 1_k \end{bmatrix} \begin{bmatrix} a & 0 \\ c & d \end{bmatrix} \begin{bmatrix} 1_n & 0 \\ d^{(1,2,4)}bf & 1_l \end{bmatrix} = \hat{P}\hat{A}\hat{Q}. \tag{2.6.13}$$

Similarly, with an application of Theorem 2.3.15, we derive the analogue of previous theorem.

**Theorem 2.6.5 ([15, Theorem 4.5])** *Let $e$, $f$, and $c$ be as in (2.6.12) and $T$ be as in (2.6.13). Assume that $c\{1, 4\} \neq \emptyset$ and let $c^{(1,4)} \in c\{1, 4\}$. Then $T\{1, 4\} \neq \emptyset$ if and only if $v = 1_l + d^{(1,2,4)}bhf(d^{(1,2,4)}b)^*$ is invertible, where $h = 1_n - c^{(1,4)}c$. In this case,*

$$T^{(1,4)} = \begin{bmatrix} a^{(1,4)} - hf(d^{(1,2,4)}b)^*\mu & hf(d^{(1,2,4)}b)^*\rho + c^{(1,4)}e \\ -\mu & \rho \end{bmatrix} \begin{bmatrix} 1_m & 0 \\ -ba^{(1,4)} & 1_k \end{bmatrix}, \tag{2.6.14}$$

*where $\rho = v^{-1}d^{(1,2,4)}(1 - bfc^{(1,4)}e)$ and $\mu = v^{-1}d^{(1,2,4)}bfa^{(1,4)}$.*

We will characterize the Moore-Penrose invertibility of $T$ when $a^\dagger$ and $d^\dagger$ exist. Set

$$e = 1_k - dd^\dagger, \quad f = 1_n - a^\dagger a, \quad c = ebf. \tag{2.6.15}$$

We next present conditions for the existence of $c^\dagger$.

**Proposition 2.6.6 ([15, Propositian 4.6])** *Let $e$, $f$ and $c$ be as in (2.6.15). If any of the following conditions hold, then $c^\dagger$ exists.*

(1) $w = cc^* + dd^*$ *is invertible.*
(2) $z = c^*c + a^*a$ *is invertible.*

*Proof*

(1) First, we prove that if $w$ is invertible then $c$ is regular. Let $x = c^*w^{-1}$. Since $ew = cc^*$ we also have $e = cc^*w^{-1} = cx$. Using this, we get $cxc = ec = c$. Hence, $x$ is a {1}-inverse of $c$.

   Now, choose $c^{(1)} = c^*w^{-1}$. Then $cc^{(1)} = e$ and by Theorem 2.2.14, $c^\dagger$ exists if and only if $v = cc^* + 1 - e$ is invertible. We only need to show that $v$ is invertible. Since $(cc^* + 1 - e)(dd^* + e) = cc^* + dd^*$ we have that $cc^* + 1 - e$ is invertible because both $w$ and $dd^* + e$ are invertible, the last one due to the fact that $d^\dagger$ exists.

(2) The proof is similar to the case (1). □

**Theorem 2.6.7 ([15, Theorem 4.7])** *Let $T$ be as in (2.6.8) and let $e$, $f$, and $c$ be as in (2.6.15). If $c^\dagger$ exists, then $T^\dagger$ exists if and only if both $u = 1_m + (ba^\dagger)^*egba^\dagger$ and $v = 1_l + d^\dagger bhf(d^\dagger b)^*$ are invertible, where $g = 1_k - cc^\dagger$ and $h = 1_n - c^\dagger c$.*

*In this case,*

$$\begin{bmatrix} a & 0 \\ b & d \end{bmatrix}^{\dagger} = \begin{bmatrix} (1 - hf(d^{\dagger}b)^* v^{-1} d^{\dagger}b)\sigma & \gamma \\ -\rho ba^{\dagger}u^{-1} & \rho(1_l - ba^{\dagger}u^{-1}(ba^{\dagger})^* eg) \end{bmatrix}, \qquad (2.6.16)$$

*where* $\rho = v^{-1} d^{\dagger}(1_l - bc^{\dagger})$, $\sigma = (1_k - c^{\dagger}b)a^{\dagger}u^{-1}$ *and*

$$\gamma = c^{\dagger} + hf(d^{\dagger}b)^* \rho(1_l \quad ba^{\dagger}u^{-1}(ba^{\dagger})^* eg) + \sigma(ba^{\dagger})^* eg. \qquad (2.6.17)$$

**Proof** We can apply Theorem 2.6.4 with $a^{(1,2,3)} = a^{\dagger}$, $d^{(1,3)} = d^{\dagger}$, and $c^{(1,3)} = c^{\dagger}$ to obtain that $T^{(1,3)}$ exists if and only if $u = 1_m + (ba^{\dagger})^* egba^{\dagger}$ is invertible and, using that $d^{\dagger}e = 0$, a {1, 3}-inverse of $T$ is of the form

$$T^{(1,3)} = \begin{bmatrix} 1_n & 0 \\ -d^{\dagger}b & 1_l \end{bmatrix} \begin{bmatrix} \sigma & \sigma(ba^{\dagger})^* eg + fc^{\dagger} \\ 0 & d^{\dagger} \end{bmatrix} = \tilde{Q}X,$$

where $\sigma = (1 - c^{\dagger}b) a^{\dagger}u^{-1}$. Similarly, we apply Theorem 2.6.5 to derive that $T^{(1,4)}$ exists if and only if $v = 1_l + d^{\dagger}bhf(d^{\dagger}b)^*$ is invertible, and a {1, 4}-inverse of $T$ is of the form

$$T^{(1,4)} = \begin{bmatrix} a^{\dagger} & hf(d^{\dagger}b)^* \rho + c^{\dagger}e \\ 0 & \rho \end{bmatrix} \begin{bmatrix} 1_m & 0 \\ -ba^{\dagger} & 1_k \end{bmatrix} = Y\tilde{P},$$

where $\rho = v^{-1}d^{\dagger}(1_l - bc^{\dagger})$. We now compute $T^{\dagger} = T^{(1,4)}TT^{(1,3)} = Y\tilde{P}T\tilde{Q}X$. One sees that

$$\tilde{P}T\tilde{Q} = \begin{bmatrix} a & 0 \\ bf - dd^{\dagger}b & d \end{bmatrix}.$$

Using that $\rho(bf - dd^{\dagger}b) = -v^{-1}d^{\dagger}b(a^{\dagger}a + c^{\dagger}c)$ and $\rho d = v^{-1}d^{\dagger}d$, we have

$$Y\tilde{P}T\tilde{Q} = \begin{bmatrix} (1 - hf(d^{\dagger}b)^* v^{-1}d^{\dagger}b)(a^{\dagger}a + c^{\dagger}c) & hf(d^{\dagger}b)^* v^{-1}d^{\dagger}d \\ -v^{-1}d^{\dagger}b(a^{\dagger}a + c^{\dagger}c) & v^{-1}d^{\dagger}d \end{bmatrix}.$$

Using $(a^{\dagger}a + c^{\dagger}c)\sigma = \sigma$, we obtain

$$T^{\dagger} = Y\tilde{P}T\tilde{Q}X = \begin{bmatrix} (1 - hf(d^{\dagger}b)^* v^{-1}d^{\dagger}b)\sigma & \gamma \\ -v^{-1}d^{\dagger}b\sigma & -v^{-1}d^{\dagger}b\sigma(ba^{\dagger})^* eg + \rho \end{bmatrix},$$

where $\gamma = \left(1 + hf\left(d^\dagger b\right)^* v^{-1} d^\dagger b\right)\left(\sigma\left(ba^\dagger\right)^* eg + c^\dagger\right) + hf\left(d^\dagger b\right)^* v^{-1} d^\dagger$.

Finally, the formula in this theorem is proved by taking into account that $v^{-1} d^\dagger b \sigma = \rho b a^\dagger u^{-1}$.  $\square$

We can rewrite $u$ and $v$ in Theorem 2.6.7 as $u = 1_m + \left(egba^\dagger\right)^*\left(egba^\dagger\right)$ and $v = 1_l + d^\dagger bhf\left(d^\dagger bhf\right)^*$. On account of this, we obtain the next corollary.

**Corollary 2.6.8 ([15, Corollary 4.8])** *Let $R$ be a GN ring. Consider $T = \begin{bmatrix} a & 0 \\ b & d \end{bmatrix}$ with $a, b, d \in R$. Let $e$, $f$, and $c$ be as in (2.6.15). If $a^\dagger$, $d^\dagger$ and $c^\dagger$ exist, then $T^\dagger$ exists and is given by (2.6.16)–(2.6.17).*

**Corollary 2.6.9 ([15, Corollary 4.9])** *Let $T = \begin{bmatrix} a & 0 \\ b & d \end{bmatrix}$ be such that $a^\dagger$ exists and $d$ be an invertible matrix of order $k \times k$. Then $T^\dagger$ exists if and only if $v = 1_k + d^{-1} bf\left(d^{-1} b\right)^*$ is invertible. In this case*

$$\begin{bmatrix} a & 0 \\ b & d \end{bmatrix}^\dagger = \begin{bmatrix} \left(1_n - f\left(d^{-1} b\right)^* v^{-1} d^{-1} b\right) a^\dagger & f\left(d^{-1} b\right)^* v^{-1} d^{-1} \\ -v^{-1} d^{-1} b a^\dagger & v^{-1} d^{-1} \end{bmatrix}.$$

**Proof** Follows from previous theorem with $d^\dagger = d^{-1}$. Then $e = 0$, $c = 0$, $g = 1_k$, and $h = 1_n$ and (2.6.16) reduces to the formula in this corollary.  $\square$

## 2.7   The Moore-Penrose Inverse of a Companion Matrix

In this section, we consider the Moore-Penrose invertibility of the $(n+1) \times (n+1)$ companion matrix $M = \begin{bmatrix} 0 & a \\ I_n & b \end{bmatrix}$ with $a \in R$ and $b \in R^n$.

Suppose $a^\dagger$ exists and consider the unit

$$u = aa^* + 1 - aa^\dagger, \quad \text{with } u^{-1} = (a^*)^\dagger a^\dagger + 1 - aa^\dagger.$$

Note that $u^{-1} a = (a^*)^\dagger$ and $a^* u^{-1} = a^\dagger$.
The matrix

$$A = \begin{bmatrix} a & 0 \\ b & I_n \end{bmatrix}$$

is Moore-Penrose invertible if and only if $U = AA^* + I_{n+1} - AA^-$ is invertible for one, and hence, all choices of inner inverses $A^-$ of $A$ by Theorem 2.2.14. We may take

$$A^- = \begin{bmatrix} a^\dagger & 0 \\ -ba^\dagger & I_n \end{bmatrix},$$

for which choice we obtain

$$AA^- = \begin{bmatrix} aa^\dagger & 0 \\ 0 & I_n \end{bmatrix}$$

and

$$U = \begin{bmatrix} aa^* + 1 - aa^\dagger & ab^* \\ ba^* & bb^* + I_n \end{bmatrix} = \begin{bmatrix} 1 & 0 \\ ba^* u^{-1} & I_n \end{bmatrix} \begin{bmatrix} u & 0 \\ 0 & Z \end{bmatrix} \begin{bmatrix} 1 & u^{-1} ab^* \\ 0 & I_n \end{bmatrix},$$

where

$$Z = bb^* + I_n - ba^* u^{-1} ab^*$$

$$= I_n + b \left( 1 - a^* u^{-1} a \right) b^*$$

$$= I_n + b \left( 1 - a^\dagger a \right) b^*.$$

Now, the invertibility of $Z$ is equivalent to $z = 1 + b^* b \left( 1 - a^\dagger a \right)$ being a unit of $R$. Writing $b = [b_1, b_2, \cdots, b_n]^{\mathrm{T}}$, this is the same as

$$z = 1 + \sum_{i=1}^{n} b_i^* b_i \left( 1 - a^\dagger a \right) \in R^{-1}.$$

**Theorem 2.7.1 ([95, Theorem 2.1])** *Let* $a \in R$ *such that* $a^\dagger$ *exists and* $b = [b_1, b_2, \cdots, b_n]^{\mathrm{T}}$. *Then the following are equivalent:*

(1) *The companion matrix* $M = \begin{bmatrix} 0 & a \\ I_n & b \end{bmatrix}$ *is Moore-Penrose invertible.*

(2) $1 + \left( 1 - a^\dagger a \right) b^* b \left( 1 - a^\dagger a \right) \in R^{-1}$.

(3) $1 + b^* b \left( 1 - a^\dagger a \right) \in R^{-1}$.

(4) $1 + \left( 1 - a^\dagger a \right) b^* b \in R^{-1}$.

We now carry out the construction of the Moore-Penrose inverse of the companion matrix, in the case $a^\dagger$ exists. Using Theorem 2.2.14

$$A^\dagger = \left( U^{-1} A \right)^*,$$

which leads to

$$\begin{bmatrix} 0 & a \\ I_n & b \end{bmatrix}^\dagger = \begin{bmatrix} 0 & I_n \\ 1 & 0 \end{bmatrix} A^\dagger = \begin{bmatrix} 0 & I_n \\ 1 & 0 \end{bmatrix} \left( U^{-1} A \right)^*.$$

Note that $u$, $U$ and $Z$ are Hermitian, and hence also are their inverses. Therefore,

$$U^{-1} = \begin{bmatrix} 1 & -(a^*)^\dagger b^* \\ 0 & I_n \end{bmatrix} \begin{bmatrix} u^{-1} & 0 \\ 0 & Z^{-1} \end{bmatrix} \begin{bmatrix} 1 & 0 \\ ba^\dagger & I_n \end{bmatrix}$$

$$= \begin{bmatrix} u^{-1} + (a^*)^\dagger b^* Z^{-1} ba^\dagger & -(a^*)^\dagger b^* Z^{-1} \\ -Z^{-1} ba^\dagger & Z^{-1} \end{bmatrix}$$

with $Z^{-1} = (I_n + b(1 - a^\dagger a)b^*)^{-1} = I_n - b(1 - a^\dagger a)z^{-1}(1 - a^\dagger a)b^*$ and $z = 1 + (1 - a^\dagger a)b^* b(1 - a^\dagger a)$. Then

$$A^\dagger = A^* \left( U^* \right)^{-1}$$

$$= \begin{bmatrix} a^* u^{-1} + a^\dagger ab^* Z^{-1} ba^\dagger - b^* Z^{-1} ba^\dagger & -a^\dagger ab^* Z^{-1} + b^* Z^{-1} \\ -Z^{-1} ba^\dagger & Z^{-1} \end{bmatrix}$$

$$= \begin{bmatrix} a^\dagger - \left(1 - a^\dagger a\right) b^* Z^{-1} ba^\dagger & \left(1 - a^\dagger a\right) b^* Z^{-1} \\ -Z^{-1} ba^\dagger & Z^{-1} \end{bmatrix}.$$

Finally,

$$\begin{bmatrix} 0 & a \\ I_n & b \end{bmatrix}^\dagger = \begin{bmatrix} 0 & I_n \\ 1 & 0 \end{bmatrix} A^\dagger$$

$$= \begin{bmatrix} -Z^{-1} ba^\dagger & Z^{-1} \\ a^\dagger - (1 - a^\dagger a)b^* Z^{-1} ba^\dagger & (1 - a^\dagger a)b^* Z^{-1} \end{bmatrix}.$$

We note that the companion matrix

$$\begin{bmatrix} 0 & a \\ I_n & b \end{bmatrix}$$

is regular if and only if $a$ is regular. This follows from the factorization

$$\begin{bmatrix} 0 & a \\ I_n & b \end{bmatrix} = \begin{bmatrix} 0 & 1 \\ I_n & 0 \end{bmatrix} \begin{bmatrix} I_n & 0 \\ 0 & a \end{bmatrix} \begin{bmatrix} I_n & b \\ 0 & 1 \end{bmatrix}.$$

Suppose $a$ is regular and let $a^+$ be any reflexive inverse of $a$. The matrix

$$A = \begin{bmatrix} a & 0 \\ b & I_n \end{bmatrix}$$

is Moore-Penrose invertible if and only if $V = A^*A + I_{n+1} - A^- A$ is invertible for one, and hence, all choices of inner inverses $A^-$ of $A$ by Theorem 2.2.14. We may take

$$A^- = \begin{bmatrix} a^+ & 0 \\ -ba^+ & I_n \end{bmatrix},$$

for which choice we obtain

$$A^- A = \begin{bmatrix} a^+a & 0 \\ -ba^+a + b & I_n \end{bmatrix}$$

and

$$V = \begin{bmatrix} a^*a + 1 - a^+a + b^*b \, b^* & \\ ba^+a & I_n \end{bmatrix} = \begin{bmatrix} 1 & b^* \\ 0 & I_n \end{bmatrix} \begin{bmatrix} \zeta & 0 \\ 0 & I_n \end{bmatrix} \begin{bmatrix} 1 & 0 \\ ba^+a & I_n \end{bmatrix},$$

where

$$\zeta = a^*a + 1 - a^+a + b^*b \left( 1 - a^+a \right)$$

$$= a^*a + 1 - a^+a + \sum_{i=1}^{n} b_i^* b_i \left( 1 - a^+a \right)$$

with $b = \left[ b_1, b_2, \cdots, b_n \right]^{\mathrm{T}}$.

**Theorem 2.7.2 ([95, Theorem 3.1])** *Let $a \in R$ and $b = \left[ b_1, \cdots, b_n \right]^{\mathrm{T}}$. Then the companion matrix $M = \begin{bmatrix} 0 & a \\ I_n & b \end{bmatrix}$ is Moore-Penrose invertible if and only if $a$ is regular and*

$$\zeta = a^*a + 1 - a^+a + \sum_{i=1}^{n} b_i^* b_i \left( 1 - a^+a \right) \in R^{-1}$$

*for some reflexive inverse $a^+$ of $a$.*

We now construct the Moore-Penrose inverse of the companion matrix, in the case $a$ is regular. Using Theorem 2.2.14, the Moore-Penrose inverse of $A$ is given by $A^\dagger = \left( AV^{-1} \right)^*$ where

$$V^{-1} = \left( A^*A + I_{n+1} - A^- A \right)^{-1}$$

$$= \begin{bmatrix} 1 & 0 \\ -ba^+a & I_n \end{bmatrix} \begin{bmatrix} \zeta^{-1} & 0 \\ 0 & I_n \end{bmatrix} \begin{bmatrix} 1 & -b^* \\ 0 & I_n \end{bmatrix}$$

and $\zeta = a^*a + 1 - a^+a + b^*b\left(1 - a^+a\right)$. Then

$$\left(V^{-1}\right)^* = \begin{bmatrix} (\zeta^*)^{-1} & -(\zeta^*)^{-1}\left(a^+a\right)^*b^* \\ -b(\zeta^*)^{-1} & I_n + b(\zeta^*)^{-1}\left(a^+a\right)^*b^* \end{bmatrix}.$$

Substituting in the expression of $A^\dagger$,

$$A^\dagger = \left(V^{-1}\right)^* A^*$$

$$= \begin{bmatrix} (\zeta^*)^{-1} & -(\zeta^*)^{-1}(a^+a)^*b^* \\ -b(\zeta^*)^{-1} & I_n + b(\zeta^*)^{-1}\left(a^+a\right)^*b^* \end{bmatrix} \begin{bmatrix} a^* & b^* \\ 0 & I_n \end{bmatrix}$$

$$= \begin{bmatrix} (\zeta^*)^{-1}a^* & (\zeta^*)^{-1}\left(1 - (a^+a)^*\right)b^* \\ -b(\zeta^*)^{-1}a^* & I_n - b(\zeta^*)^{-1}\left(1 - (a^+a)^*\right)b^* \end{bmatrix}$$

from which we deduce

$$\begin{bmatrix} 0 & a \\ I_n & b \end{bmatrix}^\dagger = \begin{bmatrix} 0 & I_n \\ 1 & 0 \end{bmatrix} A^\dagger$$

$$= \begin{bmatrix} -b(\zeta^*)^{-1}a^* & I_n - b(\zeta^*)^{-1}\left(1 - (a^+a)^*\right)b^* \\ (\zeta^*)^{-1}a^* & (\zeta^*)^{-1}\left(1 - (a^+a)^*\right)b^* \end{bmatrix}.$$

## 2.8 The Moore-Penrose Inverse of a Sum of Morphisms

In this section, $\mathscr{C}$ is an additive category with an involution $*$.

**Lemma 2.8.1 ([63, Proposition 1])** *Suppose that* $\varphi : X \to Y$ *is a morphism of* $\mathscr{C}$ *with a* $\{1, 2\}$*-inverse* $\varphi^{(1,2)} : Y \to X$ *and that* $\eta : X \to Y$ *is a morphism of* $\mathscr{C}$ *such that* $1_X + \varphi^{(1,2)}\eta$ *is invertible. Let* $\epsilon = (1_Y - \varphi\varphi^{(1,2)})\eta(1_X + \varphi^{(1,2)}\eta)^{-1}(1_X - \varphi^{(1,2)}\varphi)$. *Then* $f = \varphi + \eta - \epsilon$ *has a* $\{1\}$*-inverse and* $(1_X + \varphi^{(1,2)}\eta)^{-1}\varphi^{(1,2)} \in f\{1, 2\}$. *Moreover, if* $\tau \in (\varphi + \eta)\{1\}$, *we have* $\tau \in \epsilon\{1\}$.

**Lemma 2.8.2 ([126, Lemma 2])** *Suppose that* $\varphi : X \to Y$ *is a morphism of* $\mathscr{C}$ *with a* $\{1, 2, i\}$*-inverse* $\varphi^{(1,2,i)} : Y \to X$ *(with respect to the involution* $*$*) and that* $\eta : X \to Y$ *is a morphism of* $\mathscr{C}$ *such that* $1_X + \varphi^{(1,2,i)}\eta$ *is invertible. Let*

$$\alpha_i = (1_X + \varphi^{(1,2,i)}\eta)^{-1}, \quad \beta_i = (1_X + \eta\varphi^{(1,2,i)})^{-1}$$

*and* $\epsilon_i = (1_Y - \varphi\varphi^{(1,2,i)})\eta(1_X + \varphi^{(1,2,i)}\eta)^{-1}(1_X - \varphi^{(1,2,i)}\varphi)$, $i = 3, 4$. *We then have the following equalities:*

(1) $\alpha_i\varphi^{(1,2,i)}\eta = \varphi^{(1,2,i)}\eta\alpha_i$, $\beta_i\eta\varphi^{(1,2,i)} = \eta\varphi^{(1,2,i)}\beta_i$.
(2) $1_X - \alpha_i\varphi^{(1,2,i)}\eta = \alpha_i$, $1_Y - \beta_i\eta\varphi^{(1,2,i)} = \beta_i$.

(3) $\varphi^{(1,2,i)}\beta_i = \alpha_i\varphi^{(1,2,i)}, \ \beta_i\eta = \eta\alpha_i.$
(4) $\varphi^{(1,2,i)}\epsilon_i = 0$ and $\epsilon_i\varphi^{(1,2,i)} = 0.$

**Proof** Follows from direct calculation and Jacobson's lemma.                    □

**Proposition 2.8.3 ([126, Proposition 2])** *Suppose that $\varphi : X \to Y$ is a morphism of $\mathscr{C}$ with a $\{1, 2, 4\}$-inverse $\varphi^{(1,2,4)} : Y \to X$ and that $\eta : X \to Y$ is a morphism of $\mathscr{C}$ such that $1_X + \varphi^{(1,2,4)}\eta$ is invertible. Let $\alpha_4, \epsilon_4$ be as in Lemma 2.8.2 and*

$$\lambda = \alpha_4(1_X - \varphi^{(1,2,4)}\varphi)\eta^*(\varphi^{(1,2,4)})^*\alpha_4^*.$$

*Then the following conditions are equivalent:*

(1) $f = \varphi + \eta - \epsilon_4$ *is $\{1, 2, 4\}$-invertible.*
(2) $1_X - \lambda$ *is invertible.*
(3) $1_X - \lambda$ *is left invertible.*

*In this case*

$$(1_X - \lambda)^{-1}\alpha_4\varphi^{(1,2,4)} \in f\{1, 2, 4\}$$

*and*

$$(1_X - \lambda)^{-1} = 1_X - \varphi^{(1,2,4)}\varphi + f^{(1,2,4)}f\varphi^{(1,2,4)}\varphi.$$

**Proof** By Lemma 2.8.1, $\alpha_4\varphi^{(1,2,4)} \in f\{1, 2\}$. Writting $f_0^{(1,2)} = \alpha_4\varphi^{(1,2,4)}$, we then have the following facts:

$$\begin{aligned}
f_0^{(1,2)}f &= \alpha_4\varphi^{(1,2,4)}f \\
&= \alpha_4\varphi^{(1,2,4)}(\varphi + \eta - \epsilon_4) \\
&= \alpha_4(\varphi^{(1,2,4)}\varphi + \varphi^{(1,2,4)}\eta) \\
&= \alpha_4\varphi^{(1,2,4)}\varphi(1_X + \varphi^{(1,2,4)}\eta) \\
&= \alpha_4\varphi^{(1,2,4)}\varphi\alpha_4^{-1},
\end{aligned}$$

$$\begin{aligned}
1_X - f_0^{(1,2)}f &= \alpha_4(1_X - \varphi^{(1,2,4)}\varphi)\alpha_4^{-1} \\
&= \alpha_4(1_X - \varphi^{(1,2,4)}\varphi)(1_X + \varphi^{(1,2,4)}\eta) \\
&= \alpha_4(1_X - \varphi^{(1,2,4)}\varphi),
\end{aligned}$$

$$(1_X - f_0^{(1,2)} f)(f_0^{(1,2)} f)^* = \alpha_4(1_X - \varphi^{(1,2,4)}\varphi)(\alpha_4\varphi^{(1,2,4)}\varphi\alpha_4^{-1})^*$$

$$= \alpha_4(1_X - \varphi^{(1,2,4)}\varphi)^*(\varphi^{(1,2,4)}\varphi\alpha_4^{-1})^*\alpha_4^*$$

$$= \alpha_4[\varphi^{(1,2,4)}\varphi(1_X + \varphi^{(1,2,4)}\eta)(1_X - \varphi^{(1,2,4)}\varphi)]^*\alpha_4^*$$

$$= \alpha_4[\varphi^{(1,2,4)}\eta(1_X - \varphi^{(1,2,4)}\varphi)]^*\alpha_4^*$$

$$= \alpha_4(1_X - \varphi^{(1,2,4)}\varphi)\eta^*(\varphi^{(1,2,4)})^*\alpha_4^* = \lambda.$$

This implies

$$f\lambda = f(1_X - f_0^{(1,2)} f)(f_0^{(1,2)} f)^* = 0 \qquad (2.8.1)$$

and

$$(1_X - \lambda)f^* = [1_X - (1_X - f_0^{(1,2)} f)(f_0^{(1,2)} f)^*]f^*$$

$$= f^* - (1_X - f_0^{(1,2)} f)(f f_0^{(1,2)} f)^*$$

$$= f^* - (1_X - f_0^{(1,2)} f)f^* \qquad (2.8.2)$$

$$= f_0^{(1,2)} f f^*.$$

Now we are ready to show the equivalence of the three conditions.

(1) $\Rightarrow$ (2). First, we show that

$$1_X - \varphi^{(1,2,4)}\varphi + f^{(1,2,4)} f \varphi^{(1,2,4)}\varphi.$$

is a right inverse of $1_X - \lambda$. Note that

$$(1_X - \lambda)f^{(1,2,4)} f = (1_X - \lambda)(f^{(1,2,4)} f)^*$$

$$= (1_X - \lambda)f^*(f^{(1,2,4)})^*$$

$$= f_0^{(1,2)} f f^*(f^{(1,2,4)})^*$$

$$= f_0^{(1,2)} f$$

$$= \alpha_4\varphi^{(1,2,4)}(\varphi + \eta - \varepsilon_4)$$

$$= \alpha_4(\varphi^{(1,2,4)}\varphi + \varphi^{(1,2,4)}\eta)$$

$$= \alpha_4[\varphi^{(1,2,4)}\varphi - 1_X + (1_X + \varphi^{(1,2,4)}\eta)]$$

$$= 1_X + \alpha_4(\varphi^{(1,2,4)}\varphi - 1_X).$$

Multiplying the equality by $\varphi^{(1,2,4)}\varphi$ on the right, we obtain

$$(1_X - \lambda)f^{(1,2,4)}f\varphi^{(1,2,4)}\varphi = \varphi^{(1,2,4)}\varphi.$$

This equality together with

$$\lambda(1_X - \varphi^{(1,2,4)}\varphi) = \alpha_4(1_X - \varphi^{(1,2,4)}\varphi)\eta^*(\varphi^{(1,2,4)})^*\alpha_4^*(1_X - \varphi^{(1,2,4)}\varphi)$$
$$= \alpha_4(1_X - \varphi^{(1,2,4)}\varphi)\eta^*[(1_X - \varphi^{(1,2,4)}\varphi)\alpha_4\varphi^{(1,2,4)}]^*$$
$$= \alpha_4(1_X - \varphi^{(1,2,4)}\varphi)\eta^*[(1_X - \varphi^{(1,2,4)}\varphi)\varphi^{(1,2,4)}\beta_4]^*$$
$$= 0$$

implies

$$(1_X - \lambda)[1_X - \varphi^{(1,2,4)}\varphi + f^{(1,2,4)}f\varphi^{(1,2,4)}\varphi] = 1_X - \varphi^{(1,2,4)}\varphi + \varphi^{(1,2,4)}\varphi = 1_X.$$

Next, we show that this right inverse is also a left inverse for $1_X - \lambda$. We can write $\lambda$ as $\alpha_4(1_X - \varphi^{(1,2,4)}\varphi)(1_X - \alpha_4^*)$ since $1_X - \alpha_4^* = (\alpha_4\varphi^{(1,2,4)}\eta)^*$.

Note that

$$\varphi^{(1,2,4)}\varphi\lambda = \varphi^{(1,2,4)}\varphi\alpha_4(1_X - \varphi^{(1,2,4)}\varphi)(1_X - \alpha_4^*)$$
$$= \varphi^{(1,2,4)}\varphi(1_X - \varphi^{(1,2,4)}\beta_4\eta)(1_X - \varphi^{(1,2,4)}\varphi)(1_X - \alpha_4^*)$$
$$= (\varphi^{(1,2,4)}\varphi - \varphi^{(1,2,4)}\beta_4\eta)(1_X - \varphi^{(1,2,4)}\varphi)(1_X - \alpha_4^*)$$
$$= -\varphi^{(1,2,4)}\beta_4\eta(1_X - \varphi^{(1,2,4)}\varphi)(1_X - \alpha_4^*)$$
$$= (\alpha_4 - 1_X)(1_X - \varphi^{(1,2,4)}\varphi)(1_X - \alpha_4^*)$$
$$= \lambda - (1_X - \varphi^{(1,2,4)}\varphi)(1_X - \alpha_4^*).$$

This implies $(1_X - \varphi^{(1,2,4)}\varphi)\lambda = (1_X - \varphi^{(1,2,4)}\varphi)(1_X - \alpha_4^*)$ and

$$f^{(1,2,4)}f\varphi^{(1,2,4)}\varphi\lambda = f^{(1,2,4)}f[\lambda - (1_X - \varphi^{(1,2,4)}\varphi)(1_X - \alpha_4^*)]$$
$$= f^{(1,2,4)}f\lambda - f^{(1,2,4)}f(1_X - \varphi^{(1,2,4)}\varphi)(1_X - \alpha_4^*)$$
$$= -f^{(1,2,4)}f(1_X - \varphi^{(1,2,4)}\varphi)(1_X - \alpha_4^*).$$

On the other hand,

$$1_X - \varphi^{(1,2,4)}\varphi + \varphi^{(1,2,4)}\varphi f^{(1,2,4)}f$$
$$= f^{(1,2,4)}f + (1_X - \varphi^{(1,2,4)}\varphi)(1_X - f^{(1,2,4)}f)$$
$$= f^{(1,2,4)}f + (1_X + \varphi^{(1,2,4)}\eta - \varphi^{(1,2,4)}f)(1_X - f^{(1,2,4)}f)$$
$$= f^{(1,2,4)}f + (1_X + \varphi^{(1,2,4)}\eta)(1_X - f^{(1,2,4)}f)$$
$$= f^{(1,2,4)}f + \alpha_4^{-1}(1_X - f^{(1,2,4)}f).$$

Conjugating the two sides of the above equality, we have

$$1_X - \varphi^{(1,2,4)}\varphi + f^{(1,2,4)}f\varphi^{(1,2,4)}\varphi = f^{(1,2,4)}f + (1_X - f^{(1,2,4)}f)(\alpha_4^*)^{-1}.$$

Finally,

$$
\begin{aligned}
&(1_X - \varphi^{(1,2,4)}\varphi + f^{(1,2,4)}f\varphi^{(1,2,4)}\varphi)(1_X - \lambda) \\
&= 1_X - \varphi^{(1,2,4)}\varphi - (1_X - \varphi^{(1,2,4)}\varphi)\lambda + f^{(1,2,4)}f\varphi^{(1,2,4)}\varphi \\
&\quad - f^{(1,2,4)}f\varphi^{(1,2,4)}\varphi\lambda \\
&= 1_X - \varphi^{(1,2,4)}\varphi - (1_X - \varphi^{(1,2,4)}\varphi)(1_X - \alpha_4^*) + f^{(1,2,4)}f\varphi^{(1,2,4)}\varphi \\
&\quad + f^{(1,2,4)}f(1_X - \varphi^{(1,2,4)}\varphi)(1_X - \alpha_4^*) \\
&= (1_X - \varphi^{(1,2,4)}\varphi)\alpha_4^* + f^{(1,2,4)}f\varphi^{(1,2,4)}\varphi + f^{(1,2,4)}f(1_X - \alpha_4^*) \\
&\quad - f^{(1,2,4)}f\varphi^{(1,2,4)}\varphi + f^{(1,2,4)}f\varphi^{(1,2,4)}\varphi\alpha_4^* \\
&= f^{(1,2,4)}f(1_X - \alpha_4^*) + (1_X - \varphi^{(1,2,4)}\varphi + f^{(1,2,4)}f\varphi^{(1,2,4)}\varphi)\alpha_4^* \\
&= f^{(1,2,4)}f(1_X - \alpha_4^*) + [f^{(1,2,4)}f + (1_X - f^{(1,2,4)}f)(\alpha_4^*)^{-1}]\alpha_4^* \\
&= f^{(1,2,4)}f - f^{(1,2,4)}f\alpha_4^* + f^{(1,2,4)}f\alpha_4^* + 1_X - f^{(1,2,4)}f = 1_X.
\end{aligned}
$$

Hence $1_X - \lambda$ is invertible and

$$(1_X - \lambda)^{-1} = 1_X - \varphi^{(1,2,4)}\varphi + f^{(1,2,4)}f\varphi^{(1,2,4)}\varphi.$$

(2) $\Rightarrow$ (3). It is obvious.

(3) $\Rightarrow$ (1). Let $\omega$ be the left inverse of $1_X - \lambda$, i.e., $\omega(1_X - \lambda) = 1_X$. Since $(1_X - \lambda)f^* = f_0^{(1,2)}ff^*$ (see (2.8.2)), we have $f^* = \omega f_0^{(1,2)}ff^*$. Let $x = \omega f_0^{(1,2)}$, i.e., $f^* = xff^*$. Conjugating the two sides of the equality, we obtain that $f = ff^*x^*$. Then $xf = xff^*x^* = xf(xf)^*$ and $(xf)^* = (xf)(xf)^* = xf$. So

$$fxf = (fx)ff^*x^* = f(xff^*)x^* = ff^*x^* = f.$$

This means $x$ is a $\{1, 4\}$-inverse of $f$, so $xfx$ is a $\{1, 2, 4\}$-inverse of $f$, and $f\{1, 2, 4\} \neq \emptyset$. From (1) $\Rightarrow$ (2), we know that $1_X - \lambda$ is invertible and $\omega$ is the inverse of $1_X - \lambda$, i.e., $\omega = (1_X - \lambda)^{-1}$.

Since $f\lambda = 0$ (see (2.8.1)), $f(1_X - \lambda) = f$ and $f = f(1_X - \lambda)^{-1}$. We have

$$
\begin{aligned}
xfx &= (1_X - \lambda)^{-1}f_0^{(1,2)}f(1_X - \lambda)^{-1}f_0^{(1,2)} \\
&= (1_X - \lambda)^{-1}f_0^{(1,2)}ff_0^{(1,2)} \\
&= (1_X - \lambda)^{-1}f_0^{(1,2)} \\
&= x.
\end{aligned}
$$

So $x = (1_X - \lambda)^{-1}f_0^{(1,2)} = (1_X - \lambda)^{-1}\alpha_4\varphi^{(1,2,4)} \in f\{1, 2, 4\}$. This completes the proof.                                                                        $\square$

**Corollary 2.8.4 ([126, Corollary 2])** *Let $J(R)$ be the Jacobson radical of $R$. If an element $a$ of $R$ has a $\{1, 2, 4\}$-inverse $a^{(1,2,4)}$ and $j$ is an element of $J(R)$, then $(a + j)\{1, 2, 4\} \neq \emptyset$ if and only if*

$$(1 - aa^{(1,2,4)})j(1 + a^{(1,2,4)}j)^{-1}(1 - a^{(1,2,4)}a) = 0.$$

*In that case, $(1 - \lambda)^{-1}(1 + a^{(1,2,4)}j)^{-1}a^{(1,2,4)} \in (a + j)\{1, 2, 4\}$, where*

$$\lambda = (1 + a^{(1,2,4)}j)^{-1}(1 - a^{(1,2,4)}a)j^*(a^{(1,2,4)})^*[(1 + a^{(1,2,1)}j)^{-1}]^*.$$

**Proof** Since $j \in J(R)$, the element $1 + a^{(1,2,4)}j$ is invertible. Here we have $\alpha_4 = (1 + a^{(1,2,4)}j)^{-1}$ and $\lambda = \alpha_4(1 - a^{(1,2,4)}a)j^*(a^{(1,2,4)})^*\alpha_4^*$. Therefore

$$\lambda^* = a_4a^{(1,2,4)}j(1 - a^{(1,2,4)}a)\alpha_4^* \in J(R)$$

and $1 - \lambda^*$ is invertible, which follows that $1 - \lambda$ is invertible.

($\Leftarrow$): By hypothesis, $\epsilon_4 = 0$, so $f = a + j$. Since $1 - \lambda$ is invertible, $(a + j)\{1, 2, 4\} \neq \emptyset$ by Proposition 2.8.3. In that case, $(1 - \lambda)^{-1}\alpha_4a^{(1,2,4)} \in (a + j)\{1, 2, 4\}$.

($\Rightarrow$): Assume $\tau \in (a + j)\{1, 2, 4\}$. From Lemma 2.8.1, $\tau \in \epsilon_4\{1\}$, i.e., $\epsilon_4\tau\epsilon_4 = \epsilon_4$ and $\epsilon_4(1 - \tau\epsilon_4) = 0$. Since

$$\epsilon_4 = (1 - aa^{(1,2,4)})j(1 + a^{(1,2,4)}j)^{-1}(1 - a^{(1,2,4)}a) \in J(R),$$

$1 - \tau\epsilon_4$ is invertible and $\epsilon_4 = 0$.                                                                □

Similar to Proposition 2.8.3, we have:

**Proposition 2.8.5 ([126, Proposition 3])** *Suppose that $\varphi : X \to Y$ is a morphism of $\mathscr{C}$ with a $\{1, 2, 3\}$-inverse $\varphi^{(1,2,3)} : Y \to X$ and that $\eta : X \to Y$ is a morphism of $\mathscr{C}$ such that $1_X + \varphi^{(1,2,3)}\eta$ is invertible. Let $\beta_3$, $\epsilon_3$ be as in Lemma 2.8.2 and*

$$\mu = \beta_3^*(\varphi^{(1,2,3)})^*\eta^*(1_Y - \varphi\varphi^{(1,2,3)})\beta_3.$$

*Then the following conditions are equivalent:*

(1) $f = \varphi + \eta - \epsilon_3$ is $\{1, 2, 3\}$-invertible.
(2) $1_Y - \mu$ is invertible.
(3) $1_Y - \mu$ is right invertible.

*In this case*

$$\varphi^{(1,2,3)}\beta_3(1_Y - \mu)^{-1} \in f\{1, 2, 3\}$$

*and*

$$(1_Y - \mu)^{-1} = 1_Y - \varphi\varphi^{(1,2,3)} + \varphi\varphi^{(1,2,3)}ff^{(1,2,3)}.$$

**Corollary 2.8.6 ([126, Corollary 3])** *Let $J(R)$ be the Jacobson radical of R. If an element a of R has a $\{1, 2, 3\}$-inverse $a^{(1,2,3)}$ and j is an element of $J(R)$, then $(a + j)\{1, 2, 3\} \neq \emptyset$ if and only if*

$$(1 - aa^{(1,2,3)})j(1 + a^{(1,2,3)}j)^{-1}(1 - a^{(1,2,3)}a) = 0.$$

*In that case, $a^{(1,2,3)}(1 + ja^{(1,2,3)})^{-1}(1 - \mu)^{-1} \in (a + j)\{1, 2, 3\}$, where*

$$\mu = [(1 + ja^{(1,2,3)})^{-1}]^*(a^{(1,2,3)})^*j^*(1 - aa^{(1,2,3)})(1 + ja^{(1,2,3)})^{-1}.$$

**Proposition 2.8.7 ([126, Proposition 4])** *Suppose that $\varphi : X \to Y$ is a morphism of $\mathscr{C}$ with Moore-Penrose inverse $\varphi^\dagger : Y \to X$, and that $\eta : X \to Y$ is a morphism of $\mathscr{C}$ such that $1_X + \varphi^\dagger\eta$ is invertible. Let*

$$\epsilon = (1_Y - \varphi\varphi^\dagger)\eta(1_X + \varphi^\dagger\eta)^{-1}(1_X - \varphi^\dagger\varphi),$$

$$\lambda = (1_X + \varphi^\dagger\eta)^{-1}(1_X - \varphi^\dagger\varphi)\eta^*(\varphi^\dagger)^*(1_X + \eta^*(\varphi^\dagger)^*)^{-1}.$$

*and*

$$\mu = (1_Y + (\varphi^\dagger)^*\eta^*)^{-1}(\varphi^\dagger)^*\eta^*(1_Y - \varphi\varphi^\dagger)(1_Y + \eta\varphi^\dagger)^{-1}.$$

*Then the following conditions are equivalent:*

*(1)  $f = \varphi + \eta - \epsilon$ is Moore-Penrose invertible.*
*(2)  $1_X - \lambda$ and $1_Y - \mu$ are invertible.*
*(3)  $1_X - \lambda$ is left invertible and $1_Y - \mu$ is right invertible.*

*In this case*

$$f^\dagger = (1_X - \lambda)^{-1}(1_X + \varphi^\dagger\eta)^{-1}\varphi^\dagger(1_Y - \mu)^{-1},$$

$$(1_X - \lambda)^{-1} = 1_X - \varphi^\dagger\varphi + f^\dagger f\varphi^\dagger\varphi,$$

*and*

$$(1_Y - \mu)^{-1} = 1_Y - \varphi\varphi^\dagger + \varphi\varphi^\dagger ff^\dagger.$$

***Proof*** The equivalence of conditions (1)–(3) can be obtained from Propositions 2.8.3 and 2.8.5.

By Proposition 2.8.3, $(1_X - \lambda)^{-1}(1_X + \varphi^\dagger\eta)^{-1}\varphi^\dagger \in f\{1, 2, 4\}$.

By Proposition 2.8.5, $\varphi^\dagger(1_Y + \eta\varphi^\dagger)^{-1}(1_Y - \mu)^{-1} \in f\{1, 2, 3\}$.

Hence $f^\dagger$ exists. If we denote $(1_X + \varphi^\dagger \eta)^{-1}$ and $(1_Y + \eta\varphi^\dagger)^{-1}$ by $\alpha$ and $\beta$, respectively, then

$$
\begin{aligned}
f^\dagger &= f^{(1,2,4)} f f^{(1,2,3)} \\
&= (1_X - \lambda)^{-1} \alpha\varphi^\dagger f \varphi^\dagger \beta (1_Y - \mu)^{-1} \\
&= (1_X - \lambda)^{-1} \alpha\varphi^\dagger (\varphi + \eta - \varepsilon)\varphi^\dagger \beta (1_Y - \mu)^{-1} \\
&= (1_X - \lambda)^{-1} \alpha(\varphi^\dagger \varphi + \varphi^\dagger \eta)\psi^\dagger \beta (1_Y - \mu)^{-1} \\
&= (1_X - \lambda)^{-1} \alpha(\varphi^\dagger + \varphi^\dagger \eta\varphi^\dagger)\beta (1_Y - \mu)^{-1} \\
&= (1_X - \lambda)^{-1} \alpha\varphi^\dagger (1_Y + \eta\varphi^\dagger)\beta (1_Y - \mu)^{-1} \\
&= (1_X - \lambda)^{-1} \alpha\varphi^\dagger (1_Y - \mu)^{-1}.
\end{aligned}
$$

$\square$

**Corollary 2.8.8 ([64, Proposition 2])** *Let $J(R)$ be the Jacobson radical of R. If an element a of R has the Moore-Penrose inverse $a^\dagger$ and j is an element of $J(R)$, then $a + j$ is Moore-Penrose invertible if and only if*

$$(1 - aa^\dagger)j(1 + a^\dagger j)^{-1}(1 - a^\dagger a) = 0.$$

*In that case,*

$$(a + j)^\dagger = (1 - \lambda)^{-1}(1 + a^\dagger j)^{-1}a^\dagger (1 - \mu)^{-1},$$

*in which*

$$\lambda = (1 + a^\dagger j)^{-1}(1 - a^\dagger a)j^*(a^\dagger)^*[(1 + a^\dagger j)^{-1}]^*,$$

$$\mu = [(1 + ja^\dagger)^{-1}]^*(a^\dagger)^* j^*(1 - aa^\dagger)(1 + ja^\dagger)^{-1}.$$

# Chapter 3
# Group Inverses

Unlike the Moore-Penrose inverse whose definition requires an involution in the setting, in this chapter we discuss another generalized inverse which can be considered in an arbitrary ring or semigroup and is closely related to strong regularity of elements. Historically, Goro Azumaya [1] proved in 1954 that every strongly regular element $a$ in a ring admits a unique reflexive inverse $x$ that commutes with $a$, although he did not assign any name and symbol to this $x$. Since an element $a$ possessing an Azumaya's generalized inverse $x$ belongs to a multiplicative group with identity $ax$ (see for example [24]), in 1967 Ivan Erdélyi [42] referred to this generalized inverse $x$ as the group inverse of $a$ and denoted it by $a^{\#}$.

This chapter begins with basic examples and representations for group inverses of complex matrices with index 1. Then for an element in an arbitrary semigroup or ring, we establish various equivalent characterizations of its group invertibility and give expressions of its group inverse. Moreover, we discuss group inverses in detail for some specific elements such as a special triple product $paq$, a sum of morphisms in an additive category, sums and differences of two group invertible elements in a (Dedekind-finite) ring, as well as block matrices and companion matrices over a ring.

Throughout this chapter, unless otherwise stated, the symbols $R$ and $S$ denote a ring with identity and a semigroup, respectively.

## 3.1 Group Inverses of Complex Matrices

In this section, the group inverse of a complex matrix is introduced. Then we give several necessary and sufficient conditions which guarantee the existence of the group inverse.

**Definition 3.1.1** Let $A \in \mathbb{C}^{n \times n}$. If $X \in \mathbb{C}^{n \times n}$ satisfies

$$AXA = A, \quad XAX = X, \quad AX = XA,$$

then $X$ is called the group inverse of $A$. It is unique and denoted by $A^{\#}$.

**Proposition 3.1.2** *Let $A \in \mathbb{C}^{n \times n}$. Then the following statements are equivalent:*

(1) *$A$ is group invertible*
(2) $\operatorname{rank}(A) = \operatorname{rank}(A^2)$.
(3) $\mathcal{R}(A) = \mathcal{R}(A^2)$.
(4) $\mathcal{N}(A) = \mathcal{N}(A^2)$.

*Proof* (1)$\Rightarrow$(2). Since $A$ is group invertible, we have $A = AA^{\#}A = A^2A^{\#}$ which follows that $\operatorname{rank}(A) = \operatorname{rank}(A^2)$.

(2)$\Rightarrow$(1). Since $\operatorname{rank}(A) = \operatorname{rank}(A^2)$, there exist $Y_1$ and $Y_2$ such that $A = A^2Y_1 = Y_2A^2$. It is easy to verify that $X = AY_1^2 = Y_2^2A$ is the group inverse of $A$.

(2)$\Leftrightarrow$(3)$\Leftrightarrow$(4). See Theorem 1.1.7.

$\square$

**Theorem 3.1.3** *Let $A \in \mathbb{C}^{n \times n}$ and $A = GH$ be a full rank decomposition. Then $A$ is group invertible if and only if $HG$ is invertible. In this case, $A^{\#} = G(HG)^{-2}H$.*

*Proof* Suppose $A$ is group invertible. By Proposition 3.1.2, we know that $\operatorname{rank}(A) = \operatorname{rank}(A^2)$, which implies $\operatorname{rank}(GH) = \operatorname{rank}(A)$. Therefore $GH$ is invertible.

Conversely, suppose $HG$ is invertible. It is easy to verify that $A^{\#} = G(HG)^{-2}H$.

$\square$

**Theorem 3.1.4** *Let $A = P^{-1} \begin{bmatrix} D & 0 \\ 0 & N \end{bmatrix} P$ where $D$ is invertible and $N$ is nilpotent.*

*Then $A$ is group invertible if and only if $N = 0$. In this case, $A^{\#} = P^{-1} \begin{bmatrix} D^{-1} & 0 \\ 0 & 0 \end{bmatrix} P$.*

*Proof* Suppose $A$ is group invertible. By Proposition 3.1.2, we know that $\operatorname{rank}(A) = \operatorname{rank}(A^2)$, which implies $\operatorname{rank}(N) = \operatorname{rank}(N^2)$. Since $N$ is nilpotent, it follows that $N = 0$.

Conversely, suppose $N = 0$. It is easy to verify that $A^{\#} = P^{-1} \begin{bmatrix} D^{-1} & 0 \\ 0 & 0 \end{bmatrix} P$. $\square$

## 3.2   Characterizations of Group Inverses of Elements in Semigroups and Rings

In this section, we firstly give the definition of the group inverse of elements in semigroups and rings. Then several necessary and sufficient conditions which guarantee the existence of the group inverse of an element are presented.

**Definition 3.2.1** An element $a$ in a semigroup $S$ is said to be group invertible if there exists $x \in S$ such that

$$axa = a, \quad xax = x, \quad ax = xa.$$

In this case, such $x$ is called a group inverse of $a$. It is unique and denoted by $a^{\#}$.

Hereinafter, the set of all group invertible elements of a semigroup $S$ will be denoted by $S^{\#}$.

**Theorem 3.2.2 ([1, Lemma 1])** *An element $a \in S$ is group invertible if and only if $a \in a^2 S \cap S a^2$. Moreover, if $x, y \in S$ satisfy $a = a^2 x = y a^2$, then $a^{\#} = yax = y^2 a = ax^2$.*

**Proof** Suppose $a = a^2 x = y a^2$. By computation, we have $yax = y a^2 x^2 = ax^2 = y^2 a$. Take $t = yax = y^2 a = ax^2$. Then

$$ata = a^2 x^2 a = axa = axa = ya^2 xa = ya^2 = a,$$

$$tat = y^2 a^3 x^2 = ya^2 x^2 = yax = t,$$

$$at = a^2 x^2 = ax = ya^2 x = ya = y^2 a^2 = ta.$$

$\square$

**Proposition 3.2.3 ([52, Proposition 7])** *Let $a \in R$. Then $a \in R^{\#}$ if and only if $R = aR \oplus a^{\circ}$ if and only if $R = Ra \oplus {}^{\circ}a$.*
*In this case,*

$$a^{\#} = ax^2 = y^2 a, \tag{3.2.1}$$

*where $1 = ax + u = ya + v$ for some $x, y \in R$, $u \in a^{\circ}$ and $v \in {}^{\circ}a$.*

**Proof** We shall prove that $a \in R^{\#}$ if and only if $R = aR \oplus a^{\circ}$. Suppose $a$ is group invertible. It is easy to verify that $R = aa^{\#} R \oplus (1 - aa^{\#})R$. Since $aa^{\#} R = aR$ and $a^{\circ} = (1 - aa^{\#})R$, we have $R = aR \oplus a^{\circ}$. Conversely, if $R = aR \oplus a^{\circ}$, then there exists an idempotent $e$ such that $aR = eR$ and $a^{\circ} = (1 - e)R$. Therefore, $a$ is regular , $aR = eR$ and $Ra = Re$. There exist $x_1$ and $x_2$ such that $e = ax_1 = x_2 a$. So, $x_2 a^2 = a^2 x_1 = a$. By Theorem 3.2.2, we can easily complete the rest proof. $\square$

**Theorem 3.2.4 ([108, Theorem 2.7])** *Let $a, x \in R$. Then the following statements are equivalent:*

*(1) $a$ is group invertible and $a^{\#} = x$.*
*(2) $axa = a$, $xR = aR$ and $Rx = Ra$.*
*(3) $axa = a$, ${}^{\circ}x = {}^{\circ}a$ and $x^{\circ} = a^{\circ}$.*
*(4) $axa = a$, $xR \subseteq aR$ and $Rx \subseteq Ra$.*
*(5) $axa = a$, ${}^{\circ}a \subseteq {}^{\circ}x$ and $a^{\circ} \subseteq x^{\circ}$.*

**Proof** (1)⇒(2). Let $x = a^\#$. Then clearly $a = axa$. Also, since $a = xa^2$ and $x = ax^2$ we have $xR = aR$. Similarly, since $a = a^2x$ and $x = x^2a$ we have $Rx = Ra$.

(2)⇒(3)⇒(4)⇒(5). They are obvious by Lemma 1.2.21.

(5)⇒(1). From $axa = a$ it follows that $ax - 1 \in {}^\circ a \subseteq {}^\circ x$ and $1 - xa \in a^\circ \subseteq x^\circ$ so $(ax - 1)x = 0$ and $x(1 - xa) = 0$. Now, $x = ax^2 = x^2a$, hence $ax = ax^2a = xa$ and $xax = x^2a = x$. By the uniqueness of the group inverse, $x = a^\#$.                    □

The following theorem shows that the existence of the group inverse is closely related with existence of some idempotents.

**Theorem 3.2.5 ([8, Proposition 8.24])** *Let $a \in R$. Then $a$ is group invertible if and only if there exists an idempotent $e$ such that $ae = ea = 0$ and $a + e \in R^{-1}$. In this case, $a^\# = (a + e)^{-2}a$.*

**Proof** Suppose $a$ is group invertible. Take $e = 1 - aa^\#$. Then $ae = ea = 0$ and $(a + e)^{-1} = e + a^\#$.

Conversely, if there exists an idempotent $e$ such that $ae = ea = 0$ and $a + e \in R^{-1}$. Then $(e+a)a = a^2 = a(e+a)$. Since $a+e \in R^{-1}$, we have $a = (a+e)^{-1}a^2 = a^2(a+e)^{-1}$. By Theorem 3.2.2, we have $a$ is group invertible and $a^\# = (a+e)^{-2}a$.

□

**Theorem 3.2.6 ([108, Theorem 2.11])** *Let $a \in R$. Then the following statements are equivalent:*

(1) *$a$ is group invertible.*
(2) *There exists an idempotent $q$ such that $qR = aR$ and $Ra = Rq$.*
(3) *$a$ is regular and there exists an idempotent $q$ such that ${}^\circ a = {}^\circ q$ and $a^\circ = q^\circ$.*

*If the previous statements are valid then the statements (2) and (3) deal with the same unique idempotent $q$. Moreover, $qa^{(1)}q$ is invariant under the choice of $a^{(1)} \in a\{1\}$ and*

$$a = \begin{bmatrix} a & 0 \\ 0 & 0 \end{bmatrix}_{q \times q}, \quad a^\# = \begin{bmatrix} qa^{(1)}q & 0 \\ 0 & 0 \end{bmatrix}_{q \times q}.$$

**Proof** (1) ⇒ (2). Suppose that $a$ is group invertible and set $q = aa^\# = a^\#a$. Then $a = qa = aq$ so $qR = aR$, $Rq = Ra$.

(2) ⇒ (3). From $qR = aR$ we have $q = ax$ and $a = qz$ for some $x, z \in R$. Therefore, $qa = q^2z = qz = a$ and $axa = qa = a$, so $a \in R^{(1)}$. The rest of the proof follows by Lemma 1.2.21.

(3) ⇒ (1). Suppose that $a$ is regular and that there exists an idempotent $q$ such that $a^\circ = q^\circ$ and ${}^\circ a = {}^\circ q$. Let $a^{(1)} \in a\{1\}$ be arbitrary. Since $1 - a^{(1)}a \in a^\circ \subseteq q^\circ$ we obtain $q = qa^{(1)}a$. Also, $1 - q \in q^\circ \subseteq a^\circ$, so $a = aq$. Similarly, $q = aa^{(1)}q$ and $a = qa$. Set $x = qa^{(1)}q$. We have $x = a^\#$, because

$$ax = aqa^{(1)}q = aa^{(1)}q = q, \quad xa = qa^{(1)}qa = qa^{(1)}a = q,$$

$$axa = qa = a, \quad xax = qx = x.$$

Now the invariance of $qa^{(1)}q$ under the choice of $a^{(1)} \in a\{1\}$ follows. The uniqueness of $q$ follows by Lemma 1.2.16. Since $a = qaq$ and $a^{\#} = qa^{(1)}q$, we have also proved representations

$$a = \begin{bmatrix} a & 0 \\ 0 & 0 \end{bmatrix}_{q \times q}, \quad a^{\#} = \begin{bmatrix} qa^{(1)}q & 0 \\ 0 & 0 \end{bmatrix}_{q \times q}.$$

□

For regular elements, we have the following existence criterion of group inverses.

**Theorem 3.2.7 ([22, Proposition 4.1])** *Let $k$ be a positive integer and suppose that $a$ is regular with $a^{-} \in a\{1\}$. Then the following statements are equivalent:*

(1) *$a$ is group invertible.*
(2) *$u = a^k + 1 - aa^{-} \in R^{-1}$.*
(3) *$v = a^k + 1 - a^{-}a \in R^{-1}$.*

*In this case, $a^{\#} = u^{-1}a^{2k-1}v^{-1}$.*

**Proof** (1) $\Rightarrow$ (2). Since

$$u \left( a \left( a^{\#} \right)^k a^{-} + 1 - aa^{\#} \right) = \left( a^k + 1 - aa^{-} \right) \left( a \left( a^{\#} \right)^k a^{-} + 1 - aa^{\#} \right)$$

$$= a^{k+1} \left( a^{\#} \right)^k a^{-} + 1 - aa^{-} = aa^{-} + 1 - aa^{-} = 1,$$

it follows that $u$ is right invertible. Similarly, we can prove $\left( a \left( a^{\#} \right)^k a^{-} + 1 - aa^{\#} \right) u = 1$, i.e., $u$ is left invertible. Hence, $u = a^k + 1 - aa^{-} \in R^{-1}$.

(2) $\Rightarrow$ (3). Note that $u = 1 + a \left( a^{k-1} - a^{-} \right) \in R^{-1}$ if and only if $1 + \left( a^{k-1} - a^{-} \right) a = v \in R^{-1}$.

(3) $\Rightarrow$ (1). As $v \in R^{-1}$, then $u \in R^{-1}$. Since $ua = a^{k+1} = av$, it follows that $a = a^{k+1}v^{-1} = u^{-1}a^{k+1} \in a^2R \cap Ra^2$, i.e., $a \in R^{\#}$.

Note that $a = u^{-1}a^{k-1}a^2 = a^2a^{k-1}v^{-1} \in a^2R \cap Ra^2$. It follows from Theorem 3.2.2 that $a^{\#} = u^{-1}a^{k-1}aa^{k-1}v^{-1} = u^{-1}a^{2k-1}v^{-1}$.                □

**Corollary 3.2.8** *If $a \in R$ is regular with an inner inverse $a^{-}$, then the following conditions are equivalent:*

(1) *$a \in R^{\#}$.*
(2) *$u = a^2a^{-} + 1 - aa^{-}$ is invertible.*
(3) *$v = a^{-}a^2 + 1 - a^{-}a$ is invertible.*
(4) *$u' = a + 1 - aa^{-}$ is invertible.*
(5) *$v' = a + 1 - a^{-}a$ is invertible.*

*In this case,*

$$a^{\#} = u^{-2}a = av^{-2}.$$

**Proof** It is obvious by Jacobson's lemma and Theorem 3.2.7.                    □

At the last of this section, we present the commuting properties of the group inverse.

**Proposition 3.2.9 ([41])** *Let $a_1$, $a_2 \in R^{\#}$ and $d \in R$. If $da_1 = a_2d$, then $da_1^{\#} = a_2^{\#}d$.*

**Proof** By direct computation, we have

$$da_1^{\#} = da_1(a_1^{\#})^2 = a_2d(a_1^{\#})^2 = (a_2^{\#})^2(a_2)^3d(a_1^{\#})^2$$
$$= (a_2^{\#})^2d(a_1)^3(a_1^{\#})^2 = (a_2^{\#})^2da_1 = a_2^{\#}d.$$

□

## 3.3 The Group Inverse of a Product $paq$

Suppose $a$ is a regular element of $R$. Our aim is to give necessary and sufficient conditions for the product $t = paq$ to possess a group inverse. To do this we introduce the two elements

$$u = aqpaa^- + 1 - aa^- \quad \text{and} \quad v = a^-aqpa + 1 - a^-a.$$

We may now state:

**Theorem 3.3.1 ([105, Theorem 1])** *Suppose $a \in R$ is regular, $aa^-a = a$, $u = aqpaa^- + 1 - aa^-$, $v = a^-aqpa + 1 - a^-a$ and $t = paq$. Then the following are equivalent:*

$$\text{(i) } uR = R, \quad \text{(ii) } aqpaR = aR, \quad \text{(iii) } vR = R. \tag{3.3.1}$$

*If in addition $Ra = Rpa$, then these are also equivalent to*

$$\text{(iv) } t^2R = tR \quad \text{and} \quad aR = aqR.$$

**Proof** We first note that if $a$ is regular then $aa^-R = aR$.

(i)⇔(ii). $uR = R$ implies that

$$aqpaR = aa^-\left(aqpaa^- + 1 - aa^-\right)R = aa^-uR = aa^-R = aR.$$

Conversely, if $aqpax = a$, writing $\alpha = axa^- + 1 - aa^-$, then $uR = R$ follows from

$$u\alpha = aqpaa^-axa^- + 1 - aa^- = aa^- + 1 - aa^- = 1. \tag{3.3.2}$$

(ii)$\Leftrightarrow$(iii). If $aqpax = a$, let $\beta = a^-ax + 1 - a^-a$. Then

$$v\beta = a^-aqpax + 1 - a^-a = a^-a + 1 - a^-a = 1. \tag{3.3.3}$$

Conversely if $vR = R$ then $aqpaR = avR = aR$.

(ii)$\Leftrightarrow$(iv). Clearly, (ii) implies that $aR = aqR$. Hence $tR = paqR = paR = paqpaR = (paq)(paq)R = t^2R$. Conversely, since $aqR = aR$ we have that $(paq)(paq)R = paqR$ implies $paqpaR = paR$. Now by (iv) there exists $p'$ such that $a = p'pa$. Therefore, premultiplication by $p'$ yields $aqpaR = (p'pa)qpaR = p'paR = aR$.      $\square$

### Remark 3.3.2 ([105, p. 141, Remarks])

(1) By symmetry it follows that the following are equivalent:

$$\text{(i) } Ru = R, \text{ (ii) } Raqpa = Ra, \text{ (iii) } Rv = R. \tag{3.3.4}$$

If in addition $aR = aqR$ then these are also equivalent to

$$\text{(iv) } Rt^2 = Rt \text{ and } Ra = Rpa.$$

(2) If $p$ and $q$ are invertible then the following are equivalent:

$$\text{(i) } uR = R, \text{ (ii) } aqpaR = aR, \text{ (iii) } vR = R, \text{ (iv) } t^2R = tR$$

and

$$\text{(i) } Ru = R, \text{ (ii) } Raqpa = Ra, \text{ (iii) } Rv = R, \text{ (iv) } Rt^2 = Rt.$$

Combining (3.3.1) and (3.3.4), we have

### Corollary 3.3.3 ([105, Corollary 1]) *The following are equivalent:*

(1) *$u$ is a unit.*
(2) *$Raqpa = Ra, aqpaR = aR$.*
(3) *$v$ is a unit.*
(4) *$t^\#$ exists and $aR = aqR$, $Ra = Rpa$.*

*Moreover*

$$t^\# = pu^{-1}av^{-1}q = pu^{-2}aq = pav^{-2}q, \tag{3.3.5}$$

*where*

$$u^{-1} = axa^- + 1 - aa^- = yaa^- + 1 - aa^-,$$
$$v^{-1} = a^-ax + 1 - a^-a = a^-ya + 1 - a^-a$$

*and aqpax = a = yaqpa.*

**Proof** To compute $t^\#$ we note first that $ua = aqpa = av$ and thus

$$u^{-1}a = av^{-1} \tag{3.3.6}$$

and $a = u^{-1}(aqpa) = (aqpa)v^{-1}$. Hence $paq = pu^{-1}aqpaq = pu^{-1}p'(paq)(paq)$ or $t = ht^2$, where $h = pu^{-1}p'$. Likewise, $t^2g = t$, where $g = q'v^{-1}q$. Consequently, we may conclude that $t^\# = htg = pu^{-1}p' \cdot paq \cdot q'v^{-1}q$ which shows that

$$t^\# = pu^{-1}av^{-1}q. \tag{3.3.7}$$

By (3.3.6) we also have the asymmetric forms

$$t^\# = pav^{-2}q = pu^{-2}aq. \tag{3.3.8}$$

The expressions for $u^{-1}$ and $v^{-1}$ follow from (3.3.2) and (3.3.3).    □

**Corollary 3.3.4 ([105, Corollary 2])** *If p and q are invertible then the following are equivalent:*

(1) *u is a unit.*
(2) *$Raqpa = Ra, aqpaR = aR$.*
(3) *v is a unit.*
(4) *$t^\#$ exists.*

We may also set $aq = s, pa = r$ and $t = ra^-s$. In addition we could take $p = ra^-$ and $q = a^-s$. We then have the following theorem.

**Theorem 3.3.5 ([105, Theorem 2])** *Suppose a is regular with $aa^-a = a, aR = sR, Ra = Rr$ and*

$$u = sra^- + 1 - aa^-, \qquad v = a^-sr + 1 - a^-a.$$

*Then the following are equivalent:*

$$\text{(i) } uR = R, \text{ (ii) } srR = aR, \text{ (iii) } vR = R, \text{ (iv) } t^2R = tR$$

*and*

$$(\text{i) } Ru = R, \text{ (ii) } Rsr = Ra, \text{ (iii) } Rv = R, \text{ (iv) } Rt^2 = Rt.$$

*In which case,*

$$t^{\#} = ra^-u^{-1}av^-a^-s = rv^{-2}a^-s = ra^-u^{-2}s.$$

## 3.4 The Group Inverse of a Sum of Morphisms

In this section, the group inverse of a sum of morphisms is investigated.

**Proposition 3.4.1 ([126, Proposition 1])** *Let $\mathscr{C}$ be an additive category. Suppose that $\varphi : X \longrightarrow X$ is a morphism of $\mathscr{C}$ with group inverse $\varphi^{\#}$ and that $\eta : X \longrightarrow X$ is a morphism of $\mathscr{C}$ such that $1_X + \varphi^{\#}\eta$ is invertible. Let*

$$\alpha = (1_X + \varphi^{\#}\eta)^{-1},$$
$$\beta = (1_X + \eta\varphi^{\#})^{-1},$$
$$\varepsilon = (1_X - \varphi\varphi^{\#})\eta\alpha(1_X - \varphi^{\#}\varphi),$$
$$\gamma = \alpha(1_X - \varphi^{\#}\varphi)\eta\varphi^{\#}\beta, \quad and$$
$$\delta = \alpha\varphi^{\#}\eta(1_X - \varphi\varphi^{\#})\beta.$$

*Then the following conditions are equivalent:*

(1) $f = \varphi + \eta - \varepsilon$ *has a group inverse.*
(2) $1_X - \gamma$ *and* $1_X - \delta$ *are invertible.*
(3) $1_X - \gamma$ *is left invertible and* $1_X - \delta$ *is right invertible.*

*In that case*

$$f^{\#} = (1_X - \gamma)^{-1}\alpha\varphi^{\#}(1_X - \delta)^{-1},$$
$$(1_X - \gamma)^{-1} = 1_X - \varphi^{\#}\varphi + f^{\#}f\varphi^{\#}\varphi,$$
$$(1_X - \delta)^{-1} = 1_X - \varphi\varphi^{\#} + \varphi\varphi^{\#}ff^{\#}.$$

***Proof*** By Lemma 2.8.1, $(1_X + \varphi^{\#}\eta)^{-1}\varphi^{\#} \in f\{1, 2\}$. Let $f_0^{(1,2)} = (1_X + \varphi^{\#}\eta)^{-1}\varphi^{\#} = \alpha\varphi^{\#}$. Then

$$\varphi^{\#}f = \varphi^{\#}(\varphi + \eta - \varepsilon) = \varphi^{\#}\varphi + \varphi^{\#}\eta = \varphi^{\#}\varphi(1_X + \varphi^{\#}\eta) = \varphi^{\#}\varphi\alpha^{-1}$$

and $f\varphi^{\#} = \beta^{-1}\varphi\varphi^{\#}$. So $ff_0^{(1,2)} = f\alpha\varphi^{\#} = f\varphi^{\#}\beta = \beta^{-1}\varphi\varphi^{\#}\beta$ and

$$1_X - ff_0^{(1,2)} = \beta^{-1}(1_X - \varphi\varphi^{\#})\beta$$
$$= (1_X + \eta\varphi^{\#})(1_X - \varphi\varphi^{\#})\beta$$
$$= (1_X - \varphi\varphi^{\#})\beta.$$

Similarly, we have $f_0^{(1,2)}f = \alpha\varphi^{\#}\varphi\alpha^{-1}$ and $1_X - f_0^{(1,2)}f = \alpha(1_X - \varphi^{\#}\varphi)$. Further,

$$1_X - f_0^{(1,2)}f(1_X - ff_0^{(1,2)}) = 1_X - \alpha\varphi^{\#}\varphi\alpha^{-1}(1_X - \varphi\varphi^{\#})\beta$$
$$= 1_X - \alpha\varphi^{\#}\varphi(1_X + \varphi^{\#}\eta)(1_X - \varphi\varphi^{\#})\beta$$
$$= 1_X - \alpha(\varphi^{\#}\varphi + \varphi^{\#}\eta)(1_X - \varphi\varphi^{\#})\beta$$
$$= 1_X - \alpha\varphi^{\#}\eta(1_X - \varphi\varphi^{\#})\beta = 1_X - \delta$$

and

$$1_X - (1_X - f_0^{(1,2)}f)ff_0^{(1,2)} = 1_X - \gamma.$$

We have

$$f(1_X - \gamma) = f[1_X - (1_X - f_0^{(1,2)}f)ff_0^{(1,2)}] = f$$

(which implies $f\gamma = 0$),

$$f^2 f_0^{(1,2)} = f - f(1_X - ff_0^{(1,2)})$$
$$= f[1_X - f_0^{(1,2)}f(1_X - ff_0^{(1,2)})]$$
$$= f(1_X - \delta),$$

and

$$f_0^{(1,2)}f^2 = (1_X - \gamma)f.$$

Now we are ready to show the equivalence of three conditions.
(1) $\Rightarrow$ (2). First, we show that $1_X - \varphi^{\#}\varphi + f^{\#}f\varphi^{\#}\varphi$ is a right inverse of $1_X - \gamma$. Note that $(1_X - \gamma)ff^{\#} = f_0^{(1,2)}f^2f^{\#} = f_0^{(1,2)}f = \alpha\varphi^{\#}f = \alpha(\varphi^{\#}\varphi + \varphi^{\#}\eta) = \alpha(\alpha^{-1} + \varphi^{\#}\varphi - 1_X) = 1_X + \alpha(\varphi^{\#}\varphi - 1_X)$. Multiplying the equality by $\varphi^{\#}\varphi$ on the right, we obtain $(1_X - \gamma)ff^{\#}\varphi^{\#}\varphi = \varphi^{\#}\varphi$. Since

$$\gamma(1_X - \varphi\varphi^{\#}) = \alpha(1_X - \varphi^{\#}\varphi)\eta\varphi^{\#}\beta(1_X - \varphi\varphi^{\#})$$
$$= \alpha(1_X - \varphi^{\#}\varphi)\eta\alpha\varphi^{\#}(1_X - \varphi\varphi^{\#})$$
$$= 0,$$

we obtain

$$(1_X - \gamma)(1_X - \varphi^\#\varphi + f^\# f\varphi^\#\varphi) = 1_X - \varphi^\#\varphi - \gamma(1_X - \varphi^\#\varphi) + (1_X - \gamma)f^\# f\varphi^\#\varphi$$
$$= 1_X - \varphi^\#\varphi - \gamma(1_X - \varphi\varphi^\#) + (1_X - \gamma)ff^\#\varphi^\#\varphi$$
$$= 1_X - \varphi^\#\varphi + \varphi^\#\varphi$$
$$= 1_X.$$

Next, we show that this right inverse is also a left inverse for $1_X - \gamma$. Note first that

$$\varphi^\#\varphi\gamma = \varphi^\#\varphi\alpha(1_X - \varphi^\#\varphi)\eta\varphi^\#\beta$$
$$= \varphi^\#\varphi(1_X - \varphi^\#\eta\alpha)(1_X - \varphi^\#\varphi)\eta\varphi^\#\beta$$
$$= (\varphi^\#\varphi - \varphi^\#\eta\alpha)(1_X - \varphi^\#\varphi)\eta\varphi^\#\beta$$
$$= -\varphi^\#\eta\alpha(1_X - \varphi^\#\varphi)\eta\varphi^\#\beta$$
$$= (\alpha - 1_X)(1_X - \varphi^\#\varphi)\eta\varphi^\#\beta$$
$$= \alpha(1_X - \varphi^\#\varphi)\eta\varphi^\#\beta - (1_X - \varphi^\#\varphi)\eta\varphi^\#\beta$$
$$= \gamma - (1_X - \varphi^\#\varphi)\eta\varphi^\#\beta.$$

So $(1_X - \varphi^\#\varphi)\gamma = (1_X - \varphi^\#\varphi)\eta\varphi^\#\beta = (1_X - \varphi^\#\varphi)(1_X - \beta)$ and

$$f^\# f\varphi^\#\varphi\gamma = f^\# f[\gamma - (1_X - \varphi^\#\varphi)\eta\varphi^\#\beta]$$
$$= f^\# f\gamma - f^\# f(1_X - \varphi^\#\varphi)\eta\varphi^\#\beta$$
$$= -f^\# f(1_X - \varphi^\#\varphi)\eta\varphi^\#\beta$$
$$= -f^\# f(1_X - \varphi^\#\varphi)(1_X - \beta).$$

On the other hand,

$$1_X - \varphi^\#\varphi + f^\# f\varphi^\#\varphi = 1_X - \varphi\varphi^\# + ff^\#\varphi\varphi^\#$$
$$= (1_X + \eta\varphi^\# - f\varphi^\#) + ff^\#\varphi\varphi^\#$$
$$= 1_X + \eta\varphi^\# + ff^\#\varphi\varphi^\# - ff^\# f\varphi^\#$$
$$= 1_X + \eta\varphi^\# + ff^\#(\varphi\varphi^\# - f\varphi^\#)$$
$$= 1_X + \eta\varphi^\# - ff^\#\eta\varphi^\#$$
$$= ff^\# + (1_X - ff^\#)(1_X + \eta\varphi^\#)$$
$$= f^\# f + (1_X - f^\# f)\beta^{-1}.$$

Therefore,

$$(1_X - \varphi^{\#}\varphi + f^{\#}f\varphi^{\#}\varphi)(1_X - \gamma) = 1_X - \varphi^{\#}\varphi - (1_X - \varphi^{\#}\varphi)\gamma + f^{\#}f\varphi^{\#}\varphi - f^{\#}f\varphi^{\#}\varphi\gamma$$
$$= 1_X - \varphi^{\#}\varphi - (1_X - \varphi^{\#}\varphi)(1_X - \beta) + f^{\#}f\varphi^{\#}\varphi$$
$$+ f^{\#}f(1_X - \varphi^{\#}\varphi)(1_X - \beta)$$
$$= (1_X - \varphi^{\#}\varphi)\beta + f^{\#}f\varphi^{\#}\varphi$$
$$+ f^{\#}f(1_X - \beta) - f^{\#}f\varphi^{\#}\varphi + f^{\#}f\varphi^{\#}\psi\beta$$
$$= f^{\#}f(1_X - \beta) + (1_X - \varphi^{\#}\varphi + f^{\#}f\varphi^{\#}\varphi)\beta$$
$$= f^{\#}f(1_X - \beta) + [f^{\#}f + (1_X - f^{\#}f)\beta^{-1}]\beta$$
$$= f^{\#}f - f^{\#}f\beta + f^{\#}f\beta + 1_X - f^{\#}f$$
$$= 1_X.$$

Hence $1_X - \gamma$ is invertible and $(1_X - \gamma)^{-1} = 1_X - \varphi^{\#}\varphi + f^{\#}f\varphi^{\#}\varphi$. Similarly, we have that $1_X - \delta$ is invertible, and $(1_X - \delta)^{-1} = 1_X - \varphi\varphi^{\#} + \varphi\varphi^{\#}ff^{\#}$.

(2)$\Rightarrow$(3). Obvious.

(3)$\Rightarrow$(1). Let $\omega$ denote the left inverse of $1_X - \gamma$ , i.e., $\omega(1_X - \gamma) = 1_X$, and $t$ the right inverse of $1_X - \delta$, i.e., $(1_X - \delta)t = 1_X$. Since $f(1_X - \delta) = f^2 f_0^{(1,2)}$ and $(1_X - \gamma)f = f_0^{(1,2)}f^2$, we have

$$f = f(1_X - \delta)t = f^2 f_0^{(1,2)}t, \quad f = \omega(1_X - \gamma)f = \omega f_0^{(1,2)}f^2.$$

This means that the equations $f = f^2 x$ and $f = yf^2$ have solutions $x_0 = f_0^{(1,2)}t$ and $y_0 = \omega f_0^{(1,2)}$ , respectively. Therefore, the group inverse $f^{\#}$ of $f$ exists and $f^{\#} = y_0 f x_0 = \omega f_0^{(1,2)}ff_0^{(1,2)}t = \omega f_0^{(1,2)}t$ . From (1)$\Rightarrow$(2), we know that $1_X - \gamma$ and $1_X - \delta$ are invertible and $\omega = (1_X - \gamma)^{-1}$, $t = (1_X - \delta)^{-1}$. So $f^{\#} = (1_X - \gamma)^{-1}\alpha\varphi^{\#}(1_X - \delta)^{-1}$.                                                                                   □

**Corollary 3.4.2 ([64, Proposition 3])** *If $a \in R$ has a group inverse $a^{\#}$ and $j$ an element of $J(R)$, then $a + j$ has a group inverse if and only if $(1 - aa^{\#})j(1 + a^{\#}j)^{-1}(1 - a^{\#}a) = 0$. In that case,*

$$(a + j)^{\#} = (1 - \gamma)^{-1}(1 + a^{\#}j)^{-1}a^{\#}(1 - \delta)^{-1}, \text{ in which}$$
$$\gamma = (1 + a^{\#}j)^{-1}(1 - a^{\#}a)ja^{\#}(1 + ja^{\#})^{-1},$$
$$\delta = (1 + a^{\#}j)^{-1}a^{\#}j(1 - aa^{\#})(1 + ja^{\#})^{-1}.$$

*Proof* Since $j \in J(R)$, the element $1 + a^{\#}j$ is invertible.

"$\Leftarrow$". By hypothesis, $\varepsilon = 0$, so $f = a + j$. Since $1 - \gamma$ and $1 - \delta$ are invertible, $a + j$ has a group inverse by Proposition 3.4.1. In that case, $(a+j)^{\#} = (1-\gamma)^{-1}(1 + a^{\#}j)^{-1}a^{\#}(1 - \delta)^{-1}$.

"$\Rightarrow$". Let $\tau = (a + j)^{\#}$. Then $\tau \in \varepsilon\{1\}$, which gives $\varepsilon(1 - \tau\varepsilon) = 0$. Since

$$\varepsilon = (1 - aa^{\#})j(1 + a^{\#}j)^{-1}(1 - a^{\#}a) \in J(R),$$

it follows that $1 - \tau\varepsilon$ is invertible and so $\varepsilon = 0$. □

## 3.5 The Group Inverse of the Sum of Two Group Invertible Elements

In this section, we assume that $R$ is a Dedekind-finite ring. The group inverses of sum and difference of two group invertible elements are presented under different conditions.

**Theorem 3.5.1 ([132, Theorem 3.1])** *Let $a, b \in R^{\#}$ and $2 \in R^{-1}$. If $abb^{\#} = baa^{\#}$, then*

(1) $a + b \in R^{\#}$ *and* $(a + b)^{\#} = a^{\#} + b^{\#} - \frac{1}{2}a^{\#}bb^{\#} - \frac{1}{2}bb^{\#}a^{\#} - \frac{1}{2}aa^{\#}b^{\#}$.
(2) $a - b \in R^{\#}$ *and* $(a - b)^{\#} = a^{\#} - b^{\#} - ba^{\#}a^{\#} + ab^{\#}b^{\#}$.

**Proof** Since $abb^{\#} = baa^{\#}$, we have $abb^{\#}(b^{\#}aa^{\#})abb^{\#} = ab^{\#}abb^{\#} = ab^{\#}baa^{\#} = abb^{\#}$. So $b^{\#}aa^{\#}$ is a $\{1\}$-inverse of $abb^{\#}$. Then we have

$$(1 + b^{\#}aa^{\#} - b^{\#}abb^{\#})(1 + abb^{\#} - b^{\#}abb^{\#}) = 1 + abb^{\#}$$

$$- b^{\#}abb^{\#} + b^{\#}aa^{\#} + b^{\#}abb^{\#} - b^{\#}aa^{\#}b^{\#}abb^{\#} - b^{\#}abb^{\#} - b^{\#}babb^{\#} + b^{\#}ab^{\#}abb^{\#}$$

$$= 1 + abb^{\#} + b^{\#}aa^{\#} - b^{\#}aa^{\#}b^{\#}baa^{\#} - b^{\#}baa^{\#} - b^{\#}babb^{\#} + b^{\#}ab^{\#}baa^{\#}$$

$$= 1 + abb^{\#} + b^{\#}aa^{\#} - b^{\#}b^{\#}baa^{\#}b^{\#}baa^{\#} - b^{\#}baa^{\#} - b^{\#}babb^{\#} + b^{\#}baa^{\#}$$

$$= 1 + abb^{\#} + b^{\#}aa^{\#} - b^{\#}aa^{\#} - b^{\#}baa^{\#} - baa^{\#} + b^{\#}baa^{\#} = 1.$$

Since $R$ is a Dedekind-finite ring, $(1 + abb^{\#} - b^{\#}abb^{\#})(1 + b^{\#}aa^{\#} - b^{\#}abb^{\#}) = 1$. From Theorem 3.2.7, we know that $abb^{\#} \in R^{\#}$. Since $aa^{\#}abb^{\#} = abb^{\#} = baa^{\#}$, by Lemma 3.2.9, $aa^{\#}(abb^{\#})^{\#} = b^{\#}aa^{\#}$. Since

$$aa^{\#}(abb^{\#})^{\#} = aa^{\#}abb^{\#}(abb^{\#})^{\#}(abb^{\#})^{\#} = (abb^{\#})^{\#},$$

$(abb^{\#})^{\#} = b^{\#}aa^{\#}$. Similarly, $(baa^{\#})^{\#} = a^{\#}bb^{\#}$. So $a^{\#}bb^{\#} = b^{\#}aa^{\#}$. Furthermore, $aa^{\#}bb^{\#} = ab^{\#}aa^{\#} = abb^{\#}b^{\#}aa^{\#} = b^{\#}aa^{\#}abb^{\#} = b^{\#}abb^{\#} = b^{\#}baa^{\#}$. In addition,

$$bb^{\#}ab^{\#} = bb^{\#}abb^{\#}b^{\#} = bb^{\#}baa^{\#}b^{\#} = baa^{\#}b^{\#} = abb^{\#}b^{\#} = ab^{\#},$$

$$bb^{\#}a^{\#}b = bb^{\#}(a^{\#}bb^{\#})b = bb^{\#}b^{\#}aa^{\#}b = (b^{\#}aa^{\#})b = a^{\#}bb^{\#}b = a^{\#}b.$$

Similarly, $aa^\#ba^\# = ba^\#$ and $aa^\#b^\#a = b^\#a$. We know that $abaa^\# = abb^\#haa^\# = baa^\#abb^\# = babb^\# = bbaa^\#$. Hence $aba^\# = bba^\#$. Since $a^\#b^\#aa^\# = a^\#bb^\#b^\#aa^\# = b^\#aa^\#a^\#bb^\# = b^\#a^\#bb^\# = b^\#b^\#aa^\#$, $a^\#b^\#a = b^\#b^\#a$. Similarly, we can get $bab^\# = aab^\#$ and $b^\#a^\#b = a^\#a^\#b$. From the above discussion, we have

$$abb^\# = baa^\#, \tag{3.5.1}$$

$$a^\#bb^\# = b^\#aa^\#, \quad aa^\#bb^\# = bb^\#aa^\#, \tag{3.5.2}$$

$$aa^\#b^\#a = b^\#a, \quad bb^\#a^\#b = a^\#b, \tag{3.5.3}$$

$$bab^\# = aab^\#, \quad aba^\# = bba^\#, \tag{3.5.4}$$

$$a^\#b^\#a = b^\#b^\#a, \quad a^\#a^\#b = b^\#a^\#b, \tag{3.5.5}$$

$$aa^\#ba^\# = ba^\#, \quad bb^\#ab^\# = ab^\#. \tag{3.5.6}$$

1. Let $x = a^\# + b^\# - \frac{1}{2}a^\#bb^\# - \frac{1}{2}bb^\#a^\# - \frac{1}{2}aa^\#b^\#$. By equalities (3.5.1), (3.5.2) and (3.5.3), we have

$$(a+b)x = (a+b)(a^\# + b^\# - \frac{1}{2}a^\#bb^\# - \frac{1}{2}bb^\#a^\# - \frac{1}{2}aa^\#b^\#) = aa^\# + ab^\#$$

$$- \frac{1}{2}aa^\#bb^\# - \frac{1}{2}abb^\#a^\# - \frac{1}{2}ab^\# + ba^\# + bb^\# - \frac{1}{2}ba^\#bb^\# - \frac{1}{2}ba^\# - \frac{1}{2}baa^\#b^\#$$

$$= a^\#a + ab^\# - \frac{1}{2}aa^\#bb^\# - \frac{1}{2}ba^\# - \frac{1}{2}ab^\# + ba^\#$$

$$+ bb^\# - \frac{1}{2}aa^\#bb^\# - \frac{1}{2}ba^\# - \frac{1}{2}ab^\#$$

$$= aa^\# + bb^\# - aa^\#bb^\#$$

and

$$x(a+b) = (a^\# + b^\# - \frac{1}{2}a^\#bb^\# - \frac{1}{2}bb^\#a^\# - \frac{1}{2}aa^\#b^\#)(a+b) = a^\#a + b^\#a$$

$$- \frac{1}{2}a^\#bb^\#a - \frac{1}{2}bb^\#a^\#a - \frac{1}{2}aa^\#b^\#a + a^\#b + b^\#b - \frac{1}{2}a^\#b - \frac{1}{2}a^\#b - \frac{1}{2}aa^\#bb^\#$$

$$= a^\#a + b^\#a - \frac{1}{2}b^\#a - \frac{1}{2}aa^\#bb^\# - \frac{1}{2}b^\#a + b^\#b - \frac{1}{2}aa^\#bb^\#$$

$$= aa^\# + bb^\# - aa^\#bb^\#.$$

So $(a+b)x = x(a+b)$. By Eqs. (3.5.1) and (3.5.2), we have

$$x(a+b)x = (aa^\# + bb^\# - aa^\#bb^\#)(a^\# + b^\# - \frac{1}{2}a^\#bb^\# - \frac{1}{2}bb^\#a^\# - \frac{1}{2}aa^\#b^\#)$$

$$= a^\# + aa^\#b^\# - \frac{1}{2}a^\#bb^\# - \frac{1}{2}aa^\#bb^\#a^\#$$

$$- \frac{1}{2}aa^\#b^\# + bb^\#a^\# + b^\# - \frac{1}{2}bb^\#a^\#bb^\# - \frac{1}{2}bb^\#a^\#$$

$$- \frac{1}{2}bb^\#aa^\#b^\# - aa^\#bb^\#a^\# - aa^\#b^\# + \frac{1}{2}aa^\#bb^\#a^\#bb^\#$$

$$+ \frac{1}{2}aa^\#bb^\#a^\# + \frac{1}{2}aa^\#bb^\#aa^\#b^\#$$

$$= a^\# + aa^\#b^\# - \frac{1}{2}a^\#bb^\# - \frac{1}{2}bb^\#a^\# - \frac{1}{2}aa^\#b^\# + bb^\#a^\# + b^\#$$

$$- \frac{1}{2}a^\#bb^\# - \frac{1}{2}bb^\#a^\#$$

$$- \frac{1}{2}aa^\#b^\# - bb^\#a^\# - aa^\#b^\# + \frac{1}{2}a^\#bb^\# + \frac{1}{2}bb^\#a^\# + \frac{1}{2}aa^\#b^\#$$

$$= a^\# + b^\# - \frac{1}{2}a^\#bb^\# - \frac{1}{2}bb^\#a^\# - \frac{1}{2}aa^\#b^\# = x$$

and

$$(a+b)x(a+b) = (a+b)(aa^\# + bb^\# - aa^\#bb^\#)$$

$$= a + abb^\# - abb^\# + baa^\# + b - baa^\#bb^\#$$

$$= a + b.$$

So $(a+b)^\# = a^\# + b^\# - \frac{1}{2}a^\#bb^\# - \frac{1}{2}bb^\#a^\# - \frac{1}{2}aa^\#b^\#$.

2. Let $y = a^\# - b^\# - ba^\#a^\# + ab^\#b^\#$. By equalities (3.5.3), (3.5.4) and (3.5.5),

$$(a-b)y = (a-b)(a^\# - b^\# - ba^\#a^\# + ab^\#b^\#)$$

$$= aa^\# - ab^\# - aba^\#a^\# + aab^\#b^\# - ba^\# + bb^\# + bba^\#a^\# - bab^\#b^\#$$

$$= aa^\# - ab^\# - bba^\#a^\# + aab^\#b^\# - ba^\# + bb^\# + bba^\#a^\# - aab^\#b^\#$$

$$= aa^\# - ab^\# - ba^\# + bb^\#$$

and

$$y(a - b) = (a^{\#} - b^{\#} - ba^{\#}a^{\#} + ab^{\#}b^{\#})(a - b)$$

$$= a^{\#}a - b^{\#}a - ba^{\#}a^{\#}a + ab^{\#}b^{\#}a - a^{\#}b + b^{\#}b + ba^{\#}a^{\#}b - ab^{\#}b^{\#}b$$

$$= a^{\#}a - b^{\#}a - ba^{\#} + b^{\#}a - a^{\#}b + b^{\#}b + a^{\#}b - ab^{\#}$$

$$- aa^{\#} - ab^{\#} - ba^{\#} + bb^{\#}.$$

So $(a - b)y = y(a - b)$. By equalities (3.5.2) and (3.5.6), we have

$$ab^{\#}a^{\#} = a(b^{\#}aa^{\#})a^{\#} = aa^{\#}bb^{\#}a^{\#},$$

$$ba^{\#}b^{\#} = b(a^{\#}bb^{\#})b^{\#} = bb^{\#}aa^{\#}b^{\#},$$

$$ba^{\#}ba^{\#}a^{\#} = b(a^{\#}bb^{\#})ba^{\#}a^{\#} = (bb^{\#}aa^{\#})ba^{\#}a^{\#} = (aa^{\#}ba^{\#})a^{\#} = ba^{\#}a^{\#},$$

$$ab^{\#}ab^{\#}b^{\#} = ab^{\#}(aa^{\#}a)b^{\#}b^{\#} = a(b^{\#}aa^{\#})ab^{\#}b^{\#} = (aa^{\#}bb^{\#})ab^{\#}b^{\#}$$

$$= bb^{\#}aa^{\#}ab^{\#}b^{\#} = (bb^{\#}ab^{\#})b^{\#} = ab^{\#}b^{\#}.$$

Thus,

$$y(a - b)y = (aa^{\#} - ab^{\#} - ba^{\#} + bb^{\#})(a^{\#} - b^{\#} - ba^{\#}a^{\#} + ab^{\#}b^{\#})$$

$$= a^{\#} - ab^{\#}a^{\#} - ba^{\#}a^{\#} + bb^{\#}a^{\#} - aa^{\#}ba^{\#}a^{\#} + ab^{\#}ba^{\#}a^{\#} + ba^{\#}ba^{\#}a^{\#} - ba^{\#}a^{\#}$$

$$- aa^{\#}b^{\#} + ab^{\#}b^{\#} + ba^{\#}b^{\#} - b^{\#} + ab^{\#}b^{\#} - ab^{\#}ab^{\#}b^{\#} - ba^{\#}ab^{\#}b^{\#} + bb^{\#}ab^{\#}b^{\#}$$

$$= a^{\#} - aa^{\#}bb^{\#}a^{\#} - ba^{\#}a^{\#} + bb^{\#}a^{\#} - ba^{\#}a^{\#} + ba^{\#}a^{\#} + ba^{\#}a^{\#} - ba^{\#}a^{\#}$$

$$- aa^{\#}b^{\#} + ab^{\#}b^{\#} + bb^{\#}aa^{\#}b^{\#} - b^{\#} + ab^{\#}b^{\#} - ab^{\#}b^{\#} - ab^{\#}b^{\#} + ab^{\#}b^{\#}$$

$$= a^{\#} - b^{\#} - ba^{\#}a^{\#} + ab^{\#}b^{\#} = y.$$

Since

$$ab^{\#}a = a(b^{\#}aa^{\#})a = aa^{\#}bb^{\#}a, \quad ba^{\#}b = b(a^{\#}bb^{\#})b = bb^{\#}aa^{\#}b,$$

we have

$$(a - b)y(a - b) = (aa^{\#} - ab^{\#} - ba^{\#} + bb^{\#})(a - b)$$

$$= a - ab^{\#}a - ba^{\#}a + bb^{\#}a - aa^{\#}b + ab^{\#}b + ba^{\#}b - b$$

$$= a - aa^{\#}bb^{\#}a + bb^{\#}a - aa^{\#}b + bb^{\#}aa^{\#}b - b$$

$$= a - bb^{\#}a + bb^{\#}a - aa^{\#}b + aa^{\#}b - b = a - b.$$

Therefore, $(a - b)^{\#} = a^{\#} - b^{\#} - ba^{\#}a^{\#} + ab^{\#}b^{\#}$.

$\square$

Next, we present the expressions of $(a + b)^{\#}$ and $(a - b)^{\#}$ under the condition $bb^{\#}a = aa^{\#}b$.

**Theorem 3.5.2 ([132, Theorem 3.2])** *Let* $a$, $b \in R^{\#}$ *and* $2 \in R^{-1}$. *If* $bb^{\#}a = aa^{\#}b$, *then*

(1) $a + b \in R^{\#}$ *and* $(a + b)^{\#} = a^{\#} + b^{\#} - \frac{1}{2}bb^{\#}a^{\#} - \frac{1}{2}a^{\#}bb^{\#} - \frac{1}{2}b^{\#}aa^{\#}$.

(2) $a - b \in R^{\#}$ *and* $(a - b)^{\#} = a^{\#} - b^{\#} - a^{\#}a^{\#}b + b^{\#}b^{\#}a$.

*Proof* Since $bb^{\#}a(aa^{\#}b^{\#})bb^{\#}a = bb^{\#}ab^{\#}a = aa^{\#}bb^{\#}a = bb^{\#}a$, $aa^{\#}b^{\#}$ is a $\{1\}$-inverse of $bb^{\#}a$. Since $bb^{\#}a = aa^{\#}b$, we have

$$(1 + bb^{\#}a - bb^{\#}ab^{\#})(1 + aa^{\#}b^{\#} - bb^{\#}ab^{\#}) = 1 + bb^{\#}a - bb^{\#}ab^{\#}$$

$$+ aa^{\#}b^{\#} + bb^{\#}ab^{\#} - bb^{\#}ab^{\#}aa^{\#}b^{\#} - bb^{\#}ab^{\#} - bb^{\#}abb^{\#}ab^{\#} + bb^{\#}ab^{\#}ab^{\#}$$

$$= 1 + bb^{\#}a + aa^{\#}b^{\#} - aa^{\#}bb^{\#}aa^{\#}b^{\#} - bb^{\#}ab^{\#} - bb^{\#}abb^{\#} + aa^{\#}bb^{\#}ab^{\#}$$

$$= 1 + bb^{\#}a + aa^{\#}b^{\#} - aa^{\#}b^{\#} - aa^{\#}bb^{\#} - aa^{\#}b + aa^{\#}bb^{\#} = 1.$$

Since $R$ is a Dedekind-finite ring, $(1 + aa^{\#}b^{\#} - bb^{\#}ab^{\#})(1 + bb^{\#}a - bb^{\#}ab^{\#}) = 1$. So $1 + bb^{\#}a - bb^{\#}ab^{\#} \in R^{-1}$. From Theorem 3.2.7, we know that $bb^{\#}a \in R^{\#}$. Since $bb^{\#}aa^{\#}a = bb^{\#}a = aa^{\#}b$, $(bb^{\#}a)^{\#}aa^{\#} = aa^{\#}b^{\#}$. Then we have

$$(bb^{\#}a)^{\#}aa^{\#} = (bb^{\#}a)^{\#}(bb^{\#}a)^{\#}bb^{\#}aaa^{\#} = (bb^{\#}a)^{\#}.$$

Hence $(bb^{\#}a)^{\#} = aa^{\#}b^{\#}$. Similarly, $(aa^{\#}b)^{\#} = bb^{\#}a^{\#}$. Therefore, $bb^{\#}a^{\#} = aa^{\#}b^{\#}$. We have

$$aa^{\#}bb^{\#} = bb^{\#}ab^{\#} = (bb^{\#}a)(aa^{\#}b^{\#}) = (aa^{\#}b^{\#})(bb^{\#}a) = bb^{\#}aa^{\#}.$$

In addition,

$$ab^{\#}aa^{\#} = aaa^{\#}b^{\#}aa^{\#} = abb^{\#}a^{\#}aa^{\#} = aaa^{\#}b^{\#} = ab^{\#},$$

$$ba^{\#}bb^{\#} = b(bb^{\#}a^{\#})bb^{\#} = baa^{\#}b^{\#}bb^{\#} = baa^{\#}b^{\#} = bbb^{\#}a^{\#} = ba^{\#}.$$

Since $bb^{\#}ab = bb^{\#}aaa^{\#}b = aa^{\#}bbb^{\#}a = aa^{\#}ba = bb^{\#}aa$, $b^{\#}ab = b^{\#}aa$. Similarly, $a^{\#}ba = a^{\#}bb$. Since $bb^{\#}a^{\#}b^{\#} = bb^{\#}a^{\#}aa^{\#}b^{\#} = aa^{\#}b^{\#}bb^{\#}a^{\#} = aa^{\#}b^{\#}a^{\#} = bb^{\#}a^{\#}a^{\#}$, $ba^{\#}b^{\#} = ba^{\#}a^{\#}$. Hence, $b^{\#}a^{\#}b^{\#} = b^{\#}a^{\#}a^{\#}$. Similarly, $ab^{\#}a^{\#} = ab^{\#}b^{\#}$ and $a^{\#}b^{\#}a^{\#} = a^{\#}b^{\#}b^{\#}$. To sum up, we have the following equalities:

$$bb^{\#}a = aa^{\#}b, \tag{3.5.7}$$

$$a^{\#}b^{\#}b^{\#} = a^{\#}b^{\#}a^{\#}, \tag{3.5.8}$$

$$bb^{\#}a^{\#} = aa^{\#}b^{\#}, \quad aa^{\#}bb^{\#} = bb^{\#}aa^{\#}, \tag{3.5.9}$$

$$ab^\# aa^\# = ab^\#, \ ba^\# bb^\# = ba^\#, \tag{3.5.10}$$

$$ab^\# a^\# = ab^\# b^\#, \ ba^\# b^\# = ba^\# a^\#, \tag{3.5.11}$$

$$a^\# ba = a^\# bb, \ b^\# ab = b^\# aa. \tag{3.5.12}$$

1. Let $x = a^\# + b^\# - \frac{1}{2} bb^\# a^\# - \frac{1}{2} a^\# bb^\# - \frac{1}{2} b^\# aa^\#$. By equalities (3.5.7), (3.5.9) and (3.5.10) we obtain that

$$(a + b)x = (a + b)(a^\# + b^\# - \frac{1}{2} bb^\# a^\# - \frac{1}{2} a^\# bb^\# - \frac{1}{2} b^\# aa^\#) = aa^\# + ab^\#$$

$$- \frac{1}{2} abb^\# a^\# - \frac{1}{2} aa^\# bb^\# - \frac{1}{2} ab^\# aa^\# + ba^\# + bb^\# - \frac{1}{2} ba^\#$$

$$- \frac{1}{2} ba^\# bb^\# - \frac{1}{2} bb^\# aa^\#$$

$$= aa^\# + ab^\# - \frac{1}{2} ab^\# - \frac{1}{2} aa^\# bb^\# - \frac{1}{2} ab^\# + ba^\# + bb^\#$$

$$- \frac{1}{2} ba^\# - \frac{1}{2} ba^\# - \frac{1}{2} aa^\# bb^\#$$

$$= aa^\# + bb^\# - aa^\# bb^\#$$

and

$$x(a + b) = (a^\# + b^\# - \frac{1}{2} bb^\# a^\# - \frac{1}{2} a^\# bb^\# - \frac{1}{2} b^\# aa^\#)(a + b) = a^\# a + b^\# a$$

$$- \frac{1}{2} bb^\# a^\# a - \frac{1}{2} a^\# bb^\# a - \frac{1}{2} b^\# a + a^\# b + b^\# b - \frac{1}{2} bb^\# a^\# b - \frac{1}{2} a^\# b - \frac{1}{2} b^\# aa^\# b$$

$$= a^\# a + b^\# a - \frac{1}{2} aa^\# bb^\# - \frac{1}{2} a^\# b - \frac{1}{2} b^\# a + a^\# b + b^\# b$$

$$- \frac{1}{2} aa^\# b^\# b - \frac{1}{2} a^\# b - \frac{1}{2} b^\# a$$

$$= aa^\# + bb^\# - aa^\# bb^\#.$$

So $(a + b)x = x(a + b)$. By equalities (3.5.7) and (3.5.9), we have

$$x(a + b)x = (aa^\# + bb^\# - aa^\# bb^\#)(a^\# + b^\# - \frac{1}{2} bb^\# a^\# - \frac{1}{2} a^\# bb^\# - \frac{1}{2} b^\# aa^\#)$$

$$= a^\# + aa^\# b^\# - \frac{1}{2} aa^\# bb^\# a^\# - \frac{1}{2} a^\# bb^\# - \frac{1}{2} aa^\# b^\# aa^\# + bb^\# a^\# + b^\#$$

$$- \frac{1}{2} bb^\# a^\# - \frac{1}{2} bb^\# a^\# bb^\#$$

$$-\frac{1}{2}b^\#aa^\# - aa^\#bb^\#a^\# - aa^\#b^\# + \frac{1}{2}aa^\#bb^\#a^\# + \frac{1}{2}aa^\#bb^\#a^\#bb^\# + \frac{1}{2}aa^\#b^\#aa^\#$$

$$= a^\# + aa^\#b^\# - \frac{1}{2}aa^\#b^\# - \frac{1}{2}a^\#bb^\# - \frac{1}{2}bb^\#a^\# + bb^\#a^\# + b^\#$$

$$-\frac{1}{2}bb^\#a^\# - \frac{1}{2}aa^\#b^\#$$

$$-\frac{1}{2}b^\#aa^\# - aa^\#b^\# - aa^\#b^\# + \frac{1}{2}aa^\#b^\# + \frac{1}{2}aa^\#b^\# + \frac{1}{2}bb^\#a^\#$$

$$= a^\# + b^\# - \frac{1}{2}bb^\#a^\# - \frac{1}{2}a^\#bb^\# - \frac{1}{2}b^\#aa^\# = x,$$

and

$$(a+b)x(a+b) = (aa^\# + bb^\# - aa^\#bb^\#)(a+b)$$

$$= a + bb^\#a - aa^\#bb^\#a + aa^\#b + b - aa^\#b = a + b.$$

So $(a+b)^\# = a^\# + b^\# - \frac{1}{2}bb^\#a^\# - \frac{1}{2}a^\#bb^\# - \frac{1}{2}b^\#aa^\#$.

2. Let $y = a^\# - b^\# - a^\#a^\#b + b^\#b^\#a$. By equalities (3.5.10), (3.5.11), and (3.5.12),

$$(a-b)y = (a-b)(a^\# - b^\# - a^\#a^\#b + b^\#b^\#a)$$

$$= aa^\# - ab^\# - a^\#b + ab^\#b^\#a - ba^\# + bb^\# + ba^\#a^\#b - b^\#a$$

$$= aa^\# - ab^\# - a^\#b + ab^\#a^\#a - ba^\# + bb^\# + ba^\#b^\#b - b^\#a$$

$$= aa^\# - a^\#b - b^\#a + bb^\#$$

and

$$y(a-b) = (a^\# - b^\# - a^\#a^\#b + b^\#b^\#a)(a-b)$$

$$= a^\#a - b^\#a - a^\#a^\#ba + b^\#b^\#aa - a^\#b + b^\#b + a^\#a^\#bb - b^\#b^\#ab$$

$$= a^\#a - a^\#b - b^\#a + bb^\#.$$

Hence, $(a-b)y = y(a-b)$. Since $a^\#b^\#a = a^\#aa^\#b^\#a = a^\#bb^\#a^\#a$, by equalities (3.5.7), (3.5.8), (3.5.9) and (3.5.11), we obtain

$$y(a-b)y = (a^\#a - a^\#b - b^\#a + bb^\#)(a^\# - b^\# - a^\#a^\#b + b^\#b^\#a)$$

$$= a^\# - aa^\#b^\# - a^\#a^\#b + aa^\#b^\#b^\#a - a^\#ba^\# + a^\#bb^\# + a^\#ba^\#a^\#b - a^\#b^\#a$$

$$- b^\#aa^\# + b^\#ab^\# + b^\#a^\#b - b^\#ab^\#b^\#a + bb^\#a^\# - b^\# - bb^\#a^\#a^\#b + b^\#b^\#a$$

$$= a^\# - aa^\#b^\# - a^\#a^\#b + aa^\#b^\#a^\#a - a^\#aa^\#ba^\# + a^\#bb^\#$$

$$+ a^{\#}aa^{\#}ba^{\#}a^{\#}b - a^{\#}b^{\#}a$$

$$- b^{\#}aa^{\#} + b^{\#}bb^{\#}ab^{\#} + b^{\#}bb^{\#}a^{\#}b - b^{\#}bb^{\#}ab^{\#}b^{\#}a$$

$$+ bb^{\#}a^{\#} - b^{\#} - bb^{\#}a^{\#}b^{\#}b + b^{\#}b^{\#}a$$

$$= a^{\#} - aa^{\#}b^{\#} - a^{\#}a^{\#}b + aa^{\#}b^{\#} - a^{\#}bb^{\#}a^{\#}a + a^{\#}bb^{\#}$$

$$+ a^{\#}bb^{\#}aa^{\#}a^{\#}b - a^{\#}bb^{\#}aa^{\#}$$

$$- b^{\#}aa^{\#} + b^{\#}aa^{\#}bb^{\#} + b^{\#}aa^{\#}bb^{\#} - b^{\#}aa^{\#}bb^{\#}b^{\#}a$$

$$+ bb^{\#}a^{\#} - b^{\#} - bb^{\#}a^{\#} + b^{\#}b^{\#}a$$

$$= a^{\#} - bb^{\#}a^{\#} - a^{\#}a^{\#}b + bb^{\#}a^{\#} - a^{\#}bb^{\#} + a^{\#}bb^{\#}$$

$$+ a^{\#}aa^{\#}b^{\#}b - a^{\#}bb^{\#} - b^{\#}aa^{\#}$$

$$+ b^{\#}aa^{\#} + b^{\#}aa^{\#} - b^{\#}bb^{\#}a^{\#}a + bb^{\#}a^{\#} - b^{\#}$$

$$- bb^{\#}a^{\#} + b^{\#}b^{\#}a$$

$$= a^{\#} - b^{\#} - a^{\#}a^{\#}b + b^{\#}b^{\#}a = y.$$

By equalities (3.5.7) and (3.5.12), we have

$$(a - b)y(a - b) = (a^{\#}a - a^{\#}b - b^{\#}a + bb^{\#})(a - b)$$

$$= a - a^{\#}ba - b^{\#}aa + bb^{\#}a - aa^{\#}b + a^{\#}bb + b^{\#}ab - b$$

$$= a - a^{\#}bb - b^{\#}aa + bb^{\#}a - bb^{\#}a + a^{\#}bb + b^{\#}aa - b = a - b.$$

Thus $(a - b)^{\#} = a^{\#} - b^{\#} - a^{\#}a^{\#}b + b^{\#}b^{\#}a$.

□

According to the above two theorems, we have the following corollary.

**Corollary 3.5.3** *Let* $a$, $b \in R^{\#}$ *and* $2 \in R^{-1}$. *If* $abb^{\#} = baa^{\#}$ *and* $bb^{\#}a = aa^{\#}b$, *then*

(1) $a + b \in R^{\#}$ *and* $(a + b)^{\#} = a^{\#} + b^{\#} - \frac{3}{2}a^{\#}bb^{\#}$.
(2) $a - b \in R^{\#}$ *and* $(a - b)^{\#} = a^{\#} - b^{\#}$.

**Proof** By the proof of the above two theorems, we have

$$a^{\#}bb^{\#} = b^{\#}aa^{\#}, \ bb^{\#}a^{\#} = aa^{\#}b^{\#},$$

$$b^{\#}b^{\#}a = b^{\#}a^{\#}b = a^{\#}bb^{\#},$$

$$ab = aaa^{\#}b = abb^{\#}a = baa^{\#}a = ba,$$

$$a^{\#}a^{\#}b = a^{\#}b^{\#}a = b^{\#}a^{\#}a = a^{\#}bb^{\#},$$

$$ab^{\#} = aaa^{\#}b^{\#} = abb^{\#}a^{\#} = baa^{\#}a^{\#} = ba^{\#},$$

$$a^{\#}b = a^{\#}aa^{\#}b = a^{\#}bb^{\#}a = b^{\#}aa^{\#}a = b^{\#}a,$$

$$aa^{\#}b^{\#} = a^{\#}(ab^{\#}) = (a^{\#}b)a^{\#} = b^{\#}aa^{\#},$$

$$ba^{\#}a^{\#} = ab^{\#}a^{\#} = ab^{\#}aa^{\#}b^{\#} = aa^{\#}b^{\#} = a^2a^{\#}(a^{\#}bb^{\#})b^{\#}$$

$$= a^2(a^{\#}b^{\#}a)a^{\#}b^{\#} = a^2a^{\#}a^{\#}(ba^{\#})b^{\#} = aa^{\#}(ab^{\#})b^{\#} = ab^{\#}b^{\#}.$$

From Theorem 3.5.1, we obtain

$$(a+b)^{\#} = a^{\#} + b^{\#} - \frac{1}{2}a^{\#}bb^{\#} - \frac{1}{2}bb^{\#}a^{\#} - \frac{1}{2}aa^{\#}b^{\#}$$

$$= a^{\#} + b^{\#} - \frac{1}{2}a^{\#}bb^{\#} - aa^{\#}b^{\#}$$

$$= a^{\#} + b^{\#} - \frac{1}{2}a^{\#}bb^{\#} - b^{\#}aa^{\#}$$

$$= a^{\#} + b^{\#} - \frac{1}{2}a^{\#}bb^{\#} - a^{\#}bb^{\#}$$

$$= a^{\#} + b^{\#} - \frac{3}{2}a^{\#}bb^{\#}$$

and

$$(a-b)^{\#} = a^{\#} - b^{\#} - ba^{\#}a^{\#} + ab^{\#}b^{\#}$$

$$= a^{\#} - b^{\#}.$$

□

## 3.6   The Group Inverse of the Product of Two Regular Elements

Let $a, b$ be regular elements in $R$, with reflexive inverses $a^{+}, b^{+}$, respectively. Let also

$$w = \left(1 - bb^{+}\right)\left(1 - a^{+}a\right)$$

which we will assume to be regular in $R$. Note that the regularity of $w$ does not depend on the choices of $a^{+}$ and $b^{+}$. That is to say, if $w$ is regular for a particular choice of $a^{+}$ and of $b^{+}$, then it must be regular for all choices of $a^{+}$ and $b^{+}$. This can be easily proved by noting that $w$ being regular is equivalent to the matrix $\begin{bmatrix} a & 0 \\ 1 & b \end{bmatrix}$

being regular, using [99], which in turn is equivalent to $(1 - bb^=)(1 - a^=a)$ being regular, for any other choices of inner inverses $a^=$ and $b^=$ of $a$ and $b$. Consider the matrix $M = \begin{bmatrix} ab & a \\ 0 & 1 \end{bmatrix} = AQ$ with $A = \begin{bmatrix} a & 0 \\ 1 & -b \end{bmatrix}, Q = \begin{bmatrix} b & 1 \\ 1 & 0 \end{bmatrix}$. It is well known that $M^\#$ exists if and only if $(ab)^\#$ exists, using [59]. Furthermore, the $(1,1)$ entry of $M^\#$ equals $(ab)^\#$. Also, $M^\#$ exists if and only if $U = AQ + I - AA^-$ is invertible, see [97, 105], in which case $(AQ)^\# = U^{-2}(AQ)$.

As $AQ + I - AA^- = A(Q - A^-) + I$ then $AQ + I - AA^-$ is invertible if and only if $(Q - A^-)A + I = QA + I - A^-A$ is invertible, using Jacobson's lemma, which in turn means $(QA)^\#$ exists. Therefore, by considering the matrix $W = QA = \begin{bmatrix} ba + 1 & -b \\ a & 0 \end{bmatrix}$ then $(ab)^\#$ exists if and only if $W$ is group invertible. Using [96], the matrix $W$ is group invertible if and only if

$$z = (1 + ba)(1 - a^+a) + ba + (1 - ww^-)(1 - bb^+)(1 + ba)$$
$$= 1 - a^+a + ba + (1 - ww^-)(1 - bb^+)$$

is a unit. We have, hence, the equivalence $(ab)^\#$ exists if and only if $1 - a^+a + ba + (1 - ww^-)(1 - bb^+)$ is a unit. Using the expression presented in [96] does not give a tractable algorithm to actually compute $(ab)^\#$. We will, therefore, pursue a different strategy and compute the $(1,1)$ entry of $M^\#$. Recall that for $M = AQ$ and $Q$ invertible, the group inverse of $M$ exists if and only if $U = AQ + I - AA^-$ is invertible. For $A = \begin{bmatrix} a & 0 \\ 1 & -b \end{bmatrix}$, there exists $A^-$ for which

$$AA^- = \begin{bmatrix} aa^+ & 0 \\ -(1 - ww^-)(1 - bb^+)a^+ & bb^+ + ww^-(1 - bb^+) \end{bmatrix}$$

using [99]. The matrix $U$ then becomes

$$U = \begin{bmatrix} ab + 1 - aa^+ & a \\ (1 - ww^-)(1 - bb^+)a^+ & 2 - bb^+ - ww^+(1 - bb^+) \end{bmatrix}.$$

Multiplication on the right by $K = \begin{bmatrix} 1 & 0 \\ a^+ - b & 1 \end{bmatrix}$ gives

$$G = \begin{bmatrix} 1 & a \\ \alpha & 2 - bb^+ - ww^-(1 - bb^+) \end{bmatrix},$$

where

$$\alpha = \left(1 - ww^-\right)\left(1 - bb^+\right)a^+ + \left(2 - bb^+ - ww^-\left(1 - bb^+\right)\right)\left(a^+ - b\right)$$
$$= a^+ - b + 2\left(1 - ww^-\right)\left(1 - bb^+\right)a^+.$$

We are left with showing when $G$ is invertible. We do so using the Schur complement on the $(1,1)$ entry. This Schur complement equals

$$G/I = \left(2 - bb^+ - ww^-\left(1 - bb^+\right)\right) - \left(\left(1 - ww^-\right)\left(1 - bb^+\right)a^+\right.$$
$$+ \left(2 - bb^+ - ww^-\left(1 - bb^+\right)\right)\left(a^+ - b\right)\right)a$$
$$= \left(2 - bb^+ - ww^-\left(1 - bb^+\right)\right)\left(1 - a^+a + ba\right) - \left(1 - ww^-\right)\left(1 - bb^+\right)a^+a$$
$$= \left(1 + \left(1 - ww^-\right)\left(1 - bb^+\right)\right)\left(1 - a^+a + ba\right) - \left(1 - ww^-\right)\left(1 - bb^+\right)a^+a$$
$$= 1 - a^+a + ba + \left(1 - ww^-\right)\left(1 - bb^+\right)\left(1 - 2a^+a\right)$$
$$= 1 - a^+a + ba + \left(1 - ww^-\right)\left(1 - bb^+\right)\left(1 - a^+a\right)$$
$$+ \left(1 - ww^-\right)\left(1 - bb^+\right)a^+a$$
$$= 1 - a^+a + ba + \left(1 - ww^-\right)\left(1 - bb^+\right).$$

This gives, and as previously shown, $(ab)^{\#}$ exists if and only if $z = 1 - a^+a + ba + \left(1 - ww^-\right)\left(1 - bb^+\right)$ is a unit. As a side note, we construct another unit associated with $z$, namely we may show that $z = 1 - a^+a + ba + \left(1 - ww^-\right)\left(1 - bb^+\right)$ is a unit if and only if $z' = 1 - aa^+ + ab - a(1 - ww^-)\left(1 - bb^+\right)a^+$ is a unit. This follows by the sequence of identities $\left(1 - ww^-\right)\left(1 - bb^+\right) = \left(1 - ww^-\right)\left(1 - bb^+\right)\left(1 - a^+a + a^+a\right) = \left(1 - ww^-\right)\left(1 - bb^+\right)a^+a$ together with Jacobson's Lemma. We remark that given a reflexive inverse $w^+$ of $w$, the element $\tilde{w} = \left(1 - a^+a\right)w^+(1 - bb^+)$ is an idempotent reflexive inverse of $w$. As such $z$ and $z'$ simplify to

$$1 - a^+a + ba + 1 - bb^+ - w\tilde{w} \text{ and } 1 + ab - abb^+a^+ - aw\tilde{w}a^+, \text{ respectively.}$$

We know, using [97, Corollary 3.3(4)], that $(AQ)^{\#}$ exists if and only if $U$ is invertible, in which case $(AQ)^{\#} = U^{-2}(AQ)$. The matrices $U$ and $G$ are equivalent, and we are able to relate their inverses by means of the matrix $K$. Indeed, since $G = UK$, we have $U^{-1} = KG^{-1}$. Firstly, we need to compute the inverse of $G$, for which we will use the following known result:

**Lemma 3.6.1**

$$\begin{bmatrix} 1 & y \\ x & z \end{bmatrix}^{-1} = \begin{bmatrix} 1 + y\zeta^{-1}x & -y\zeta^{-1} \\ -\zeta^{-1}x & \zeta^{-1} \end{bmatrix}$$

*where $\zeta = z - xy$ is the Schur complement.*

Our purpose is to derive an expression for $(ab)^{\#}$, which equals the $(1,1)$ entry of $M^{\#}$. The $(1,1)$ entry of $M^{\#}$ is obtained by multiplying the first row of $U^{-2}$ by the first column of $AQ$, which is $\begin{bmatrix} ab \\ 0 \end{bmatrix}$. So, in fact we just need the $(1,1)$ entry of $U^{-2}$, which is then multiplied on the right by $ab$ to give $(ab)^{\#}$.

We recall that $G = UK$, where $K = \begin{bmatrix} 1 & 0 \\ a^+ & -b \ 1 \end{bmatrix}$, which gives $U^{-1} = KG^{-1}$ and $U^{-2} = KG^{-1}KG^{-1}$. Pre-multiplication with $K$ does not affect the first row, and so we just need the $(1,1)$ element of $G^{-1}KG^{-1}$. Calculations show that

$$U^{-2} = \left(KG^{-1}\right)^{-2} = \begin{bmatrix} (1 + az^{-1}\alpha)^2 - az^{-1}(a^+ - b)(1 + az^{-1}\alpha) + az^{-2}\alpha & ? \\ ? & ? \end{bmatrix}.$$

We will need the simplification

$$b - zb = a^+ab - bab, \tag{3.6.1}$$

from where we obtain

$$\alpha ab = \left(a^+ - b\right)ab = b - zb,$$
$$\left(1 + az^{-1}\alpha\right)ab = az^{-1}b.$$

Indeed, $\alpha ab = aa^+ab - bab + 2\left(1 - ww^-\right)\left(1 - bb^+\right)a^+ab$ whose last summand can be expressed as

$$2\left(1 - ww^-\right)\left(1 - bb^+\right)a^+ab = -2\left(1 - ww^-\right)\left(1 - bb^+\right)\left(1 - a^+a - 1\right)b$$

$$= -2\left(1 - ww^-\right)w + 2\left(1 - ww^-\right)\left(1 - bb^+\right)b = 0,$$

and therefore $\alpha ab = a^+ab - bab$. Therefore,

$$(AQ)^{\#} = \begin{bmatrix} (ab)^{\#} & ? \\ 0 & ? \end{bmatrix}$$

$$= \begin{bmatrix} \left((1 + az^{-1}\alpha)^2 - az^{-1}(a^+ - b)(1 + az^{-1}\alpha) + az^{-2}\alpha\right)ab & ? \\ 0 & ? \end{bmatrix},$$

from which we obtain the general formula

$$(ab)^{\#} = \left(\left(1 + az^{-1}\alpha\right)^2 - az^{-1}\left(a^+ - b\right)\left(1 + az^{-1}\alpha\right) + az^{-2}\alpha\right) ab$$

$$=ab + 2\left(az^{-1}b - ab\right) + az^{-1}\alpha\left(az^{-1}b - ab\right)$$

$$\quad - az^{-1}\left(a^+ - b\right)az^{-1}b + az^{-1}\left(z^{-1}b - b\right)$$

$$=\left(az^{-1}\alpha az^{-1}b - az^{-1}\left(a^+ - b\right)az^{-1}b\right) + az^{-2}b$$

$$=2az^{-1}\left(1 - ww^-\right)\left(1 - bb^+\right)z^{-1}b + az^{-2}b$$

$$=2az^{-1}b - 2\left(az^{-1}b\right)^2 + az^{-2}b.$$

From $1 - a^+a = z^{-1} - z^{-1}a^+a$ we obtain, by postmultiplying by $b$,

$$b - a^+ab = z^{-1}b - z^{-1}a^+ab \qquad (3.6.2)$$

which implies

$$az^{-1}b = az^{-1}a^+ab. \qquad (3.6.3)$$

Now, from (3.6.1) we have $z^{-1}b = b + z^{-1}a^+ab - z^{-1}bab$ which implies, using (3.6.2), that

$$z^{-1}bab = a^+ab \qquad (3.6.4)$$

which it turns delivers

$$ab = az^{-1}bab. \qquad (3.6.5)$$

Using (3.6.4) and (3.6.5), together with $(ab)^{\#} = 2az^{-1}b - 2\left(az^{-1}b\right)^2 + az^{-2}b$, we write the idempotent $(ab)^{\#}(ab)$ as

$$(ab)^{\#}ab = \left(2az^{-1}b - 2\left(az^{-1}b\right)^2 + az^{-2}b\right)ab$$

$$= 2az^{-1}bab - 2az^{-1}bab + az^{-1}z^{-1}bab$$

$$= az^{-1}a^+ab.$$

Using (3.6.3), this equals $az^{-1}b$ and therefore $az^{-1}b$ is an idempotent, the unit of the group generated by $ab$. This simplifies the expression of $(ab)^\#$ to

$$(ab)^\# = az^{-2}b.$$

It comes with no surprise that the expression of $(ab)^\#$ is of the form $aXb$, for a suitable $X$. We have, from the above, our main result:

**Theorem 3.6.2 ([91, Theorem 2.2])** *Let $a, b \subset R$ be regular with reflexive inverses $a^+$ and $b^+$, respectively. Assume that $w = \left(1 - bb^+\right)\left(1 - a^+a\right)$ is regular. Then $(ab)^\#$ exists if and only if $z = 1 - a^+a + ba + \left(1 - ww^-\right)\left(1 - bb^+\right)$ is a unit. In this case,*

$$(ab)^\# = az^{-2}b.$$

## 3.7  Group Inverses of Block Matrices

In this section, we consider group inverses of block matrices over a ring. First, we introduce two useful lemmas.

**Lemma 3.7.1 ([14, Lemma 1.1])** *Let $a \in R$ be regular, $b \in R$, and assume that there exists a reflexive inverse $a^+$ such that $a^+b = ba^+ = 0$. If $a + b$ is regular then $b$ is regular and $b^- = (a + b)^-$ is a $\{1\}$-inverse of $b$ for any $(a + b)^-$.*

**Lemma 3.7.2 ([14, Lemma 1.2])** *Let $P, Q, A \in R^{2\times 2}$. If $M = PAQ$ where $P$ and $Q$ are units and $A$ is regular, then the group inverse of $M$ exists if and only if $S = AQP + I - AA^-$ is a unit of $R^{2\times 2}$, independent of the choice of $A^-$, or equivalently, $T = QPA + I - A^-A$ is a unit, in which case*

$$M^\# = PS^{-2}AQ = PAT^{-2}Q.$$

**Proof** It is obvious by Corollary 3.3.3.                                     □

From now on we assume that $d \in R$ is regular and that $d^+$ is a fixed but arbitrary $\{1, 2\}$-inverse of $d$. Let us introduce the notation

$$e = 1 - dd^+, \quad f = 1 - d^+d, \quad s = a - cd^+b. \tag{3.7.1}$$

With this notation, we have the following decomposition of the matrix $M \in R^{2\times 2}$

$$M = \begin{bmatrix} 1 & cd^+ \\ 0 & 1 \end{bmatrix} \begin{bmatrix} s & cf \\ eb & d \end{bmatrix} \begin{bmatrix} 1 & 0 \\ d^+b & 1 \end{bmatrix} := PAQ. \tag{3.7.2}$$

In the notation of (3.7.1), we assume both $eb$ and $cf$ to be regular elements in $R$. Set

$$p = 1 - (eb)^+ eb, \quad q = 1 - cf(cf)^+, \quad w = qsp, \tag{3.7.3}$$

for fixed but arbitrary $(eb)^+$ and $(cf)^+$.

**Lemma 3.7.3 ([14, Lemma 2.1])** *Let $e, f, s, p, q$ and $w$ be as in (3.7.1) and (3.7.3). We have that $A = \begin{bmatrix} s & cf \\ eb & d \end{bmatrix}$ is regular in $R^{2\times2}$ if and only if $w$ is regular in $R$. In this case,*

$$A^- = \begin{bmatrix} 1 & 0 \\ -f(cf)^+ s & 1 \end{bmatrix} \begin{bmatrix} pw^-q & (eb)^+ e \\ f(cf)^+ & d^+ \end{bmatrix} \begin{bmatrix} 1 & -qs(eb)^+ e \\ 0 & 1 \end{bmatrix} \tag{3.7.4}$$

*is a {1}-inverse of $A$ and*

$$I - AA^- = \begin{bmatrix} (1 - ww^-)q & -(1 - ww^-)qs(eb)^+ e \\ 0 & (1 - eb(eb)^+)e \end{bmatrix}. \tag{3.7.5}$$

**Proof** Let us first observe that the following relations hold:

$$ed = df = 0, \quad fd^+ = d^+ e = 0, \quad qcf = (cf)^+ q = 0.$$

Now, consider

$$\begin{bmatrix} 1 & -qs(eb)^+ e \\ 0 & 1 \end{bmatrix} \begin{bmatrix} s & cf \\ eb & d \end{bmatrix} \begin{bmatrix} 1 & 0 \\ -f(cf)^+ s & 1 \end{bmatrix} = \begin{bmatrix} qsp & cf \\ eb & d \end{bmatrix} := \tilde{A}. \tag{3.7.6}$$

Hence, $A$ is regular if and only if $\tilde{A}$ is regular. Using $w = qsp$, we can write

$$\tilde{A} = \begin{bmatrix} 0 & cf \\ eb & d \end{bmatrix} + \begin{bmatrix} w & 0 \\ 0 & 0 \end{bmatrix} := Y + W.$$

Next, we will prove that $\tilde{A}$ is regular if and only if $w$ is regular. We have that $Y$ is regular and $Y^+ = \begin{bmatrix} 0 & (ed)^+ e \\ f(cf)^+ & d^+ \end{bmatrix}$ is a {1,2}-inverse of $Y$ such that $WY^+ = Y^+W = 0$. Now, if $\tilde{A}$ is regular, then $W$ is regular, by Lemma 3.7.1. This implies that $w$ is regular. Conversely, assume that $w$ is regular. Let $X = \begin{bmatrix} pw^-q & (eb)^+ e \\ f(cf)^+ & d^+ \end{bmatrix}$. We claim that $X$ is a {1} -inverse of $\tilde{A}$. Indeed.

$$\tilde{A}X\tilde{A} = \begin{bmatrix} w & cf \\ eb & d \end{bmatrix} \begin{bmatrix} pw^-w + 1 - p & 0 \\ 0 & d^+d + f(cf)^+cf \end{bmatrix} = \tilde{A}.$$

In view of (3.7.6) we conclude that a {1}-inverse of $A$ is given by (3.7.4) and, thus, $A$ is regular. It remains to prove (3.7.5) but the proof of this is straightforward.  □

We can now formulate our main result. It is required that $d$ has the group inverse. Accordingly, we can set $d^+ = d^\#$. In this case, in the notation of (3.7.1) we have $e = 1 - dd^\# = 1 - d^\# d = f$. It follows that $de = ed^\# = 0$. Moreover $d + e$ is a unit of $R$ and $(d + e)^{-1} = d^\# + e$.

**Theorem 3.7.4 ([14, Theorem 2.2])** *Let $d$ be group invertible. With the notation (3.7.1) and under the assumptions of (3.7.3), with $d^\dagger$ replaced by $d^\#$, if $w$ is regular in $R$ then the group inverse of the matrix $M = \begin{bmatrix} a & c \\ b & d \end{bmatrix}$ exists if and only if*

$$u = sp - ceb + \left(1 - ww^-\right)q\left[s + \left(1 + c\left(d^\#\right)^2 b\right)p\right] \tag{3.7.7}$$

*is a unit of $R$. In this case, $M^\# = \begin{bmatrix} \gamma_1 & \gamma_3 \\ \gamma_2 & \gamma_4 \end{bmatrix}$, where*

$$
\begin{aligned}
\gamma_1 &= \alpha - p\left(\delta\alpha - u^{-1}\right), \\
\gamma_2 &= -d^\#\left(b - d^\# bp\right)\alpha + \left(d^\# bp + eb\right)\left(\delta\alpha - u^{-1}\right), \\
\gamma_3 &= -\beta + p\left(\alpha c\left(d^\#\right)^2 + \delta\beta - u^{-1}cd^\#\right), \\
\gamma_4 &= d^\# + d^\#\left(b - d^\# bp\right)\beta - \left(d^\# bp + eb\right)\left(\alpha c\left(d^\#\right)^2 + \delta\beta - u^{-1}cd^\#\right),
\end{aligned}
$$
$$\tag{3.7.8}$$

*with*

$$
\begin{aligned}
&\alpha = u^{-1}\left(1 - ww^-\right)q, \quad \beta = \alpha cd^\# + u^{-1}ce, \\
&\delta = \alpha\left(1 + c\left(d^\#\right)^2\left(b - d^\# bp\right)\right) + u^{-1}\left(s + \left(1 + c\left(d^\#\right)^2 b\right)p\right).
\end{aligned}
\tag{3.7.9}
$$

*Proof* Write $M = PAQ$ as in (3.7.2). By Lemma 3.7.2, the group inverse of $M$ exists if and only if $S = AQP + I - AA^-$ is a unit, independent of the choice of $A^-$. For the {1}-inverse provided in Lemma 3.7.3, we have

$$S = AQP + I - AA^- = \begin{bmatrix} s + \left(1 - ww^-\right)q & scd^\# + ce - \left(1 - ww^-\right)qs(eb)^+ e \\ b & d + bcd^\# + \left(1 - eb(eb)^+\right)e \end{bmatrix}.$$

Now, let us introduce the matrix $F = \begin{bmatrix} 1 & (eb)^+ e - cd^\# \\ 0 & 1 \end{bmatrix}$. We have

$$SF = \begin{bmatrix} s + \left(1 - ww^-\right)q & x \\ b & h \end{bmatrix}, \tag{3.7.10}$$

where

$$h = d + e + dd^\# b(eb)^+ e = (d + e)\left(1 + d^\# b(eb)^+ e\right),$$
$$x = s(eb)^+ e + ce - \left(1 - ww^-\right)q\left[(s - 1)(eb)^+ e + cd^\#\right]. \tag{3.7.11}$$

Since the element $d^\# b(eb)^+ e$ is 2-nilpotent, it follows that $1 + d^\# b(eb)^+ e$ is a unit of $R$. Moreover, $d + e$ is a unit because $d$ has group inverse. Thus, $h$ is a unit and

$$g := h^{-1} = \left(1 - d^\# b(eb)^+ e\right)\left(d^\# + e\right). \tag{3.7.12}$$

On account that the element (2,2) of the matrix $SF$ is a unit, it follows that $SF$ is a unit of $R^{2\times2}$ if and only if the Schur complement is a unit of $R$. Therefore, the matrix $S$ is a unit if and only if

$$u = s + \left(1 - ww^-\right)q - xgb \tag{3.7.13}$$

is a unit of $R$. From (3.7.12) we get $gb = eb + d^\# bp$. Further, using the last relation of (3.7.11) we obtain

$$xgb = s(1 - p) + ceb - \left(1 - ww^-\right)q\left[(s - 1) + \left(1 + c\left(d^\#\right)^2 b\right)p\right].$$

By substituting this expression in (3.7.13) we conclude that $u$ has the form given in (3.7.7). By Lemma 3.7.2, $M^\# = PS^{-2}AQ$. From (3.7.10) it follows that

$$(SF)^{-1} = \begin{bmatrix} u^{-1} & -u^{-1}xg \\ -gbu^{-1} & g + gbu^{-1}xg \end{bmatrix}.$$

Next, we compute

$$PS^{-1} = PF(SF)^{-1} = \begin{bmatrix} 1 & cd^\# \\ 0 & 1 \end{bmatrix}\begin{bmatrix} 1 & (eb)^+ e - cd^\# \\ 0 & 1 \end{bmatrix}(SF)^{-1}$$

$$= \begin{bmatrix} pu^{-1} & (eb)^+ e - pu^{-1}xg \\ -gbu^{-1} & g + gbu^{-1}xg \end{bmatrix}, \tag{3.7.14}$$

the last equality is due to the fact that $eg = e$. In the sequel, we denote $\alpha = u^{-1}\left(1 - ww^-\right)q$ and $\beta = \alpha cd^\# + u^{-1}ce$. From (3.7.12), (3.7.11) and (3.7.13), it follows that

$$gd = d^\# d,$$
$$u^{-1}xd^\# d = -\alpha cd^\#, \tag{3.7.15}$$
$$u^{-1}xgb = u^{-1}\left(s + \left(1 - ww^-\right)q\right) - 1 = u^{-1}s + \alpha - 1,$$

respectively. In deriving the last equality wc have multiplied on the left expression (3.7.13) by $u^{-1}$. Then

$$S^{-1}AQ = F(SF)^{-1}AQ = F(SF)^{-1}\begin{bmatrix} s & ce \\ eb & d \end{bmatrix}\begin{bmatrix} 1 & 0 \\ d^{\#}b & 1 \end{bmatrix}$$

$$= F\begin{bmatrix} u^{-1} & -u^{-1}xg \\ -gbu^{-1} & g+gbu^{-1}xg \end{bmatrix}\begin{bmatrix} s & ce \\ b & d \end{bmatrix}$$

$$= \begin{bmatrix} 1 & (eb)^{+}e-cd^{\#} \\ 0 & 1 \end{bmatrix}\begin{bmatrix} 1-\alpha & \beta \\ gb\alpha & d^{\#}d-gb\beta \end{bmatrix}$$

$$= \begin{bmatrix} 1-\left(1+c\left(d^{\#}\right)^{2}b\right)p\alpha & -cd^{\#}+\left(1+c\left(d^{\#}\right)^{2}b\right)p\beta \\ gb\alpha & d^{\#}d-gb\beta \end{bmatrix}.$$

$$(3.7.16)$$

Using (3.7.14) and (3.7.16) we obtain

$$PS^{-2}AQ = \begin{bmatrix} pu^{-1} & (eb)^{+}e-pu^{-1}xg \\ -gbu^{-1} & g+gbu^{-1}xg \end{bmatrix}$$

$$\cdot \begin{bmatrix} 1-\left(1+c\left(d^{\#}\right)^{2}b\right)p\alpha & -cd^{\#}+\left(1+c\left(d^{\#}\right)^{2}b\right)p\beta \\ gb\alpha & d^{\#}d-gb\beta \end{bmatrix}$$

$$= \begin{bmatrix} \gamma_{1} & \gamma_{3} \\ \gamma_{2} & \gamma_{4} \end{bmatrix},$$

where

$$\gamma_{1} = (1-p)\alpha - pu^{-1}\left(xg^{2}b\alpha - 1 + \left(1+c\left(d^{\#}\right)^{2}b\right)p\alpha\right),$$
$$\gamma_{2} = g^{2}b\alpha + gbu^{-1}\left(xg^{2}b\alpha - 1 + \left(1+c\left(d^{\#}\right)^{2}b\right)p\alpha\right),$$
$$\gamma_{3} = (p-1)\beta + pu^{-1}\left(xg^{2}b\beta - xd^{\#} - cd^{\#} + \left(1+c\left(d^{\#}\right)^{2}b\right)p\beta\right),$$
$$\gamma_{4} = d^{\#} - g^{2}b\beta - gbu^{-1}\left(xg^{2}b\beta - xd^{\#} - cd^{\#} + \left(1+c\left(d^{\#}\right)^{2}b\right)p\beta\right).$$

$$(3.7.17)$$

From (3.7.12) it follows that $g^{2}b = gb - d^{\#}\left(b - d^{\#}bp\right)$. Using this latter expression and (3.7.15) we get

$$u^{-1}xg^{2}b = u^{-1}xgb - u^{-1}xd^{\#}\left(b - d^{\#}bp\right) = u^{-1}s + \alpha - 1 + \alpha c\left(d^{\#}\right)^{2}\left(b - d^{\#}bp\right).$$

By substituting this into (3.7.17), using $u^{-1}xd^{\#} = -\alpha c\left(d^{\#}\right)^{2}$, and regrouping terms we get (3.7.8).                                                                                          □

Next, some applications are indicated. We begin with the case that $d$ is a unit of $R$.

**Corollary 3.7.5 ([14, Corollary 3.1])** *Let* $M = \begin{bmatrix} a & c \\ b & d \end{bmatrix}$*, where $d$ is a unit of $R$, and let $s = a - cd^{-1}b$ be regular. Then the group inverse of $M$ exists if and only if*

$$u = s + \left(1 - ss^-\right)\left(1 + cd^{-2}b\right) \tag{3.7.18}$$

*is a unit of $R$. In this case,* $M^\# = \begin{bmatrix} \gamma_1 & \gamma_3 \\ \gamma_2 & \gamma_4 \end{bmatrix}$*, where*

$$\gamma_1 = u^{-1} + (1 - \delta)\alpha,$$

$$\gamma_2 = -d^{-1}bu^{-1} - d^{-1}\left(b - d^{-1}b - b\delta\right)\alpha,$$

$$\gamma_3 = \alpha cd^{-2} - u^{-1}cd^{-1} - (1 - \delta)\beta, \tag{3.7.19}$$

$$\gamma_4 = d^{-1} - d^{-1}b\left(\alpha cd^{-2} - u^{-1}cd^{-1}\right) + d^{-1}\left(b - d^{-1}b - b\delta\right)\beta,$$

*with*

$$\alpha = u^{-1}\left(1 - ss^-\right), \quad \beta = \alpha cd^{-1}, \tag{3.7.20}$$

$$\delta = \alpha\left(1 + cd^{-2}\left(b - d^{-1}b\right)\right) + u^{-1}\left(s + 1 + cd^{-2}b\right).$$

In addition, let $s$ be group invertible. Denote $t = 1 - ss^\#$. Then the following statements are equivalent:

(1) $M^\#$ exists.
(2) $u = s + t\left(1 + tcd^{-2}b\right)$ is a unit of $R$.
(3) $1 + tcd^{-2}b$ is a unit of $R$.
(4) $d^2 + btc$ is a unit of $R$.

**Proof** Note that $d$ is a unit, we have $e = 1 - dd^\# = 0$ and so $p = q = 1$. Thus $w = s$. The rest of the proof follows from Theorem 3.7.4. □

The next corollary provides a compact representation for the group inverse of $M$.

**Corollary 3.7.6 ([14, Corollary 3.2])** *Let* $M = \begin{bmatrix} a & c \\ b & 0 \end{bmatrix}$*, where $b, c$ and $w = \left(1 - cc^+\right)a\left(1 - b^+b\right)$ are regular elements in $R$. Then the group inverse of $M$ exists if and only if*

$$u = a\left(1 - b^+b\right) - cb + \left(1 - ww^-\right)\left(1 - cc^+\right)\left(a + 1 - b^+b\right)$$

*is a unit of R. In this case, $M^{\#} = \begin{bmatrix} \gamma_1 & \gamma_3 \\ \gamma_2 & \gamma_4 \end{bmatrix}$, where*

$$\gamma_1 = \alpha - \left(1 - b^+ b\right)\left(\delta\alpha - u^{-1}\right),$$
$$\gamma_2 = b\left(\delta\alpha - u^{-1}\right),$$
$$\gamma_3 = -u^{-1}c + \left(1 - b^+ b\right)\delta u^{-1}c,$$
$$\gamma_4 = -b\delta u^{-1}c,$$

*with $\alpha = u^{-1}\left(1 - ww^-\right)\left(1 - cc^+\right)$ and $\delta = \alpha + u^{-1}\left(a + 1 - b^+ b\right)$.*

**Proof** This result follows from Theorem 3.7.4 since, in this case, $d = 0$ and so $e = 1, s = a, p = 1 - b^+ b, q = 1 - cc^+$ and $w = \left(1 - cc^+\right)a\left(1 - b^+ b\right)$.     □

## 3.8   Group Inverses of Companion Matrices Over a Ring

One of the fundamental building block in the theory and applications of matrices, is the lower companion matrix associated with a monic polynomial $f(x) = p_0 + p_1 x + \cdots + x^n$. It is defined by

$$L = L[f(x)] = \begin{bmatrix} 0 & 0 & \cdots & 0 & 0 & -p_0 \\ 1 & 0 & \cdots & 0 & 0 & -p_1 \\ 0 & 1 & \cdots & 0 & 0 & -p_2 \\ \cdot & \cdot & \cdot & \cdot & \cdot & \cdot \\ \cdot & \cdot & \cdot & \cdot & \cdot & \cdot \\ 0 & 0 & \cdots & 1 & 0 & -p_{n-2} \\ 0 & 0 & \cdots & 0 & 1 & -p_{n-1} \end{bmatrix}$$

and is the simplest example of an $n \times n$ non-derogatory matrix. Throughout this section $n$ will be fixed.

It was shown in [59] that over an arbitrary ring, $L^{\#}$ with $n \geq 2$, exists exactly when there exist solutions $x$ and $y$ in the ring $R$, such that
(i)

$$[p_0, p_1]\begin{bmatrix} x \\ y \end{bmatrix} = -1 = [x, y]\begin{bmatrix} p_0 \\ p_1 \end{bmatrix}, \tag{3.8.1}$$

(ii)

$$p_0 y = 0,$$

(iii)

$$[x, y] \begin{bmatrix} p_1 \\ p_2 \end{bmatrix} y = 0.$$

In which case $L^{\#}$ has the form

$$L^{\#} = \begin{bmatrix} \underline{x} \ \underline{y} \ 0 \cdots 0 \\ \quad I_{n-2} \\ x \ y \ 0 \cdots 0 \end{bmatrix}, \tag{3.8.2}$$

where $\underline{x} = [x_1, \ldots, x_{n-1}]^T$, $\underline{y} = [y_1, \ldots, y_{n-1}]^T$, and

$$y_i = \delta_{i1} + p_i y \quad \text{and} \quad x_i = y_{i+1} + p_i x, \quad i = 1, \ldots, n - 1.$$

The case $n = 2$ is included in this if we set $p_2 = 1$, $y_2 = y$ and drop $I_{n-2}$ from (3.8.2). This gives

$$y_1 = 1 + p_1 y \quad \text{and} \quad x_1 = y + p_1 x.$$

Consider the $n \times n$ companion matrix $L = \begin{bmatrix} 0 \ a \\ I \ \underline{k} \end{bmatrix}$ over a general ring with 1, where $a = a_0$ and $\underline{k} = [a_1, a_2, \ldots, a_{n-1}]^T$.

**Theorem 3.8.1 ([105, Theorem 3])** *The following are equivalent:*

(1) *The group inverse $L^{\#}$ of $L$ exists.*
(2) *$a$ is regular and $h = a - (1 - aa^-) a_1$ is invertible.*
(3) *$a$ is regular and $k = a - a_1 (1 - a^- a)$ is invertible.*

**Proof** For convenience we set $a_1 = b$ and $a_2 = c$. Consider the companion matrix $L$ and its factorization

$$L = \begin{bmatrix} 0 \ a \\ I \ \underline{k} \end{bmatrix} = \begin{bmatrix} 0 \ 1 \\ I \ 0 \end{bmatrix} \begin{bmatrix} I \ 0 \\ 0 \ a \end{bmatrix} \begin{bmatrix} I \ \underline{k} \\ 0 \ 1 \end{bmatrix} = PAQ.$$

Because $P$ and $Q$ are invertible, it is necessary in order for $L^{\#}$ to exist, that $A$ and hence $a$ be regular. Let us now assume that $a$ is a regular and apply Corollary 3.3.4 to simplify the conditions for $L^{\#}$ to exist, and then use (3.3.5) to compute the actual group inverse. From (3.8.2) we recall that it suffices to compute $x = (L^{\#})_{n1}$ and $y = (L^{\#})_{n2}$. In order to construct the matrices $U$ and $V$, we must consider

$$PA = \begin{bmatrix} 0 \ a \\ I \ 0 \end{bmatrix}, \ AQ = \begin{bmatrix} I \ \underline{k} \\ 0 \ a \end{bmatrix} \text{ and } (AQ)(PA) = \begin{bmatrix} 0 \ b \ a \\ I_{n-2} \ \underline{r} \ 0 \\ 0 \ a \ 0 \end{bmatrix}$$

where $r = [c, a_3, \ldots, a_{n-1}]^T$ is $(n-2) \times 1$. Since $a$ is regular we may take $\Lambda^- = \begin{bmatrix} I & 0 \\ 0 & a^- \end{bmatrix}$. This yields

$$U = (AQ)\left(PAA^-\right) + \left(I - AA^-\right) = \begin{bmatrix} 0 & b & 1-e \\ I_{n-2} & r & 0 \\ 0 & a & e \end{bmatrix} \tag{3.8.3}$$

in which $e = 1 - aa^-$. Similarly

$$V = A^-(AQ)(PA) + \left(I - A^-A\right) = \begin{bmatrix} 0 & b & a \\ I_{n-2} & r & 0 \\ 0 & 1-f & f \end{bmatrix}, \tag{3.8.4}$$

where $f = 1 - a^-a$. Next we define

$$F = \begin{bmatrix} 0 & 1 & 0 \\ I_{n-2} & 0 & 0 \\ 0 & 0 & 1 \end{bmatrix} \text{ and } W(X) = \begin{bmatrix} I & B \\ 0 & X \end{bmatrix}, \qquad \text{where } B = [r, 0].$$

Then $U = FW_1$ and $V = FW_2$, where $W_i = W(X_i)$ and

$$X_1 = \begin{bmatrix} b & 1-e \\ a & e \end{bmatrix}, \quad X_2 = \begin{bmatrix} b & a \\ 1-f & f \end{bmatrix}. \tag{3.8.5}$$

We note that if $n = 2$, then $U = X_1$ and $V = X_2$. As such let us first turn to the case where $n > 2$. We shall need the following facts:

(i)

$$Qe_1 = e_1, \ Qe_2 = e_2, \ e_n^T P = e_{n-1}. \tag{3.8.6}$$

(ii) $F^{-1} = \begin{bmatrix} 0 & I_{n-2} & 0 \\ 1 & 0 & 0 \\ 0 & 0 & 1 \end{bmatrix}$, and thus $F^{-1}e_1 = e_{n-1}$ and $F^{-1}e_2 = e_1$.

(iii) $W_i^{-1} = \begin{bmatrix} I_{n-2} & -BX_i^{-1} \\ 0 & X_i^{-1} \end{bmatrix}$ in which $X_i^{-1} = \begin{bmatrix} \alpha_i & \gamma_i \\ \beta_i & \delta_i \end{bmatrix}$, and $-BX_i^{-1} = [r\alpha_i, r\gamma_i]$.

(iv) $e_{n-1}^T W_i^{-1} = [0, \ldots, 0, \alpha_i, \gamma_i]$.

(v) $e_{n-1}^T W_i^{-1} F^{-1} = [\alpha_i, 0, \ldots, 0, \gamma_i]$.

(vi) $s_i^T = e_n^T P W_i^{-1} F^{-1} = e_{n-1}^T W_i^{-1} F^{-1} = [\alpha_i, 0, \ldots, 0, \gamma_i] = e_n^T P A W_i^{-1} F^{-1}$.

(vii) $\underline{z}_i = W_i^{-1} F^{-1} A Q \underline{e}_1 = W_i^{-1} F^{-1} A \underline{e}_1 = W_i^{-1} F^{-1} \underline{e}_1 = W_i^{-1} \underline{e}_{n-1} =$
$$\begin{bmatrix} -\underline{r}\alpha_i \\ \alpha_i \\ \beta_i \end{bmatrix}.$$

(viii) $\underline{d}_i = W_i^{-1} F^{-1} A Q \underline{e}_2 = W_i^{-1} F^{-1} \underline{e}_2 = W_i^{-1} \underline{e}_1 = \underline{e}_1.$

Using these identities we see that

$$\underline{s}_i^T \underline{z}_i = [\alpha_i, 0, \ldots, 0, \gamma_i] \begin{bmatrix} -\underline{r}\alpha_i \\ \alpha_i \\ \beta_i \end{bmatrix} = -\alpha_i c \alpha_i + \gamma_i \beta_i \qquad (3.8.7)$$

and

$$\underline{s}_i^T \underline{d}_i = [\alpha_i, 0, \ldots, 0, \gamma_i] \underline{e}_1 = \alpha_i. \qquad (3.8.8)$$

Let us now apply these to the three formulae for $L^{\#}$.

**Case (I)** $L^{\#} = PU^{-2}AQ$. Now

$$x = \left( L^{\#} \right)_{n,1} = \underline{e}_n^T PU^{-1}U^{-1}A Q \underline{e}_1 = \left( \underline{e}_n^T P W_1^{-1} F^{-1} \right) \left( W_1^{-1} F^{-1} A Q \underline{e}_1 \right) = \underline{s}_1^T \underline{z}_1.$$

From (3.8.7) we thus see that for $n > 2$,

$$x = -\alpha_1 c \alpha_1 + \gamma_1 \beta_1. \qquad (3.8.9)$$

Likewise

$$y = \left( L^{\#} \right)_{n,2} = \left( \underline{e}_n^T P W_1^{-1} F^{-1} \right) \left( W_1^{-1} F^{-1} A Q \underline{e}_2 \right) = \underline{s}_1^T \underline{d}_1 = \alpha_1.$$

**Case (II)** $L^{\#} = PAV^{-2}Q$. Now

$$x = \left( L^{\#} \right)_{n,1} = \underline{e}_n^T PAV^{-1}V^{-1}Q \underline{e}_1 = \left( \underline{e}_n^T P A W_2^{-1} F^{-1} \right) \left( W_2^{-1} F^{-1} Q \underline{e}_1 \right) = \underline{s}_2^T \underline{z}_2,$$

which gives

$$x = -\alpha_2 c \alpha_2 + \gamma_2 \beta_2. \qquad (3.8.10)$$

Likewise

$$y = \left( L^{\#} \right)_{n,2} = \underline{e}_n^T PAV^{-1}V^{-1}Q \underline{e}_2$$

$$= \left( \underline{e}_n^T P A W_2^{-1} F^{-1} \right) \left( W_2^{-1} F^{-1} Q \underline{e}_2 \right) = \underline{s}_2^T \underline{d}_2 = \alpha_2.$$

**Case (III)** $L^{\#} = PU^{-1}AV^{-1}Q$. Then

$$x = \underline{e}_n^T PU^{-1}AV^{-1}Q\underline{e}_1 = \left(\underline{e}_n^T PW_1^{-1}F^{-1}\right) A \left(W_2^{-1}F^{-1}Q\underline{e}_1\right)$$

$$= s_1^T \underline{A}\underline{z}_2 = -\alpha_1 c\alpha_2 + \gamma_1 a\beta_2$$

and

$$y = \left(\underline{e}_n^T PW_1^{-1}F^{-1}\right) A \left(W_2^{-1}F^{-1}Q\underline{e}_2\right) = \underline{s}_1^T A\underline{d}_2 = \alpha_1.$$

Thus equating the three cases we arrive for $n > 2$ at the identities

$$y = \alpha_1 = \alpha_2, \tag{3.8.11}$$

$$x = -\alpha_1 c\alpha_1 + \gamma_1\beta_1 = -\alpha_2 c\alpha_2 + \gamma_2\beta_2 = -\alpha_1 c\alpha_2 + \gamma_1 a\beta_2. \tag{3.8.12}$$

This shows that we must have

$$\gamma_1\beta_1 = \gamma_2\beta_2 = \gamma_1 a\beta_2. \tag{3.8.13}$$

In order to verify these we need the explicit expressions for $X_i^{-1}$. This we now pursue.

From (3.8.5) we may factor the $X_i$ as

$$X_1 = \begin{bmatrix} b & 1-e \\ a & e \end{bmatrix} = \begin{bmatrix} 1 & 1-e \\ -1 & e \end{bmatrix} \begin{bmatrix} -h & 0 \\ a+b & 1 \end{bmatrix}, \tag{3.8.14}$$

where $h = a - eb$, and by symmetry

$$X_2 = \begin{bmatrix} b & a \\ 1-f & f \end{bmatrix} = \begin{bmatrix} -k & a+b \\ 0 & 1 \end{bmatrix} \begin{bmatrix} 1 & -1 \\ 1-f & f \end{bmatrix}, \tag{3.8.15}$$

where $k = a - bf$.

This shows that $U$ and $V$ are invertible in $R_{n \times n}$ with $n \geq 2$, exactly when $h = a - eb$ and $k = a - bf$ are respectively invertible in $R$. Again this is independent of the choice of $a^-$. Inverting the $X_i$ we see that

$$X_1^{-1} = \begin{bmatrix} -h^{-1} & 0 \\ (a+b)h^{-1} & 1 \end{bmatrix} \begin{bmatrix} e & e-1 \\ 1 & 1 \end{bmatrix}$$

$$= \begin{bmatrix} -h^{-1}e & h^{-1}(1-e) \\ 1+(a+b)h^{-1}e & 1-(a+b)h^{-1}(1-e) \end{bmatrix} = \begin{bmatrix} \alpha_1 & \gamma_1 \\ \beta_1 & \delta_1 \end{bmatrix}. \tag{3.8.16}$$

Likewise

$$X_2^{-1} = \begin{bmatrix} f & 1 \\ f-1 & 1 \end{bmatrix} \begin{bmatrix} -k^{-1} & k^{-1}(a+b) \\ 0 & 1 \end{bmatrix}$$

$$= \begin{bmatrix} -fk^{-1} & 1+fk^{-1}(a+b) \\ (1-f)k^{-1} & 1-(1-f)k^{-1}(a+b) \end{bmatrix} = \begin{bmatrix} \alpha_2 & \gamma_2 \\ \beta_2 & \delta_2 \end{bmatrix}.$$

$$(3.8.17)$$

Thus

$$\begin{aligned}
\alpha_1 &= -h^{-1}e, & \alpha_2 &= -fk^{-1}, \\
\beta_1 &= 1+(a+b)h^{-1}e, & \beta_2 &= (1-f)k^{-1}, \\
\gamma_1 &= h^{-1}(1-e), & \gamma_2 &= 1+fk^{-1}(a+b), \\
\delta_1 &= 1-(a+b)h^{-1}(1-e), & \delta_2 &= 1-(1-f)k^{-1}(a+b).
\end{aligned}$$

Let us now check these identities.
First $hf = -ebf = ek$, and thus

$$h^{-1}e = fk^{-1}, \qquad (3.8.18)$$

which ensures that $\alpha_1 = \alpha_2$. Next, we observe that $fk^{-1}a = h^{-1}ea = 0$, and hence

$$a = a(1-f) = (h+eb)(1-f) = h\left(1+h^{-1}eb\right)(1-f)$$

$$= h\left(1+fk^{-1}b\right)(1-f)$$

which says that

$$\gamma_1 a\beta_2 = h^{-1}ak^{-1} = \left(1+fk^{-1}b\right)(1-f)k^{-1} = \gamma_2\beta_2. \qquad (3.8.19)$$

The remaining identity follows by symmetry, i.e., $\gamma_1 a\beta_2 = \gamma_1\beta_1$. In conclusion, let us compute the actual expression for $L^\#$. $\qquad\qquad\qquad \square$

**Theorem 3.8.2 ([105, Theorem 4])** *Suppose* $L = \begin{bmatrix} 0 & a \\ I & \underline{k} \end{bmatrix}$ *with* $\underline{k} = [b, a_2, \ldots,$
$a_{n-1}]^T$. *Define the constants*

$$h = a - \left(1-aa^-\right)b,$$
$$\alpha_1 = -h^{-1}\left(1-aa^-\right), \quad \gamma_1 = h^{-1}aa^-,$$
$$\beta_1 = 1+(a+b)h^{-1}\left(1-aa^-\right) \quad \text{and} \quad \delta_1 = 1-(a+b)h^{-1}aa^-.$$

*If $L^{\#}$ exists then the constants $x$ and $y$ from (3.8.2) are given by*

$$(i)\ for\ n > 2 : x = -\alpha_1 a_2 \alpha_1 + \gamma_1 \beta_1, \quad y = \alpha_1;$$

$$(ii)\ for\ n = 2 : x = \alpha_1^2 + \gamma_1 \beta_1, \quad y = [\alpha_1, \gamma_1] \begin{bmatrix} \alpha_1 & \gamma_1 \\ \beta_1 & \delta_1 \end{bmatrix} \begin{bmatrix} b \\ a \end{bmatrix}.$$

**Proof** All that remains is to prove the case where $n = 2$. In this case we again only need compute $x$ and $y$ after which we set $x_1 = y - bx$ and $y_1 = 1 - by$. It follows from (3.8.3)–(3.8.4) that $U = X_1$ and $V = X_2$. Moreover the values of $\underline{s}_i$, $\underline{z}_i$ and $\underline{d}_i$ change to

$$\underline{s}_i^T = [\alpha_i, \beta_i], \quad \underline{z}_i = \begin{bmatrix} \alpha_i \\ \beta_i \end{bmatrix} \text{ and } \underline{d}_i = X_i^{-1} \begin{bmatrix} b \\ a \end{bmatrix}.$$

This gives

$$x = \left(L^{\#}\right)_{2,1} = \underline{e}_2^T P X_1^{-1} X_1^{-1} A Q \underline{e}_1 = \left(\underline{e}_1^T X_1^{-1}\right) \left(X_1^{-1} \underline{e}_1\right) = \alpha_1^2 + \gamma_1 \beta_1.$$

Likewise

$$y = \left(L^{\#}\right)_{2,2} = \underline{e}_2^T P X_1^{-1} X_1^{-1} A Q \underline{e}_2 = \left(\underline{e}_1^T X_1^{-1}\right) \left(X_1^{-1} \begin{bmatrix} b \\ a \end{bmatrix}\right)$$

$$= [\alpha_1, \gamma_1] \begin{bmatrix} \alpha_1 & \gamma_1 \\ \beta_1 & \delta_1 \end{bmatrix} \begin{bmatrix} b \\ a \end{bmatrix},$$

completing the proof.                                                                                                  □

## 3.9    EP Elements

Recall that $A \in \mathbb{C}^{n \times n}$ is called EP if $\mathcal{R}(A) = \mathcal{R}(A^*)$. In this case, $A = U \begin{bmatrix} D & 0 \\ 0 & 0 \end{bmatrix} U^*$ for some unitary matrix $U$ and non-singular matrix $D$ (see Theorem 1.1.10). In this section we introduce the studies that extend the notion of EP matrices to EP elements in a $*$-ring. Throughout this section, $R$ denotes a $*$-ring unless otherwise stated.

**Definition 3.9.1** Let $a \in R$ be Moore-Penrose invertible and group invertible. If $a^\dagger = a^\#$, then $a$ is called an EP element.

**Proposition 3.9.2** *Let $a \in R$. Then the following statements are equivalent:*

(1) *$a$ is EP.*
(2) *$a^\dagger$ exists and $aa^\dagger = a^\dagger a$.*
(3) *$a^\#$ exists and $(aa^\#)^* = aa^\#$.*

**Proof** (1) $\Rightarrow$ (2). Since $a$ is EP, we have $a^\dagger = a^\#$. Therefore, $aa^\dagger = aa^\# = a^\#a = a^\dagger a$.

(2) $\Rightarrow$ (3). Because $aa^\dagger = a^\dagger a$, $a$ is group invertible and $a^\# = a^\dagger$. So, $(aa^\#)^* = (aa^\dagger)^* = aa^\dagger = aa^\#$.

(3) $\Rightarrow$ (1). Since $(aa^\#)^* = aa^\#$ and $aa^\# = a^\#a$, we have $(a^\#a)^* = a^\#a$. Therefore, $a$ is Moore-Penrose invertible and $a^\dagger = a^\#$. That is, $a$ is EP. $\square$

**Theorem 3.9.3 ([100, Proposition 2])** *Let $a \in R$. Then the following statements are equivalent:*

(1) *$a$ is EP.*
(2) *$a^\dagger$ exists and $aR = a^*R$.*
(3) *$a^\#$ exists and $aR = a^*R$.*

**Proof** (1) $\Rightarrow$ (2). By Proposition 3.9.2, we have $aa^\dagger = a^\dagger a$, which implies $aa^\dagger R = a^\dagger aR$. Since $aa^\dagger a = a$ and $(a^\dagger a)^* = a^\dagger a$, we have $aR = aa^\dagger R$ and $a^*R = a^\dagger aR$. Therefore, $aR = a^*R$.

(2) $\Rightarrow$ (3). Since $aa^\dagger R = aR = a^*R$, $aa^\dagger a^* = a^*$ which implies $a^2 a^\dagger = a$. Because $aR = a^*R = a^\dagger aR$, we have $a^\dagger a^2 = a$. From Theorem 3.2.2, $a^\#$ exists.

(3) $\Rightarrow$ (1). Since $a^\#$ exists and $aR = a^*R$, $aa^\# R = aR = a^*R = (aa^\#)^*R$ which implies $(aa^\#)^* = aa^\#(aa^\#)^*$. Therefore, $aa^\# = (aa^\#)^*$. Then $a$ is EP by Proposition 3.9.2. $\square$

In particular, for $*$-regular rings we have the following.

**Corollary 3.9.4** *Let $R$ be a $*$-regular ring and $a \in R$. Then $a$ is EP if and only if $aR = a^*R$.*

The next theorem characterizes EP elements by the invertibility of some elements.

**Theorem 3.9.5 ([100, Theorem 4])** *Let $R$ be a $*$-ring. If $a \in R$ is regular with a $\{1\}$-inverse $a^-$, then the following are equivalent and independent from the choice of $a^-$:*

(1) *$a$ is EP.*
(2) *$aa^*aa^- + 1 - aa^-$ and $a^2aa^- + 1 - aa^-$ are invertible and*

$$\left[ \left( aa^*aa^- + 1 - aa^- \right)^{-1} a \right]^* = \left( a^2aa^- + 1 - aa^- \right)^{-1} a.$$

(3) *$a^-aa^*a + 1 - a^-a$ and $a^-aa^2 + 1 - a^-a$ are invertible and*

$$\left[ a \left( a^-aa^*a + 1 - a^-a \right)^{-1} \right]^* = a \left( a^-aa^2 + 1 - a^-a \right)^{-1}.$$

*Moreover, if* $u = a^2aa^- + 1 - aa^-, \quad v = a^-aa^2 + 1 - a^-a, \tilde{u} = aa^*aa^- + 1$
*$-aa^-$ and $\tilde{v} = a^-aa^*a + 1 - a^-a$ then*

$$a^\# = a^\dagger = u^{-1}a = av^{-1} = \left(\tilde{u}^{-1}a\right)^* = \left(a\tilde{v}^{-1}\right)^*$$

*and equals a* $\left(a^2\right)^- a \left(a^2\right)^- a$.

**Proof** Follows directly from the results in [94, 105] if we can replace $a^2a^- + 1 - aa^-$ by $a^2aa^- + 1 - aa^-$, and analogously $a^- u^? + 1 - a^-a$ by $a^-aa^2 + 1 - a^-a$. Indeed,

$$a^2a^- + 1 - aa^-$$

is invertible iff

$$\left(a^2a^- + 1 - aa^-\right)^2 = \left(a^2a^- + 1 - aa^-\right)\left(a^2a^- + 1 - aa^-\right)$$

$$= a^2a^-a^2a^- + 1 - aa^-$$

$$= a^3a^- + 1 - aa^-$$

is invertible. Then

$$\left(a^2a^- + 1 - aa^-\right)^{-2} = \left[\left(a^2a^- + 1 - aa^-\right)^2\right]^{-1}$$

$$= \left(a^3a^- + 1 - aa^-\right)^{-1}.$$

The remaining fact to prove is that $a^\# = a^\dagger = a\left(a^2\right)^- a \left(a^2\right)^- a$. Indeed, if $a^\#$ exists then $a^2$ is von Neumann regular and

$$\left(a^2a^- + 1 - aa^-\right)^{-1} = a\left(a^2\right)^- aa^- + 1 - aa^-$$

since

$$\left(a^2a^- + 1 - aa^-\right)\left(a\left(a^2\right)^- aa^- + 1 - aa^-\right) = a^2a^-a\left(a^2\right)^- aa^- + 1 - aa^-$$

$$= a^2\left(a^2\right)^- aa^- + 1 - aa^-$$

$$= a^2\left(a^2\right)^- a^2a^\#a^- + 1 - aa^-$$

$$= a^2a^\#a^- + 1 - aa^-$$

$$= 1$$

and

$$\left(a\left(a^2\right)^- aa^- + 1 - aa^-\right)\left(a^2a^- + 1 - aa^-\right) = a\left(a^2\right)^- aa^- a^2a^- + 1 - aa^-$$
$$= a\left(a^2\right)^- a^2a^- + 1 - aa^-$$
$$= a^{\#}a^2\left(a^2\right)^- a^2a^- + 1 - aa^-$$
$$= a^{\#}a^2a^- + 1 - aa^-$$
$$= 1.$$

Therefore,

$$\left(a^3a^- + 1 - aa^-\right)^{-1} = \left(a^2a^- + 1 - aa^-\right)^{-2}$$
$$= \left(a\left(a^2\right)^- aa^- + 1 - aa^-\right)^2$$

and

$$a^{\#} = a^{\dagger} = \left(\left(a\left(a^2\right)^-\right)^2 aa^- + 1 - aa^-\right)a = a\left(a^2\right)^- a\left(a^2\right)^- a.$$

$\square$

**Remark 3.9.6 ([100, Remark])** A regular element $a$ in a $*$-ring $R$ has a group inverse $a^{\#}$ and a Moore-Penrose inverse $a^{\dagger}$ with respect to $*$ such that $a^{\#} = a^{\dagger}$ if and only if

$$\left(a^3a^- + 1 - aa^-\right)^{-1} \quad \text{and} \quad \left(a^- aa^*a + 1 - a^- a\right)^{-1} \text{ exist}$$

and

$$a^* = \left[\left(a^- aa^*a + 1 - a^- a\right)^* a\left(a^2\right)^- a\left(a^2\right)^-\right]a$$

for any choice of $a^-$, since

$$a\left(a^- a^3 + 1 - a^- a\right)^{-1} = \left(a^3a^- + 1 - aa^-\right)^{-1}a = a\left(a^2\right)^- a\left(a^2\right)^- a.$$

For a complex matrix $A$, from a result of Katz (see also [4, p. 159, Exercise 17]), it can be seen that $A^{\dagger} = A^{\#}$ if and only if there is $Y$ such that $A^* = YA$. The next result extends this fact from the $n \times n$ full complex matrix ring to any general Dedekind-finite ring.

**Proposition 3.9.7 ([100, Fact 5])** *If $R$ is a Dedekind-finite $*$-ring and $a \in R$ is Moore-Penrose invertible, then $a$ is EP if and only if $a^* = ya$ for some $y \in R$.*

**Proof** If $a^\dagger$ exists, then also $\left(a^\dagger\right)^*$ exists and equals $(a^*)^\dagger$. Since $a^* = ya$, we have $a = a^* y^*$ and hence $aR \subseteq a^* R$

Moreover, $aR \cong a^* R$ since $\phi : aR \to a^* R$, with $\phi(ax) = a^\dagger ax$, is a $R$ - module isomorphism. Then, also $aa^\dagger R \cong a^\dagger aR$, which implies $aR = a^* R$. By Theorem 3.9.3, $a^\#$ exists and $a^\dagger = a^\#$. Conversely, if $a^\#$ exists and $a^\dagger = a^\#$ then $a^* = \left(aa^\dagger a\right)^* = a^* aa^\dagger = a^* aa^\# = a^* a^\# a$. It suffices to take $y = a^* a^\#$.  □

The following theorem shows that the equality $aR = a^* R$ in Theorem 3.9.3 can be replaced by the weaker inclusions $aR \subseteq a^* R$ or $a^* R \subseteq aR$.

**Theorem 3.9.8 ([123, Theorem 3.9])** *Let $a \in R$. Then the following statements are equivalent:*

(1) *$a$ is EP.*
(2) *$a \in R^\#$ and $aR \subseteq a^* R$.*
(3) *$a \in R^\#$ and $Ra \subseteq Ra^*$.*

**Proof** (1) $\Rightarrow$ (2). It is obvious by Theorem 3.9.3.

(2) $\Rightarrow$ (1). By $aR \subseteq a^* R$, we have $a = a^* r$ for some $r \in R$, then $a = \left(aa^\# a\right)^* r = \left(a^\# a\right)^* a^* r = \left(a^\# a\right)^* a$. Thus $a^\# a = aa^\# = \left(a^\# a\right)^* aa^\# = \left(a^\# a\right)^* a^\# a$, which gives $\left(a^\# a\right)^* = a^\# a$. Therefore, $a$ is EP.

(1) $\Leftrightarrow$ (3). It is similar to the proof of (1) $\Leftrightarrow$ (2).  □

Recall that an infinite matrix $M$ is said to be bi-finite if it is both row-finite and column-finite.

**Example 3.9.9** Let $R$ be the ring of all bi-finite real matrices with transposing of matrices as its involution and let $e_{i,j}$ be the matrix in $R$ with 1 in the $(i, j)$ position and 0 elsewhere. Let $A = \sum_{i=1}^\infty e_{i+1,i}$ and $B = A^*$, now $AB = \sum_{i=2}^\infty e_{i,i}$, $BA = I$. So $A^\dagger = B$ and $A^\dagger = A^\dagger BA = B^2 A$. It is easy to check that $B^2$ is not left invertible and $A$ is not EP (since $AB \neq BA$). In addition, $A$ is not group invertible. (If $A \in R^\#$, then $AA^\# = A^\# A = BAA^\# A = BA = I$; thus, $A$ is invertible, which is not possible.) This example also shows that the equality $aR = a^* R$ in the equivalence ($a$ is EP $\Leftrightarrow a \in R^\dagger, aR = a^* R$) cannot be replaced by the inclusions $aR \subseteq a^* R$ or $a^* R \subseteq aR$.

**Theorem 3.9.10 ([123, Theorem 4.4])** *Let $a \in R^\dagger$. Then the following statements are equivalent:*

(1) *$a$ is EP.*
(2) *There exists a unit $u \in R$ such that $a^\dagger = ua$.*
(3) *There exists a left invertible element $v \in R$ such that $a^\dagger = va$.*

***Proof*** (1) $\Rightarrow$ (2). If $a$ is EP, then $a \in R^\dagger$ and $a^\dagger = a^\#$. Let $u = \left(a^\#\right)^2 + 1 - aa^\#$. Since $u\left(a^2 + 1 - aa^\#\right) = \left(a^2 + 1 - aa^\#\right)u = 1$ we get that $u$ is a unit. Furthermore, $ua = \left(\left(a^\#\right)^2 + 1 - aa^\#\right)a = a^\# = a^\dagger$.

(2) $\Rightarrow$ (3). It is clear.

(3) $\Rightarrow$ (1). Suppose that there exists a left invertible element $v \in R$ such that $a^\dagger = va$. Then $1 = tv$ for some $t \in R$ and $ta^\dagger = tva = a$. Thus $Ra^\dagger \subseteq Ra$ and $Ra \subseteq Ra^\dagger$. Since $Ra^\dagger = Ra^*$, we deduce that $Ra^* = Ra$, showing that $a$ is an EP element. $\qquad\square$

In the following theorem, we show that an EP element in a ring can be described by three equations.

**Theorem 3.9.11 ([123, Theorem 2.2])** *Let $a \in R$. Then $a$ is EP if and only if there exists $x \in R$ such that*

$$(xa)^* = xa, \quad xa^2 = a \text{ and } ax^2 = x.$$

***Proof*** Suppose $a$ is EP. Let $x = a^\dagger = a^\#$. Then $(xa)^* = \left(a^\dagger a\right)^* = a^\dagger a = xa$, $xa^2 = a^\# a^2 = a$ and $ax^2 = a\left(a^\#\right)^2 = a^\# = x$. Conversely, if there exists $x \in R$ such that $(xa)^* = xa, xa^2 = a$ and $ax^2 = x$, then $a\left(x^2 a\right) = \left(ax^2\right)a = xa = x\left(xa^2\right) = \left(x^2 a\right)a, a\left(x^2 a\right)a = (xa)a = xa^2 = a$, and $\left(x^2 a\right)a\left(x^2 a\right) = (xa)\left(x^2 a\right) = x\left(ax^2\right)a = x^2 a$. These three equalities prove that $a \in R^\#, a^\# = x^2 a$, and $aa^\# = xa$. By Proposition 3.9.2, we get $a$ is EP. $\qquad\square$

We will characterize when $a \in R$ is EP by another three equations.

**Theorem 3.9.12 ([123, Theorem 2.11])** *Let $a \in R$. Then $a$ is EP if and only if there exists $x \in R$ such that*

$$a^2 x = a, \quad ax = xa \text{ and } (ax)^* = ax.$$

***Proof*** If $a$ is EP, by taking $x = a^\dagger = a^\#$, we get the above three equations. Conversely, assume that there exists $x \in R$ such that the above three equations are satisfied. We shall show that $a$ is EP and $a^\# = ax^2$. Since $ax = xa$, we get $a\left(ax^2\right) = \left(ax^2\right)a$, but in addition, $a\left(ax^2\right) = \left(a^2 x\right)x = ax$, which leads to $a\left(ax^2\right)a = a^2 x = a$ and $\left(ax^2\right)a\left(ax^2\right) = \left(ax^2\right)ax = \left(a^2 x\right)x^2 = ax^2$. Since $aa^\# = a^2 x^2 = ax$ is Hermitian, the conclusion follows from Proposition 3.9.2. $\qquad\square$

The following theorem gives a necessary and sufficient condition of an EP element in C*-algebra, which is also true in a ring with involution.

**Theorem 3.9.13 ([5, Theorem 2.1])** *Let $\mathcal{A}$ be a C*-algebra with unity $1$ and $a \in \mathcal{A}$. Then the following statements are equivalent:*

(1) *There exists a unique projection $p$ such that $a + p \in \mathcal{A}^{-1}$ and $ap = pa = 0$.*

(2) *$a$ is EP.*

**Proof** (1) $\Rightarrow$ (2). Since $ap = 0$ and $p^2 = p$, we get $(a+p)p = p$. The invertibility of $a + p$ entails $p = (a + p)^{-1}p$, and in a similar way we get $p = p(a + p)^{-1}$. Now, we claim

$$a\left[(a + p)^{-1} - p\right] = 1 - p. \tag{3.9.1}$$

In fact:

$$a\left[(a + p)^{-1} - p\right] = [(a + p) - p]\left[(a + p)^{-1} - p\right]$$
$$= 1 - (a + p)p - p(a + p)^{-1} + p$$
$$= 1 - p - p + p = 1 - p.$$

This proves the claim. Analogously we can prove $\left[(a + p)^{-1} - p\right]a = 1 - p$. Now, we are going to prove that $a^\dagger = (a + p)^{-1} - p$. We have

$$a\left[(a + p)^{-1} - p\right]a = (1 - p)a = a - pa = a$$

and

$$\left[(a + p)^{-1} - p\right]a\left[(a + p)^{-1} - p\right] = \left[(a + p)^{-1} - p\right](1 - p)$$
$$= (a + p)^{-1} - p - (a + p)^{-1}p + p$$
$$= (a + p)^{-1} - p - p + p$$
$$= (a + p)^{-1} - p.$$

Evidently $\left[(a + p)^{-1} - p\right]a = a\left[(a + p)^{-1} - p\right] = 1 - p$ is self-adjoint. This proves that $a^\dagger = (a + p)^{-1} - p$. Since $aa^\dagger = 1 - p = a^\dagger a$, we get $a$ is EP.

(2) $\Rightarrow$ (1). Let $p$ be the projection defined by $p = 1 - aa^\dagger$. Evidently, we have $ap = pa = 0$. Now, $pa^\dagger = pa^\dagger aa^\dagger = paa^\dagger a^\dagger = 0$, and similarly $a^\dagger p = 0$ holds. Let us prove $a + p \in \mathcal{A}^{-1}$ : We have $(a + p)(a^\dagger + p) = aa^\dagger + ap + pa^\dagger + p = aa^\dagger + p = 1$ and analogously we have also $(a^\dagger + p)(a + p) = 1$. Now, we shall prove the uniqueness. Assume that $q$ is another projection such that $aq = qa = 0$ and $a + q \in \mathcal{A}^{-1}$. The computations made in (1) $\Rightarrow$ (2) show that $a^\dagger = (a + p)^{-1} - p = (a + q)^{-1} - q$. Premultiplying by $a$ we get $a(a + p)^{-1} = a(a + q)^{-1}$. Now, (3.9.1) implies $a(a + p)^{-1} = 1 - p$ and $a(a + q)^{-1} = 1 - q$. Therefore, $1 - p = 1 - q$ and the uniqueness is proved. $\square$

Recall that $a \in R$ is called *-strongly regular if there exist a projection $p$ and $u \in R^{-1}$ such that $a = pu = up$. The following theorem can be seen as an application of Theorem 3.9.13.

**Theorem 3.9.14 ([125, Theorem 3.3])** *Let $a \in R$. Then $a$ is EP if and only if $a$ is $*$-strongly regular.*

*Proof* Suppose that $a$ is EP. Then, by Theorem 3.9.13, there exists a unique projection $p$ such that $a + p \in R^{-1}$ and $ap = pa = 0$. Write $a + p = u \in R^{-1}$. Then $a = a(1 - p) = u(1 - p) = (1 - p)u$. Therefore, $a$ is $*$-strongly regular.

Conversely, assume that $a$ is $*$-strongly regular. Then there exist a projection $p$ and $u \in R^{-1}$ such that $a = pu = up$. Since $(a + 1 - p)\left(u^{-1}p + 1 - p\right) = \left(u^{-1}p + 1 - p\right)(a+1-p) = 1$, $a+1-p \in R^{-1}$. Because $a(1-p) = (1-p)a = 0$, $a$ is EP by Theorem 3.9.13. $\qquad \square$

# Chapter 4
# Drazin Inverses

For complex matrices, we know from Chap. 3 that the group inverse only exists for matrices with index no more than 1, so it is not suitable for the more common class of matrices whose index exceed 1. In this chapter, we introduce the Drazin inverse which is a slight generalization of the group inverse.

The notion of Drazin inverses was first introduced in the context of rings and semigroups by Drazin [39] in 1958 who proved that an element $a$ in a ring is strongly $\pi$-regular if and only if there exists a (unique) element $a^D$ (which is called the pseudo-inverse of $a$ by Drazin himself) such that

$$a^D a^{k+1} = a^k \text{ for some integer } k, a(a^D)^2 = a^D \text{ and } aa^D = a^D a.$$

Noting that every square matrix over a field (or more generally over a semisimple ring) is strongly $\pi$-regular, one can deduce that every square complex matrix possesses a Drazin inverse. In particular, the Drazin inverse $A^D$ reduces to the group inverse $A^\#$ when the matrix $A$ is of index 1. Also, in contrast to the Moore-Penrose inverse, the Drazin inverse is closely related with the matrix index and has some of the spectral properties (i.e., properties relating to eigenvalues and eigenvectors).

## 4.1 Drazin Inverses of Complex Matrices

In this section, we introduce the Drazin inverse for complex matrices, and show its relationship with several matrix decompositions.

**Definition 4.1.1** Let $A \in \mathbb{C}^{n \times n}$ be of index $k$. Then $X \in \mathbb{C}^{n \times n}$ is called the Drazin inverse of $A$ if

$$XA^{k+1} = A^k, \ X = AX^2, \ AX = XA.$$

© The Author(s), under exclusive license to Springer Nature Singapore Pte Ltd. 2024     143
J. Chen, X. Zhang, *Algebraic Theory of Generalized Inverses*,
https://doi.org/10.1007/978-981-99-8285-1_4

The existence and uniqueness of the Drazin inverse are given in the following theorem.

**Theorem 4.1.2** *For any $A \in \mathbb{C}^{n \times n}$, the Drazin inverse of $A$ exists and is unique (denoted by $A^D$).*

**Proof** Let $\text{ind}(A) = k$. We first prove the uniqueness of the Drazin inverse. Let $X$, $Y$ are the Drazin inverse of $A$. Set $E = AX = XA$, $F = AY = YA$. It is clear that $E^2 = E$, $F^2 = F$. Thus

$$E = E^k = (AX)^k = A^k X^k = YA^{k+1} X^k = YA(A^k X^k) = YA(AX) = FE,$$

$$F = F^k = (YA)^k = Y^k A^k = Y^k A^{k+1} X = (Y^k A^k)AX = (YA)AX = FE.$$

Then $E = F$ and

$$X = AX^2 = (AX)X = (YA)X = Y(AX) = YAY = Y,$$

meaning that the Drazin inverse is unique.

Next we prove the existence of the Drazin inverse. Here we give three methods.

Method I:   Jordan Form

Suppose that $A$ has the Jordan form $A = P \begin{bmatrix} D & 0 \\ 0 & N \end{bmatrix} P^{-1}$, where $D$ is invertible and $N$ is nilpotent. Then we have $\text{ind}(A)=\text{ind}(N)$. Let $\text{ind}(A)=k$. Then $N^k = 0$, it follows that

$$A^k = P \begin{bmatrix} D^k & 0 \\ 0 & N^k \end{bmatrix} P^{-1} = P \begin{bmatrix} D^k & 0 \\ 0 & 0 \end{bmatrix} P^{-1}.$$

Let $X = P \begin{bmatrix} D^{-1} & 0 \\ 0 & 0 \end{bmatrix} P^{-1}$. Then clearly $X$ satisfies the equations $XA^{k+1} = A^k$, $X = AX^2$ and $AX = XA$, which follows that $A^D = X = P \begin{bmatrix} D^{-1} & 0 \\ 0 & 0 \end{bmatrix} P^{-1}$.

Method II:   Matrix Polynomial

Suppose that $A$ has the Jordan normal form $A = P \begin{bmatrix} D & 0 \\ 0 & N \end{bmatrix} P^{-1}$, where $D$ is invertible and $N$ is nilpotent. Since $D$ is invertible, there exists polynomial function $q(x)$ such that $D^{-1} = q(D)$. Then

$$A^k (q(A))^{k+1} = P \begin{bmatrix} D^k & 0 \\ 0 & 0 \end{bmatrix} P^{-1} P \begin{bmatrix} (q(D))^{k+1} & 0 \\ 0 & q(N)^{k+1} \end{bmatrix} P^{-1}$$

$$= P \begin{bmatrix} D^k (q(D))^{k+1} & 0 \\ 0 & 0 \end{bmatrix} P^{-1}.$$

It is easy to check $A^k (q(A))^{k+1} = P \begin{bmatrix} D^k (q(D))^{k+1} & 0 \\ 0 & 0 \end{bmatrix} P^{-1} = A^D.$

Method III:    Using $\{1\}$-inverses

Suppose that $A$ has the Jordan form $A = P \begin{bmatrix} D & 0 \\ 0 & N \end{bmatrix} P^{-1}$, where $D$ is invertible and $N$ is nilpotent. Then $\text{ind}(N)=\text{ind}(A)=k$, it follows that $N^k = 0$. Hence, we have $A^{2t+1} = P \begin{bmatrix} D^{2t+1} & 0 \\ 0 & 0 \end{bmatrix} P^{-1}$ for arbitrary $t \geq k$. Let $X \in (A^{2t+1})\{1\}$.

Then $X = P \begin{bmatrix} D^{-(2t+1)} & X_1 \\ X_2 & X_3 \end{bmatrix} P^{-1}$. Hence it is easy to obtain $A^t X A^t = P \begin{bmatrix} D^{-1} & 0 \\ 0 & 0 \end{bmatrix} P^{-1} = A^D$. Let $t = k$. Then $A^D = A^k (A^{2k+1})^\dagger A^k$.

$\square$

By the proof of existence of the Drazin inverse, the Drazin inverse has a simple representation in terms of the Jordan form.

**Theorem 4.1.3** *Let $A \in \mathbb{C}^{n \times n}$ have the Jordan form $A = P^{-1} \begin{bmatrix} D & 0 \\ 0 & N \end{bmatrix} P$, where $D$ is invertible and $N$ is nilpotent. Then $A^D = P^{-1} \begin{bmatrix} D^{-1} & 0 \\ 0 & 0 \end{bmatrix} P$.*

The Drazin inverse of a complex matrix can be obtained by applying the full rank factorization.

**Theorem 4.1.4** *Let $A \in \mathbb{C}^{n \times n}$. We perform a sequence of full rank factorizations:*

$$A = G_1 H_1, \ H_1 G_1 = G_2 H_2, \ H_2 G_2 = G_3 H_3, \ \cdots$$

*so that $G_i H_i$ are full rank factorizations of $H_{i-1} G_{i-1}$, for $i = 2, \ 3, \ \ldots$ Then there exists integer $k$ such that $H_k G_k = 0$ or $H_k G_k$ is invertible. In this case,*

$$A^D = \begin{cases} 0, & \text{if } H_k G_k = 0; \\ G_1 \cdots G_k (H_k G_k)^{-(k+1)} H_k \cdots H_1, & \text{if } H_k G_k \text{ is invertible.} \end{cases}$$

***Proof*** If $H_i G_i$ is $p \times p$ matrix and has rank $q < p$, then we have $H_i G_i = 0$ when $q = 0$. When $q = p$, we have $H_i G_i$ is invertible. When $0 < q < p$, we have $H_i G_i$ is not invertible. Since $H_i G_i = G_{i+1} H_{i+1}$ is full rank factorization, we know that $r(G_{i+1}) = r(H_{i+1}) = q < p$. If $H_{i+1} G_{i+1} = 0$ or $H_{i+1} G_{i+1}$ is invertible, then the conclusion holds. Othersize, the full rank factorization is continued. It follows that the conclusion holds since $p$ is finite.

Let $k$ be the smallest integer such that $H_k G_k = 0$ or $H_k G_k$ is invertible. (1) Assume that $H_k G_k$ is invertible. If $G_k \in \mathbb{C}^{p \times r}$, $H_k \in \mathbb{C}^{r \times p}$, then $r(H_{k-1} G_{k-1}) = r(G_k H_k) = r$ according to that $H_{k-1} G_{k-1} = G_k H_k$ is full rank factorizations. Since $G_1(\cdots G_k x) = 0$ implies $G_2(\cdots G_k x) = 0$, $G_2(\cdots G_k x) = 0$

implies $G_3(\cdots G_k x) = 0, \cdots, G_k x = 0$ implies $x = 0$, we have that $G_1 \cdots G_k$ is of full column rank. Similarly, we can obtain $H_k \cdots H_1$ is of full row rank. Hence $G_1 \cdots G_k (H_k G_k)$ is of full column rank. Since

$$A^k = (G_1 H_1)^k = G_1 (H_1 G_1)^{k-1} H_1$$

$$= G_1 (G_2 H_2)^{k-1} H_1 = G_1 G_2 (H_2 G_2)^{k-1} H_2 H_1$$

$$- \cdots$$

$$= G_1 \cdots G_{k-1} (H_{k-1} G_{k-1}) H_{k-1} \cdots H_1 = (G_1 \cdots G_k)(H_k \cdots H_1)$$

is full rank factorization, we have $r(A^k) = r(G_1 \cdots G_k) = r$. Also,

$$A^{k+1} = (G_1 \cdots G_k)(H_k G_k) H_k \cdots H_1$$

$$= (G_1 \cdots G_k (H_k G_k)) (H_k \cdots H_1)$$

is full rank factorization, we have $r(A^{k+1}) = r$. Since $k$ is the smallest integer, we have ind$(A)=k$. Let $X = G_1 \cdots G_k (H_k G_k)^{-k-1} H_k \cdots H_1$. Then $X = A^D$.

(2) Assume that $H_k G_k = 0$. Then $A^{k+1} = 0$, it follows that ind$(A)=k+1$. Hence $X = 0 = A^D$. □

## 4.2 Drazin Inverses of Elements in Semigroups and Rings

In this section, we first give the definition of the Drazin inverse in semigroups or rings. Then we present some basic properties of the Drazin inverse. Finally, we study the core-nilpotent decomposition.

**Definition 4.2.1 ([39])** Let $S$ be a semigroup and $a, x \in S$. Then $x$ is called the Drazin inverse of $a$ if there exists a positive integer $k$ such that

$$x a^{k+1} = a^k, \ x = ax^2, \ ax = xa.$$

The smallest integer $k$ satisfying the above equations is called the Drazin index of $a$, denoted by ind$(a)$. Clearly, the Drazin inverse of $a$ reduces to the group inverse $a^\#$ when ind$(a) = 1$.

From now on, the sets of all Drazin invertible elements of a semigroup/ring $S$ is denoted by $S^D$.

The next theorem shows that the Drazin inverse in semigroups or rings is unique if it exists.

**Theorem 4.2.2 ([39])** *Let $a \in S$. Then $a$ has at most one Drazin inverse. If the Drazin inverse of $a$ exists, we denote the unique Drazin inverse of $a$ by $a^D$.*

***Proof*** Suppose $x$ and $y$ satisfy conditions of the definition of Drazin inverse, with $m$, $n$ as Drazin index, respectively. Then we get $x = x^2a$ and $y = ay^2$. Let $k = max\{m, n\}$. Then we can obtain $xa^{k+1} = a^k = a^{k+1}y$. Hence we have

$$x = x^{k+1}a^k = x^{k+1}a^{k+1}y = xay = xa^{k+1}y^{k+1} = a^k y^{k+1} = y.$$

Therefore, $a$ has at most one Drazin inverse. □

Let us recall that an element $a \in S$ is called strongly $\pi$-regular if there exist $x, y \in R$ and positive integers $p, q$ such that

$$a^p = a^{p+1}x, \ a^q = ya^{q+1}.$$

The next result shows that an element is Drazin invertible if and only if it is strongly $\pi$-regular.

**Theorem 4.2.3 ([39])** *Let $a \in S$. Then $a$ is Drazin invertible if and only if there exist $x, y \in S$ and positive integers $p, q$ satisfying $a^p = a^{p+1}x$ and $a^q = ya^{q+1}$. In this case,*

$$a^D = a^p x^{p+1} = y^{q+1}a^q.$$

***Proof*** It is obvious to prove the necessity. Conversely, suppose that there exist $x, y \in S$ and positive integers $p, q$ satisfying $a^p = a^{p+1}x$ and $a^q = ya^{q+1}$. Let $k = max\{p, q\}$. Then, we have $a^k x = ya^{k+1}x = ya^k$. whence, by induction, $a^k x^m = y^m a^k$ ($m = 1, 2, \cdots$). Thus our choice $z = a^k x^{k+1}$ can equivalently be written $z = y^{k+1}a^k$, and we have:
$$az = aa^k x^{k+1} = a^{k+1}xx^k = a^k x^k = y^k a^k = \cdots = za \text{ by symmetry.}$$
By another induction, $a^k = a^{m+k}x^m$ ($m = 1, 2, \cdots$) and so $a^{k+1}z = a^{k+1}a^k x^{k+1} = a^{k+(k+1)}x^{k+1} = a^k$.
$$z^2a = z(az) = za^{k+1}x^{k+1} = a^{k+1}zx^{k+1} = a^k x^{k+1} = z.$$
Hence $z$ is the Drazin inverse of $a$. In this case, $a^D = a^k x^{k+1} = y^{k+1}a^k$. □

The next two theorems show that the relationships between Drazin invertibility of $a$ and Drazin invertibility of $a^n$. Moreover, the explicit expressions are given.

**Theorem 4.2.4 ([39])** *Let $a \in S$ and $n \geq 1$ be a positive integer. Then $a^D$ exists if and only if $(a^n)^D$ exists. In this case,*

$$(a^n)^D = (a^D)^n \ a^D = a^{n-1}(a^n)^D.$$

***Proof*** Suppose that $a^D = x$ with ind($a$)=$k$. Then, for arbitrary positive integer $j$, we have $a^k = a^{k+1}x = a(a^{k+1}x)x = a^{k+2}x^2 = \cdots = a^{k+j}x^j$ and $x = x^2a = x(x^2a)a = x^3a^2 = \cdots = x^{j+1}a^j$. Let $q$ be the unique integer satisfying

$0 \leq nq - k < n$. Then

$$(a^n)^q = a^{nq} = a^{nq-k}a^k = a^{nq-k}a^{k+n}x^n = (a^n)^{q+1}x^n,$$

$$x^n = x^{n-1}x = x^{n-1}x^{n+1}a^n = x^{2n}a^n,$$

$$a^n x^n = x^n a^n.$$

Thus $(a^n)^D = x^n = (a^D)^n$ with ind$(a^n) \leq q$. Finally ind$(a^n) < q$ would mean $(a^n)^{q-1} = (a^n)^q(a^D)^n$. Since $a^D = a^{n-1}(a^D)^n$, this in turn implies that $a^{n(q-1)} = a^{nq-(n-1)}a^D = a^{n(q-1)+1}a^D$, whence, by the definition of ind$(a)$, we should have $n(q-1) > $ ind$(a)$, contrary to our definition of $q$. Hence ind$(a^n) = q$.

Conversely, suppose $(a^n)^D = y$ with ind$(a^n)=m$. Then we have $y(a^n)^{m+1} = (a^n)^m$, $a^n y^2 = y$, $a^n y = ya^n$. By induction, we get $y = y^{i+1}a^{in}$ and $a^{mn+1} = a(a^n)^m = a(a^n)^{m+1}y = a^{n+1}(a^n)^{m+1}y^2 = \cdots = a^{in+1}(a^n)^{m+1}y^{i+1}$ for arbitrary positive integer $i$.

Set $x = a^{n-1}y$. In what follows, we prove $a^D = x$.

$$xa = a^{n-1}ya = a^{n-1}y^{m+1}a^{mn+1}$$

$$= a^{n-1}y^{m+1}a^{mn+1}(a^n)^{m+1}y^{m+1} = a^{n-1}a^{mn+1}y^{m+1}$$

$$= aa^{n-1}y = ax,$$

$$ax^2 = xax = a^{n-1}yaa^{n-1}y = a^{n-1}a^n y^2 = a^{n-1}y = x,$$

$$a^{mn+1}x = a^{mn+1}a^{n-1}y = (a^n)^{m+1}y = a^{mn}.$$

Hence $a^D$ exists with ind$(a) \leq mn$.                                                  □

**Theorem 4.2.5 ([70, Proposition 4.5])** *Let $a \in S$. Then $a^D$ exists if and only if there exists a positive integer $n \geq 1$ such that $(a^n)^\#$ exists. In this case,*

$$a^D = a^{n-1}(a^n)^\#, \quad (a^n)^\# = (a^D)^n.$$

***Proof*** Suppose that $x = a^D$ with ind$(a)=n$. Set $y = x^n$. Then we have

$$y(a^n)^2 = x^n(a^n)^2 = (x^n a^n)a^n = (x^{n+1}a^{n+1})a^n = (x^{n+1}a^n)a^{n+1} = xa^{n+1} = a^n,$$

$$a^n y^2 = a^n(x^n)^2 = (a^n x^n)x^n = axx^n = ax^{n+1} = x^n = y,$$

$$a^n y = a^n x^n = x^n a^n = ya^n.$$

Therefore $(a^n)^\# = y = (a^D)^n$.

On the contrary. Let $x = a^{n-1}(a^n)^\#$. We prove $x$ is the Drazin inverse of $a$. In fact,

$$xa = a^{n-1}(a^n)^\# a = a^{n-1}\left((a^n)^\#\right)^2 a^{n+1}$$

$$= a^{n-1}\left((a^n)^\#\right)^2 a^{3n+1}\left((a^n)^\#\right)^2 = a^{n-1}a^{n+1}\left((a^n)^\#\right)^2$$

$$= a^n(a^n)^\# = ax,$$

$$ax^2 = xax = a^{n-1}(a^n)^\# a^n(a^n)^\# = a^{n-1}(a^n)^\# = x,$$

$$a^{n+1}x = a^{n+1}a^{n-1}(a^n)^\# = (a^n)^2(a^n)^\# = a^n.$$

Hence $a^D$ exists and $a^D = a^{n-1}(a^n)^\#$. □

The core-nilpotent decomposition in a ring was introduced by Patrício and Puystjens [100].

**Definition 4.2.6 ([100])** Let $a \in R$. Then the sum $a = a_1 + a_2$ is called the core-nilpotent decomposition of $a$, if the following conditions hold:

(1) $a_1$ is group invertible;
(2) $a_2^k = 0$ for some $k \in \mathbb{N}^+$ ;
(3) $a_1 a_2 = a_2 a_1 = 0$.

where $a_1$ is called the core part of $a$ and $a_2$ is called the nilpotent part of $a$.

**Theorem 4.2.7 ([100])** *Let $a \in R$. Then $a^D$ exists if and only if $a$ has a core-nilpotent decomposition.*

**Proof** Suppose that $a^D$ exists with ind$(a)=k$. Set $a_1 = aa^D a$ and $a_2 = a - aa^D a = (1 - aa^D)a$. Then we have $a_1 a_2 = (aa^D a)(1 - aa^D)a = (aa^D a - aa^D a^2 a^D)a = 0$ and $a_2 a_1 = (1 - aa^D)a(aa^D a) = a^2 a^D a - aa^D a^2 a^D a = 0$. Hence $a_1 a_2 = a_2 a_1 = 0$. By induction, we obtain $a_2^k = \left((1 - aa^D)a\right)^k = (1 - aa^D)a^k = 0$. Next, we prove that $a_1^\#$ exists. Since $a^D a_1 = a^D(aa^D a) = (aa^D a)a^D = a_1 a^D$, $a_1(a^D)^2 = aa^D a(a^D)^2 = a^D$ and $a^D(a_1)^2 = a^D(aa^D a)^2 = a^4(a^D)^3 = aa^D a = a_1$, we have $a_1^\#$ exists with $a_1^\# = a^D$.

Conversely, let $a = a_1 + a_2$ be core-nilpotent decomposition of $a$. Then there exists $k$ such that $a_1^\#$ exists, $a_2^k = 0$ and $a_1 a_2 = a_2 a_1 = 0$. By induction, $a^k = (a_1 + a_2)^k = \sum_{i=0}^{k} a_1^i a_2^{k-i}$. Then we have

$$a_1^\# a^{k+1} = a_1^\#(a_1 + a_2)a^k$$

$$= a_1^\# a_1 \sum_{i=0}^{k} a_1^i a_2^{k-i} + a_1^\# a_2 \sum_{i=0}^{k} a_1^i a_2^{k-i}$$

$$= a_1^\# a_1 \sum_{i=0}^{k} a_1^i a_2^{k-i} = a^k,$$

$$a(a_1^\#)^2 = (a_1 + a_2)(a_1^\#)^2 = a_1^\#$$

and $a_1^\# a = a_1^\#(a_1 + a_2) = a_1^\# a_1 = a_1 a_1^\# = (a_1 + a_2)a_1^\# = aa_1^\#$. Hence $a^D$ exists with $a^D = a_1^\#$.                                                                                                     □

**Remark 4.2.8** The core-nilpotent decomposition is unique. In fact, let $a \in R$. Suppose that $a = a_1 + a_2$ and $a = b_1 + b_2$ are core-nilpotent decompositions of $a$. By Theorem 4.2.7, we know that $a_1^\# = a^D = b_1^\#$. So $a_1 = (a_1^\#)^\# = (b_1^\#)^\# = b_1$. Therefore $a_2 = a - a_1 = a - b_1 = b_2$, the uniqueness of core-nilpotent decomposition is proved.

## 4.3   Drazin Invertibility in Two Semigroups of a Ring

In this section, $e$ is an idempotent of a ring $R$ and we denote $eRe + 1 - e = \{exe + 1 - e : x \in R\}$. It should be stressed that this set is a semigroup. The subrings of the form $eRe$ are called corner rings.

**Lemma 4.3.1 ([100, Theorem 1])** *Let $e \in R$ be an idempotent. Then for all $x \in R$, the following statements hold:*

(1) *$exe + 1 - e$ is invertible in $R$ if and only if $exe$ is invertible in the ring $eRe$. In this case,*

$$(exe)^{-1} = e(exe + 1 - e)^{-1}e \in eRe$$

*and*

$$(exe + 1 - e)^{-1} = (exe)^{-1} + 1 - e \in eRe + 1 - e.$$

(2) *$exe + 1 - e$ is regular in $R$ if and only if $exe$ is regular in the ring $eRe$. In this case,*

$$e(exe + 1 - e)^- e \in exe\{1\}$$

*and*

$$(exe)^- + 1 - e \in (exe + 1 - e)\{1\} \cap eRe + 1 - e.$$

(3) $exe + 1 - e$ is group invertible in $R$ if and only if $exe$ is group invertible in the ring $eRe$. In this case,

$$(exe)^{\#} = e(exe + 1 - e)^{\#}e \in eRe$$

and

$$(exe + 1 - e)^{\#} = (exe)^{\#} + 1 - e \in eRe + 1 - e.$$

(4) $exe + 1 - e$ is Drazin invertible with index $k$ in $R$ if and only if $exe$ is Drazin invertible with index $k$ in the ring $eRe$. In this case,

$$(exe)^{D} = e(exe + 1 - e)^{D}e \in eRe$$

and

$$(exe + 1 - e)^{D} = (exe)^{D} + 1 - e \in eRe + 1 - e.$$

If $A \in R^{m \times n}$ is regular with $A^{-}$, $A^{=} \in A\{1\}$, then $AA^{-}$ and $A^{=}A$ are two idempotents. The next result relates Drazin inverses and the ordinary inverse between the two semigroups

$$AA^{-}R^{m \times m}AA^{-} + I_{m} - AA^{-}$$

and

$$A^{=}AR^{n \times n}A^{=}A + I_{n} - A^{=}A$$

using Lemma 4.3.1.

**Proposition 4.3.2 ([100, Proposition 3])** *Let $A \in R^{m \times n}$ be a regular matrix with inner inverses $A^{-}$ and $A^{=}$, and $B \in R^{m \times m}$. Then the following conditions are equivalent:*

(1) $\Gamma = AA^{-}BAA^{-} + I_{m} - AA^{-}$ *is an invertible matrix.*
(2) $\Omega = A^{=}AA^{-}BA + I_{n} - A^{=}A$ *is an invertible matrix.*

*Moreover,*

$$\Omega^{-1} = A^{=}AA^{-}\Gamma^{-1}A + I_{n} - A^{=}A$$

*and also*

$$\Gamma^{-1} = A\Omega^{-1}A^{=}AA^{-} + I_{m} - AA^{-}.$$

**Proof** If $AA^-BAA^- + I_m - AA^-$ is invertible in $R^{m\times m}$, then it follows from
Lemma 4.3.1 (1) that $AA^-BAA^-$ is invertible in the ring $AA^-R^{m\times m}AA^-$.
Therefore, there exists an $X \in AA^-R^{m\times m}AA^-$ such that $AA^-BAA^-X = XAA^-BAA^- = AA^-$. Multiplying on the left by $A^=$ and on the right by $A$, and as
$AA^-X = XAA^- = X$, then

$$\big((A^=A)\,A^-BA\,(A^=A)\big)\big((A^=A)\,A^-XA\,(A^=A)\big) = A^=A$$

and

$$\big((A^=A)\,A^-XA\,(A^=A)\big)\big((A^=A)\,A^-BA\,(A^=A)\big) = A^=A.$$

Hence, $(A^=A)\,A^-BA\,(A^=A)$ is invertible in the ring $A^=AR^{n\times n}A^=A$ and thus
$A^=AA^-BA + I_n - A^=A$ is an invertible matrix.

The converse is analogous. To prove $A^=AA^-\Gamma^{-1}A + I_n - A^=A$ is the inverse
of $\Omega$, we remark that

$$\Omega\left(A^=AA^-\Gamma^{-1}A + I_n - A^=A\right) = A^=AA^-BAA^-\Gamma^{-1}A + I_n - A^=A$$

$$= A^=AA^-\Gamma\Gamma^{-1}A + I_n - A^=A$$

$$= I_n$$

$$= \left(A^=AA^-\Gamma^{-1}A + I_n - A^=A\right)\Omega.$$

The expression of the inverse of $\Gamma$ can be verified analogously.                    □

**Proposition 4.3.3 ([100, Proposition 4])** *Let $A \in R^{m\times n}$ be a regular matrix with
inner inverses $A^-$ and $A^=$, and $B \in R^{m\times m}$. Then the following conditions are
equivalent:*

(1) $\Gamma = AA^-BAA^- + I_m - AA^-$ *is a regular matrix.*
(1) $\Omega = A^=AA^-BA + I_n - A^=A$ *is a regular matrix.*

*Moreover,*

$$A^=AA^-\Gamma^-A + I_n - A^=A \in \Omega\{1\}$$

*and also*

$$A\Omega^-A^=AA^- + I_m - AA^- \in \Gamma\{1\}.$$

**Proof** If $\Gamma$ is regular, then

$$\Gamma^- \in \Gamma\{1\} \Rightarrow AA^-\Gamma^-AA^- \in AA^-BAA^-\{1\}$$

$$\Rightarrow AA^-BAA^-\Gamma^-AA^-BAA^- = AA^-BAA^-$$

$$\Rightarrow A^=AA^-BA\left(A^-\Gamma^-A\right)A^=AA^-BA = A^=AA^-BA$$

$$\Rightarrow A^=AA^-\Gamma^-A \in A^=AA^-BA\{1\} = A^=A\Omega A^=A\{1\}$$

$$\Rightarrow A^=AA^-\Gamma^-A + I_n - A^=A \in \Omega\{1\}.$$

Conversely, if $\Omega$ is regular, then

$$\Omega^- \in \Omega\{1\} \Rightarrow A^=A\Omega^-A^=A \in A^=AA^-BA\{1\}$$

$$\Rightarrow A^=AA^-BA\Omega^-A^=AA^-BA = A^=AA^-BA$$

$$\Rightarrow AA^-BAA^-\left(A\Omega^-A^=\right)AA^-BAA^- = AA^-BAA^-$$

$$\Rightarrow A\Omega^-A^=AA^- \in AA^-\Gamma AA^-\{1\}$$

$$\Rightarrow A\Omega^-A^=AA^- + I_m - AA^- \in \Gamma\{1\}.$$

□

**Theorem 4.3.4** ([100, Proposition 5]) *Let* $A \in R^{m \times n}$ *be a regular matrix with inner inverses* $A^-$ *and* $A^=$, *and* $B \in R^{m \times m}$. *Then the following conditions are equivalent:*

1. $\Gamma = AA^-BAA^- + I_m - AA^-$ *is Drazin invertible with index k (group invertible if* $k = 1$).
2. $\Omega = A^=AA^-BA + I_n - A^=A$ *is Drazin invertible with index k (group invertible if* $k = 1$).

*Moreover,*

$$\Omega^D = A^=AA^-\Gamma^D A + I_n - A^=A$$

*and also*

$$\Gamma^D = A\Omega^D A^=AA^- + I_m - AA^-.$$

**Proof** Let us first consider the case $k = 1$, i.e., the group invertibility case. If $\Gamma^\#$ exists, then by Lemma 4.3.1 and Proposition 4.3.3

$$A^=AA^-\Gamma^\#A \in A^=A\Omega A^=A\{1\} = A^=AA^-BA\{1\},$$

and furthermore

$$A^= A\Omega A^= A \left( A^= AA^- \Gamma^\# A \right)$$

$$= A^= AA^- BA \left( A^= AA^- \Gamma^\# A \right)$$

$$= A^= AA^- \Gamma \Gamma^\# A$$

$$= A^= AA^- \Gamma^\# \Gamma A$$

$$= \left( A^= AA^- \Gamma^\# A \right) A^= AA^- BA$$

$$= \left( A^= AA^- \Gamma^\# A \right) A^= A\Omega A^= A.$$

Thus

$$(A^= A\Omega A^= A)^\# = A^= AA^- \Gamma^\# A\Omega A^= AA^- \Gamma^\# A$$

$$= A^= AA^- \Gamma^\# \Gamma AA^- \Gamma^\# A$$

$$= A^= AA^- \Gamma^\# A$$

since $AA^- \Gamma^\# = \Gamma^\# AA^-$. In fact, using Lemma 4.3.1 (3), it follows that

$$\Gamma^\# \in AA^- R^{m\times m} AA^- + I_m - AA^-,$$

and hence $AA^- \Gamma^\# = \Gamma^\# AA^-$. Therefore,

$$\Omega^\# = A^= AA^- \Gamma^\# A + I_n - A^= A.$$

Conversely, if $\Omega^\#$ exists then

$$A\Omega^\# A^= AA^- \in AA^- \Gamma AA^- \{1\}$$

and also

$$\left( AA^- \Gamma AA^- \right) \left( A\Omega^\# A^= AA^- \right)$$

$$= \left( AA^- BAA^- \right) \left( A\Omega^\# A^= AA^- \right)$$

$$= AA^- BA\Omega^\# A^= AA^-$$

$$= A\Omega\Omega^\# A^= AA^-$$

$$= A\Omega^\# \Omega A^= AA^-$$

$$= A\Omega^\# A^= AA^- BAA^-$$

$$= \left( A\Omega^\# A^= AA^- \right) \left( AA^- \Gamma AA^- \right).$$

So,

$$\left(AA^{-}\Gamma AA^{-}\right)^{\#} = A\Omega^{\#}\Omega A^{=}A\Omega^{\#}A^{=}AA^{-}$$
$$= A\Omega^{\#}A^{=}AA^{-}$$

since $A^{=}A\Omega^{\#} = \Omega^{\#}A^{=}A$, using $\Omega^{\#} \in A^{=}AR^{n\times n}A^{=}A + I_n - A^{=}A$ by Lemma 4.3.1 (3). Therefore,

$$\Gamma^{\#} = A\Omega^{\#}A^{=}AA^{-} + I_m - AA^{-}.$$

For the general case, suppose $\Gamma$ has index $k$, i.e., $\Gamma^{D}$ exists. Then

$$\left(\Gamma^{k}\right)^{\#} = \left(AA^{-}\left(BAA^{-}\right)^{k} + I_m - AA^{-}\right)^{\#}$$

exists. Using the first part of the proof and keeping in mind that $B$ is arbitrary,

$$\Omega^{k} = A^{=}AA^{-}\left(BAA^{-}\right)^{k}A + I_n - A^{=}A$$

is group invertible. Thus, $\Omega^{D}$ exists. Moreover,

$$\Omega^{D} = \Omega^{k-1}\left(\Omega^{k}\right)^{\#}$$
$$= \Omega^{k-1}\left(A^{=}AA^{-}\left(BAA^{-}\right)^{k}A + I_n - A^{=}A\right)^{\#}$$
$$= \Omega^{k-1}\left(A^{=}AA^{-}\left(\Gamma^{k}\right)^{\#}A + I_n - A^{=}A\right)$$
$$= A^{=}AA^{-}\left(BAA^{-}\right)^{k-1}AA^{-}\left(\Gamma^{k}\right)^{\#}A + I_n - A^{=}A$$
$$= A^{=}AA^{-}\Gamma^{k-1}\left(\Gamma^{k}\right)^{\#}A + I_n - A^{=}A$$
$$= A^{=}AA^{-}\Gamma^{D}A + I_n - A^{=}A.$$

The converse is analogous. For the expression of $\Gamma^{D}$,

$$\Gamma^{D} = \Gamma^{k-1}\left(\Gamma^{k}\right)^{\#}$$
$$= \Gamma^{k-1}\left(AA^{-}\left(BAA^{-}\right)^{k}AA^{-} + I_m - AA^{-}\right)^{\#}$$
$$= \Gamma^{k-1}\left(A\left(\Omega^{k}\right)^{\#}A^{=}AA^{-} + I_m - AA^{-}\right)$$

$$= AA^- \left(BAA^-\right)^{k-1} A \left(\Omega^k\right)^{\#} A^= AA^- + I_m - A\Lambda^-$$

$$= A\Omega^{k-1} \left(\Omega^k\right)^{\#} A^= AA^- + I_m - AA^-$$

$$= A\Omega^D A^= AA^- + I_m - AA^-.$$

$$\square$$

## 4.4  Jacobson's Lemma and Cline's Formula for Drazin Inverses

This section concerns Jacobson's lemma and Cline's formula for Drazin inverses.

**Lemma 4.4.1 ([13, Lemma 2.2])** *Let a be a regular element. Then, for any positive integer n,*

$$(a + 1 - aa^-)^n = (a^2 a^- + 1 - aa^-)^n + \sum_{i=1}^{n} a^i (1 - aa^-). \tag{4.4.1}$$

*Proof* The proof is by induction on $n$. Denote

$$z = a + 1 - aa^- \text{ and } x = a^2 a^- + 1 - aa^-.$$

It is clear that $z = x + a(1 - aa^-)$. Assume (4.4.1) to hold for $k$, we will prove it for $k + 1$. We note that

$$zx = x^2 + a(1 - aa^-) \text{ and } za = a^2.$$

Now, by the induction step,

$$z^{k+1} = z \left(x^k + \sum_{i=1}^{k} a^i (1 - aa^-)\right)$$

$$= x^{k+1} + a(1 - aa^-) + \sum_{i=1}^{k} a^{i+1} (1 - aa^-)$$

$$= x^{k+1} + \sum_{i=1}^{k+1} a^i (1 - aa^-).$$

$$\square$$

**Lemma 4.4.2 ([13])** *Let a, b ∈ R. Then, for any positive integer n,*

$$(1 - ba)^n = 1 - bra \quad and \quad (1 - ab)^n = 1 - rab,$$

*where* $r = \sum_{j=0}^{n-1}(1 - ab)^j.$

***Proof*** This can be proved by induction on $n$.                                                  □

The following result is an answer to a question raised by Patrício and Veloso Da Costa in [101] about the equivalence between the conditions ind($a^2a^- + 1 - aa^-$)=k and ind($a + 1 - aa^-$)=k, and provides a new characterization of the Drazin index.

**Lemma 4.4.3 ([13, 101])** *Let a be Drazin invertible and regular. Then, for any inner inverse $a^-$ of a integer $k \geq 0$, the following conditions are equivalent:*

*(1) ind(a) = k + 1.*
*(2) ind($a^2a^- + 1 - aa^-$) = k.*
*(3) ind($a + 1 - aa^-$) = k.*

***Proof***

(1)⟺(2). It follows by [101, Theorem 2.1].
(2)⟺(3). Denote

$$x = a^2a^- + 1 - aa^- \text{ and } z = a + 1 - aa^-.$$

Assume that ind($x$)=k, or equivalently, ind($a$)=k+1. Then $x^k = x^{k+1}R$ and $a^{k+1} = a^{k+2}w$ for some $w \in R$. By (4.4.1),

$$z^k R = \left(1 + \sum_{i=1}^{k} a^i(1 - aa^-)\right)x^k R$$

$$= \left(1 + \sum_{i=1}^{k} a^i(1 - aa^-)\right)x^{k+1} R$$

$$= \left(z^{k+1} - \sum_{i=1}^{k+1} a^i(1 - aa^-) + \sum_{i=1}^{k} a^i(1 - aa^-)\right) R$$

$$= \left(z^{k+1} - a^{k+1}(1 - aa^-)\right) R$$

$$= \left(z^{k+1} - a^{k+2}w(1 - aa^-)\right) R$$

$$= z^{k+1}\left(1 - aw(1 - aa^-)\right) R \subseteq z^{k+1} R.$$

This gives $z^k R = a^{k+1} R$. On the other hand, since $\text{ind}(x)=k$ we also have $x^k = ux^{k+1}$ for some $u \in R$. By (4.4.1),

$$Rz^k = R\left(x^k + \sum_{i=1}^{k} a^i(1 - aa^-)\right)$$

$$= R\left(ux^{k+1} + \sum_{i=1}^{k} a^i(1 - aa^-)\right)$$

$$= R\left(u - u\sum_{i=1}^{k+1} a^i(1 - aa^-) + \sum_{i=1}^{k} a^i(1 - aa^-)\right)z^{k+1} \subseteq Rz^{k+1}.$$

From this we conclude that $Rz^k = Ra^{k+1}$. Consequently, $\text{ind}(z) \le k$.

By symmetrical arguments, we can show that $\text{ind}(z)=k$ implies that $\text{ind}(x) \le k$. Further, if $\text{ind}(z) < k$, having $\text{ind}(x) = k$, then we would get that $\text{ind}(x) \le k - 1$, and we would arrive at a contradiction. Therefore $\text{ind}(z) = k$.                              □

**Lemma 4.4.4 ([13])**  Let $a$, $b \in R$. If $(1-ab)^+$ is a reflexive inverse of $1-ab \in R$, then a reflexive inverse of $1 - ba$ is given by

$$(1 - ba)^+ = 1 + b\left((1 - ab)^+ - pq\right)a,$$

where $p = 1 - (1 - ab)^+(1 - ab)$ and $q = 1 - (1 - ab)(1 - ab)^+$.

**Proof**  Let $x = 1 + b\left((1 - ab)^+ - pq\right)a$. Then $(1 - ba)x = 1 - bqa$. Further, $(1 - ba)x(1 - ba) = 1 - ba - bqa(1 - ba)a = 1 - ba$ and

$$x(1 - ba)x = x - xbqa$$

$$= x - bqa - b\left((1 - ab)^+ - pq\right)abqa$$

$$= x,$$

where we have simplified by writing $ab = 1 - (1 - ab)$ and using the relations $(1 - ab)(1 - ab)^+(1 - ab) = 1 - ab$ and $(1 - ab)^+(1 - ab)(1 - ab)^+ = (1 - ab)^+$.                                                                          □

**Theorem 4.4.5 ([13])**  Let $a$, $b \in R$. Then $1 - ab \in R^\#$ if and only if $1 - ba \in R^\#$. In this case,

$$(1 - ba)^\# = 1 + b\left((1 - ab)^\# - (1 - ab)^\pi\right)a,$$

where $(1 - ab)^\pi = 1 - (1 - ab)(1 - ab)^\#$.

**Proof**  Let $x = 1 + b\left((1 - ab)^\# - (1 - ab)^\pi\right)a$. First, we note that $(1 - ab)^\#$ is a reflexive inverse that commutes with $1 - ab$. In view of the preceding Lemma 4.4.4

we have that $x$ is a reflexive inverse of $1 - ba$. Next, we will prove that $x$ commutes with $1 - ba$. We have

$$x(1 - ba) = 1 - ba + b(1 - ab)^\#(1 - ab)a = 1 - b(1 - ab)^\pi a$$

and, similarly, $(1 - ba)x = 1 - b(1 - ab)^\pi a$, which gives $(1 - ba)x = x(1 - ba)$. Therefore $x$ satisfies the three equations involved in the definition of group inverse.

$\square$

**Theorem 4.4.6 ([13])** *Let $a, b \in R$. Then $1 - ab \in R^D$ if and only if $1 - ba \in R^D$. In this case,* $\mathrm{ind}(1 - ab) = \mathrm{ind}(1 - ba)$ *and*

$$(1 - ba)^D = 1 + b[(1 - ab)^D - (1 - ab)^\pi r]a,$$

*where* $r = \sum_{j=0}^{k-1}(1 - ab)^j$, $k = \mathrm{ind}(1 - ab) = \mathrm{ind}(1 - ba)$ *and* $(1 - ab)^\pi = 1 - (1 - ab)(1 - ab)^D$.

**Proof** Assume that $\mathrm{ind}(1 - ab) = k \geq 2$. Then $(1 - ab)^k$ is group invertible and Lemma 4.4.3 leads to $\mathrm{ind}\left(\left(1 - \left(1 - (1 - ab)^k\right)\right)(1 - ab)^k\left((1 - ab)^k\right)^\#\right) = 0$. By Lemma 4.4.2, we have

$$1 - (1 - ab)^k = rab \quad \text{and} \quad 1 - (1 - ba)^k = bra, \tag{4.4.2}$$

where $r = \sum_{j=0}^{k-1}(1 - ab)^j$. From the above relations, $1 - rab(1 - ab)^k((1 - ab)^k)^\#$ is invertible and by Jacobson's lemma we have that $1 - b(1 - ab)(1 - ab)^D ra$ is invertible. Further,

$$\left(1 - b(1 - ab)(1 - ab)^D ra\right)(1 - ba)^k$$

$$= (1 - ba)^k - b(1 - ab)(1 - ab)^D ra(1 - ba)^k$$

$$= (1 - ba)^k - b(1 - ab)^k ra$$

$$= (1 - bra)(1 - ba)^k = (1 - ba)^{2k}.$$

From this it follows that

$$(1 - ba)^k = \left(1 - b(1 - ab)(1 - ab)^D ra\right)^{-1}(1 - ba)^{2k} \in R(1 - ba)^{k+1}.$$

On the other hand,

$$(1-ba)^k \left(1 - b(1-ab)(1-ab)^D ra\right)$$

$$= (1-ba)^k - (1-ba)^k b(1-ab)(1-ab)^D ra$$

$$= (1-ba)^k - b(1-ba)^k ra = (1-ba)^{2k}$$

and hence $(1-ba)^k = (1-ba)^{2k} \left(1 - b(1-ab)(1-ab)^D ra\right)^{-1} \in (1-ba)^{k+1} R.$ Therefore $(1 - ba)^k \in R(1 - ba)^{k+1} \bigcap (1 - ba)^{k+1} R$, which implies that $\text{ind}((1-ba)) \le k$.

Further, analysis similar to that of the last part of the proof of Lemma 4.4.3 shows that $\text{ind}(1-ab)=k$. Now, $(1-ba)^D = \left((1-ba)^k\right)^{\#} (1-ba)^{k-1}$. In view of Lemma 4.4.2 and applying Theorem 4.4.5, it follows that

$$\left((1-ba)^k\right)^{\#} = (1-bra)^{\#}$$

$$= 1 + b\left((1-rab)^{\#} - (1-rab)^{\pi}\right)ra$$

$$= 1 + b\left(\left((1-ab)^k\right)^{\#} - \left((1-ab)^k\right)^{\pi}\right)ra$$

$$= 1 + b\left(\left((1-ab)^D\right)^k - (1-ab)^{\pi}\right)ra.$$

Hence,

$$(1-ba)^D = \left(1 + b\left(\left((1-ab)^D\right)^k - \left((1-ab)^k\right)^{\pi}\right)ra\right)(1-ba)^{k-1}$$

$$= (1-ba)^{k-1} + b\left(\left((1-ab)^D\right)^k - (1-ab)^{\pi}\right)(1-ab)^{k-1}ra$$

$$= 1 - br'a + b\left((1-ab)^D r - (1-ab)^{\pi}(1-ab)^{k-1}\right)a$$

$$= 1 + b\left((1-ab)^D - (1-ab)^{\pi}r' - (1-ab)^{\pi}(1-ab)^{k-1}\right)a$$

$$= 1 + b\left((1-ab)^D - (1-ab)^{\pi}r\right)a,$$

where $r' = \sum_{j=0}^{k-2}(1-ab)^j$, completing the proof.                                  □

In 1965, Cline [25] proved that if $ab$ is Drazin invertible, then $ba$ is Drazin invertible. In this case, $(ba)^D = b[(ab)^D]^2 a$. This equality is called Cline's formula.

**Theorem 4.4.7 ([25])**  *Let $a, b \in R$. Then $ab \in R^D$ if and only if $ba \in R^D$. In this case,*

$$(ba)^D = b\left((ab)^D\right)^2 a \text{ and } \text{ind}(ba) \leq \text{ind}(ab) + 1.$$

**Proof**  The sufficiency of the theorem is easy to obtain by the symmetry of $a$ and $b$. We only need to prove the necessity. Suppose that $ab \in R^D$ with $\text{ind}(ab) = k$. Then we have $ab(ab)^D = (ab)^D ab$, $ab\left((ab)^D\right)^2 = (ab)^D$, $(ab)^{k+1}(ab)^D = (ab)^k$. Then we prove that $ba \in R^D$. Let $x = b\left((ab)^D\right)^2 a$. Then

$$(ba)x = (ba)b\left((ab)^D\right)^2 a = b(ab)\left((ab)^D\right)^2 a = b(ab)^D a,$$

$$x(ba) = b\left((ab)^D\right)^2 a(ba) = b\left((ab)^D\right)^2 (ab)a = b(ab)^D a.$$

Hence $(ba)x = x(ba)$.

$$(ba)x^2 = (ba)xx = b(ab)^D ab\left((ab)^D\right)^2 a = b\left((ab)^D\right)^2 a = x,$$

$$(ba)^{k+2}x = (ba)^{k+2}b\left((ab)^D\right)^2 a = b(ab)^{k+2}\left((ab)^D\right)^2 a$$

$$= b(ab)^{k+1}(ab)^D a = b(ab)^k a = (ba)^{k+1}.$$

Hence $ba \in R^D$ with $\text{ind}(ba) \leq k + 1$, and $(ba)^D = b\left((ab)^D\right)^2 a$.  □

## 4.5  Additive Properties of Drazin Inverses of Elements

In this section, we first start with some lemmas which will be useful in proving our main results. The notation $a^\pi$ means $1 - aa^D$ for any Drazin invertible element $a \in R$.

**Lemma 4.5.1**  *Let $a, b \in R$.*

*(1) If $a$ is nilpotent and $ab = ba$, then $ab$ is nilpotent.*
*(2) If $a, b$ are nilpotent and $ab = ba$, then $a + b$ is nilpotent.*

**Lemma 4.5.2**  ([139]) *Let $a, b \in R$ be Drazin invertible and $ab = ba$. Then*

*(1) $a, b, a^D$ and $b^D$ commute.*
*(2) $ab$ is Drazin invertible with $(ab)^D = b^D a^D$.*

## Proof

(1). Note that $a^D \in comm^2(a)$. From $ab = ba$ it follows that $a^D b = b a^D$. So is $a^D b^D = b^D a^D$. Hence $a$, $b$, $a^D$ and $b^D$ commute each other.

(2). Let $x = b^D a^D$. The commutativity of $ab$ and $x$ is obvious, and it is easily verified that $xabx = x$. Finally, $ab - (ab)^2 x$ is nilpotent by using Lemma 4.5.1 and the splitting $ab - (ab)^2 x = abb^\pi + b^2 b^D aa^\pi$, where $aa^\pi$ and $bb^\pi$ are both nilpotent.

□

The next theorem characterizes the relationships of the Drazin inverse between of $a + b$ and $1 + a^D b$, and gives the explicit expressions of the Drazin inverse.

**Theorem 4.5.3** ([139]) *Let $a$, $b \in R$ be Drazin invertible and $ab = ba$. Then $1 + a^D b$ is Drazin invertible if and only if $a + b$ is Drazin invertible. In this case, we have*

$$(a + b)^D = (1 + a^D b)^D a^D + b^D (1 + aa^\pi b^D)^{-1} a^\pi$$

$$= a^D (1 + a^D b)^D bb^D + b^\pi (1 + bb^\pi a^D)^{-1} a^D + b^D (1 + aa^\pi b^D)^{-1} a^\pi,$$

*and*

$$(1 + a^D b)^D = a^\pi + a^2 a^D (a + b)^D.$$

**Proof** Suppose $\xi = 1 + a^D b$ is Drazin invertible. Then by Lemma 4.5.2, $a$, $b$, $a^D$, $b^D$, $\xi$ and $\xi^D$ commute each other. Since $aa^\pi$ is nilpotent, we get $1 + aa^\pi b^D$ is invertible. In a similar way, we conclude that $1 + bb^\pi a^D$ is invertible.

Let $x = \xi^D a^D + b^D (1 + aa^\pi b^D)^{-1} a^\pi$. In what follows, we show that $x$ is the Drazin inverse of $a + b$, i.e., the following conditions hold: $(a)$ $x \in comm(a + b)$; $(b)$ $x(a + b)x = x$ and $(c)$ $(a + b) - (a + b)^2 x$ is nilpotent.

$(a)$  By Lemma 4.5.2, $a + b$ and $x$ commute.

$(b)$  After a calculation we obtain

$$(1 + aa^\pi b^D)^{-1} = 1 - aa^\pi b^D (1 + aa^\pi b^D)^{-1}.$$

Then we can simplify

$$x(a + b) = \xi^D a^D (a + b) + b^D (1 + aa^\pi b^D)^{-1} a^\pi (a + b)$$

$$= \xi^D a^D (a + b) + aa^\pi b^D (1 + aa^\pi b^D)^{-1} + a^\pi bb^D (1 + aa^\pi b^D)^{-1}$$

$$= \xi^D a^D (a + b) + aa^\pi b^D (1 + aa^\pi b^D)^{-1}$$

$$\quad + a^\pi bb^D \left( 1 - aa^\pi b^D (1 + aa^\pi b^D)^{-1} \right)$$

$$= \xi^D a^D (a + b) + a^\pi bb^D.$$

Note that $a^D a^\pi = 0$. Then we have

$$x(a+b)x = \left(\xi^D a^D (a+b) + a^\pi b b^D\right)\left(\xi^D a^D + b^D(1+aa^\pi b^D)^{-1}a^\pi\right)$$

$$= (\xi^D)^2 (a^D)^2(a+b) + a^\pi b^D(1+aa^\pi b^D)^{-1}$$

$$= (\xi^D)^2 \xi a^D + a^\pi b^D(1+aa^\pi b^D)^{-1}$$

$$= \xi^D a^D + a^\pi b^D(1+aa^\pi b^D)^{-1}$$

$$= x.$$

(c) We have

$$(a+b) - (a+b)^2 x = (a+b) - (a+b)\xi^D a^D(a+b) - (a+b)a^\pi b b^D$$

$$= (a+b) - \xi^D(a^D(a+b))^2 a - (a+b)(1-aa^D)bb^D$$

$$= (a+b) - \xi^D(\xi - a^\pi)^2 a - a(1-aa^D)bb^D - b^2 b^D$$

$$\quad + aa^D b^2 b^D$$

$$= (a+b) - \xi^D \xi^2 a + \xi^D aa^\pi - aa^\pi bb^D - b^2 b^D$$

$$\quad + aa^D b^2 b^D$$

$$= bb^\pi - aa^\pi bb^D + a + aa^D b^2 b^D + \xi^D aa^\pi - \xi^D \xi^2 a$$

$$= bb^\pi - aa^\pi bb^D + aa^\pi \xi^D + \xi a - aa^D bb^\pi - \xi^D \xi^2 a$$

$$= a^\pi bb^\pi + aa^\pi(\xi^D - bb^D) + \xi\xi^\pi a.$$

Since $aa^\pi$, $bb^\pi$ and $\xi\xi^D$ are nilpotent, it follows that $(a+b) - (a+b)^2 x$ is nilpotent by Lemma 4.5.1.

In order to show another expression $(a+b)^D$, it is sufficient to prove

$$\xi^D a^D = a^D \xi^D bb^D + b^\pi(1+bb^\pi a^D)^{-1}a^D.$$

Noting the commutativity and

$$a^D b^\pi(1+bb^\pi a^D) = a^D b^\pi + a^D bb^\pi a^D = \xi a^D b^\pi,$$

we derive

$$\left(1 - (1+bb^\pi a^D)^{-1}\xi\xi^\pi\right)\xi^\pi a^D b^\pi = 0.$$

Observing that $1 - (1 + bb^\pi a^D)^{-1}\xi\xi^\pi$ is invertible for $(1 + bb^\pi a^D)^{-1}\xi\xi^\pi$ is nilpotent, we know

$$\xi^\pi a^D b^\pi = 0.$$

Thus

$$a^D b^\pi = \xi\xi^D a^D b^\pi = \xi^D \xi a^D b^\pi = \xi^D a^D b^\pi (1 + bb^\pi a^D).$$

Now we get

$$\xi^D a^D b^\pi = a^D b^\pi (1 + bb^\pi a^D)^{-1} = b^\pi (1 + bb^\pi a^D)^{-1} a^D.$$

Hence

$$\xi^D a^D = a^D \xi^D bb^D + b^\pi (1 + bb^\pi a^D)^{-1} a^D.$$

Conversely, if $a + b$ is Drazin invertible, we can rewrite $1 + a^D b = a_1 + b_1$, where $a_1 = a^\pi$ and $b_1 = a^D(a + b)$. Note that $a_1$ is idempotent and $a^D$ is group invertible with $(a^D)^\# = a^2 a^D$. Then by Lemma 4.5.2(2), we know that $b_1$ is Drazin invertible and

$$b_1^D = \left(a^D(a + b)\right)^D = a^2 a^D (a + b)^D.$$

In addition,

$$1 + a_1^D b_1 = 1 + a_1 b_1 = 1.$$

Therefore we can derive $(a_1 + b_1)^D$, where $a_1$ and $b_1$ satisfy the sufficient conditions. That is, $1 + a^D b$ is Drazin invertible and the formula is presented as follows:

$$(1 + a^D b)^D = (a_1 + b_1)^D$$
$$= (1 + a_1^D b_1)^D a_1^D + b_1^D (1 + a_1 a_1^\pi b_1^D)^{-1} a_1^\pi$$
$$= a^\pi + a^2 a^D (a + b)^D.$$

The proof is complete.

□

**Corollary 4.5.4 ([139])** *Let $a$, $b \in R$ be Drazin invertible with $\mathrm{ind}(a) = k$, $\mathrm{ind}(b) = l$, $ab = ba$. If $1 + a^D b$ is Drazin invertible, then $a + b$ is Drazin* *invertible and*

$$(a+b)^D = (1 + a^D b)^D a^D + \sum_{i=0}^{k-1} (b^D)^{i+1} (-a)^i a^\pi$$

$$= a^D (1 + a^D b)^D b b^D + b^\pi \sum_{i=0}^{l-1} (-b)^i (a^D)^{i+1} + \sum_{i=0}^{k-1} (b^D)^{i+1} (-a)^i a^\pi.$$

**Proof** From ind$(a)=k$, it follows that $(aa^\pi b^D)^k = 0$. Thus

$$(1 + aa^\pi b^D)^{-1} a^\pi = \left( 1 + \sum_{i=0}^{k-1} (b^D)^i (-a)^i a^\pi \right) a^\pi = \sum_{i=0}^{k-1} (b^D)^i (-a)^i a^\pi.$$

Similarly,

$$b^\pi (1 + bb^\pi a^D)^{-1} = b^\pi \sum_{i=0}^{l-1} (a^D)^i (-b)^i.$$

Using Theorem 4.5.3, we obtain the above formulae.    □

**Proposition 4.5.5 ([139])** *Let $a, b \in R$ be group invertible with $ab = ba$. Then $a + b$ is group invertible if and only if $1 + a^\# b$ is group invertible. In this case,*

$$(a+b)^\# = (1 + a^\# b)^\# a^\# + b^\# a^\pi$$

$$= a^\# (1 + a^\# b)^\# b b^\# + a^\# b^\pi + b^\# a^\pi,$$

*and*

$$(1 + a^\# b)^\# = a^\pi + a(a+b)^\#,$$

*where $a^\pi = 1 - aa^\#$, $b^\pi = 1 - bb^\#$.*

**Proof** It is similar to the proof of Theorem 4.5.3.    □

## 4.6   Drazin Inverses of Products and Differences of Idempotents

Let $p$ and $q$ be two idempotents of a ring $R$. In this section, we give some equivalent conditions for the Drazin invertibility of $p - q$, $pq$, $pq - qp$ and $pq + qp$.

**Lemma 4.6.1 ([39])**

(1) *Let $a \in R^D$. If $ab = ba$, then $a^D b = ba^D$.*
(2) *Let $a, b \in R^D$ and $ab = ba = 0$. Then $(a + b)^D = a^D + b^D$.*

**Lemma 4.6.2 ([20])** *Let $a, b \in R^D$ and $p^2 = p \in R$. If $ap = pa$ and $bp = pb$, then $ap + b(1 - p) \in R^D$ and*

$$(ap + b(1 - p))^D = a^D p + b^D(1 - p).$$

**Proof** Since $p^2 = p$, we have $p^D = p$. Thus, $a, b, p \in R^D$. Note that $ap = pa$ and $bp = pb$. We obtain $(ap)^D = a^D p$ and $(b(1 - p))^D = b^D(1 - p)$ by Lemma 4.5.2(2). As $apb(1 - p) = b(1 - p)ap = 0$, according to Lemma 4.6.1(2), it follows that $(ap + b(1 - p))^D = a^D p + b^D(1 - p)$. □

**Lemma 4.6.3 ([20])** *Let $a \in R$, $p^2 = p \in R$, $b = pa(1 - p)$ and $c = (1 - p)ap$. The following statements are equivalent:*

(1) $b + c \in R^D$.
(2) $bc \in R^D$.
(3) $b - c \in R^D$.

**Proof**

(1) $\Rightarrow$ (2). Since $(b + c)^2 = (bc + cb)$ and $b + c \in R^D$, we have $bc + cb \in R^D$. Let $x = (bc + cb)^D$. As $p(bc + cb) = (bc + cb)p$, we obtain that $px = xp$ by Lemma 4.6.1(1). Next, we prove that $(bc)^D = pxp$. Since $x = (bc + cb)x^2$, we get

$$pxp = p(bc + cb)x^2 p = bcx^2 p = bc(px^2 p) = bc(pxp)^2.$$

By $(bc + cb)x = x(bc + cb)$, we obtain $p(bc + cb)xp = px(bc + cb)p$. It follows that $bc(pxp) = (pxp)bc$.
Because $(bc + cb)^{n+1}x = (bc + cb)^n$ for some $n$, we have

$$\left((bc)^{n+1} + (cb)^{n+1}\right)x = (bc)^n + (cb)^n.$$

Multiplying the equation above by $p$ on two sides yields

$$p(bc)^{n+1}xp = p(bc)^n p,$$

i.e., $(bc)^{n+1}pxp = (bc)^n$. So, $bc$ is Drazin invertible and $(bc)^D = pxp$.
(2) $\Rightarrow$ (1). According to Lemma 4.4.7, $bc \in R^D$ is equivalent to $cb \in R^D$. Note that $bc \cdot cb = cb \cdot bc = 0$. We have $(b+c)^2 = (bc+cb) \in R^D$ by Lemma 4.6.1(2). It follows that $b + c \in R^D$.
(2) $\Leftrightarrow$ (3). Its proof is similar to (1) $\Leftrightarrow$ (2).

□

**Lemma 4.6.4 ([20])**  *Let $a \in R$ with $a - a^2 \in R^D$ or $a + a^2 \in R^D$. Then $a \in R^D$.*

***Proof***  We only need to prove the situation when $a - a^2 \in R^D$ with $x = (a - a^2)^D$.
By Lemma 4.6.1(1), it is clear $ax = xa$ since $a(a - a^2) = (a - a^2)a$.
Since $a - a^2 \in R^D$, we get $(a - a^2)^n = (a - a^2)^{n+1}x$ for some integer $n \geq 1$,
that is,

$$a^n(1 - a)^n = a^{n+1}(1 - a)^{n+1}x.$$

Note that

$$a^n(1 - a)^n = a^n \left( 1 + \sum_{i=1}^{n} C_n^i(-a)^i \right).$$

It follows that

$$a^n = a^{n+1} \left( (1 - a)^{n+1}x + \sum_{i=1}^{n} C_n^i(-a)^{i-1} \right)$$

$$= \left( (1 - a)^{n+1}x + \sum_{i=1}^{n} C_n^i(-a)^{i-1} \right) a^{n+1}.$$

This shows $a^n \in a^{n+1}R \cap Ra^{n+1}$. Hence, $a \in R^D$.                    □

**Proposition 4.6.5 ([20])**  *The following statements are equivalent:*

(1) $1 - pq \in R^D$.
(2) $p - pq \in R^D$.
(3) $p - qp \in R^D$.
(4) $1 - pqp \in R^D$.
(5) $p - pqp \in R^D$.
(6) $1 - qp \in R^D$.
(7) $q - qp \in R^D$.
(8) $q - pq \in R^D$.
(9) $1 - qpq \in R^D$.
(10) $q - qpq \in R^D$.

***Proof***

(1) $\Leftrightarrow$ (6).   It is obvious by Lemma 4.4.6. We only need to prove that (1)–(5) are
equivalent.
(1) $\Leftrightarrow$ (4).   It is clear that $1 - pq = 1 - p(pq)$. Thus, $1 - pq$ is Drazin invertible
if and only if $1 - pqp$ is Drazin invertible by Lemma 4.4.6.
(4) $\Rightarrow$ (5).   Since $p \in R^D$ and $p(1 - pqp) = (1 - pqp)p = p - pqp$, we obtain
that $p - pqp$ is Drazin invertible according to Lemma 4.5.2(2).

(5) $\Rightarrow$ (4).    Suppose $a = p - pqp$, $b = 1$. Then $a$ and $b$ are Drazin invertible. Since $1 - pqp = (p - pqp) + 1 - p$, it follows that $1 - pqp = ap + b(1 - p)$ is Drazin invertible in view of Lemma 4.6.2.

(2) $\Leftrightarrow$ (5).    Since $p - pq = pp(1 - q)$ and $p - pqp = p(1 - q)p$, the result follows by Theorem 4.4.7.

(2) $\Leftrightarrow$ (3).    It is evident according to Theorem 4.4.7.

$\square$

By replacing $p$ and $q$ with $1 - p$ and $1 - q$ in Proposition 4.6.5, respectively, We get the following result immediately.

**Corollary 4.6.6 ([20])**  *The following statements are equivalent:*

(11)  $p + q - pq \in R^D$.
(12)  $q - pq \in R^D$.
(13)  $q - qp \in R^D$.
(14)  $p + (1 - p)(q - qp) \in R^D$.
(15)  $(1 - p)q(1 - p) \in R^D$.
(16)  $p + q - qp \in R^D$.
(17)  $p - qp \in R^D$.
(18)  $p - pq \in R^D$.
(19)  $q + (1 - q)(p - pq) \in R^D$.
(20)  $(1 - q)p(1 - q) \in R^D$.

Finally, as $p - pq$ appears in both Proposition 4.6.5(2) and Corollary 4.6.6(18), we get

**Corollary 4.6.7 ([20])**  *Statements (1)–(10) of Proposition 4.6.5 are equivalent with statements (11)–(20) of Corollary 4.6.6.*

In 2012, Koliha, Cvetković-Ilić and Deng [75] proved that $p - q \in \mathscr{A}^D$ if and only if $1 - pq \in \mathscr{A}^D$ if and only if $p + q - pq \in \mathscr{A}^D$ where $p, q$ are idempotents in a Banach algebra $\mathscr{A}$. It is natural to consider whether the same property can be generalized to the Drazin inverse of ring versions. The following result illustrates the possibility.

**Theorem 4.6.8 ([20])**  *The following statements are equivalent:*

(1)  $p - q \in R^D$.
(2)  $1 - pq \in R^D$.
(3)  $p + q - pq \in R^D$.

**Proof**

(1) $\Rightarrow$ (2).    It is easy to check that $p(p - q)^2 = (p - q)^2 p = p - pqp$. Hence,

$$1 - pqp = (p - q)^2 p + 1 - p.$$

Let $a = (p - q)^2$ and $b = 1$. Then $ap = pa$, $bp = pb$. Since $p - q \in R^D$, we obtain that $a = (p - q)^2 \in R^D$. In view of Lemma 4.6.2, $1 - pqp = ap + b(1 - p) \in R^D$. Therefore, $1 - pq$ is Drazin invertible by Lemma 4.4.6.

(2) $\Leftrightarrow$ (3).    It is clear by Proposition 4.6.5 and Corollary 4.6.6.

(3) $\Rightarrow$ (1).    Let $a = 1 - pqp$, $b = 1 - (1 - p)(1 - q)(1 - p)$. Then we get $ap = pa$ and $bp = pb$. Since

$$1 - (1 - p)(1 - p)(1 - q) = p + q - pq \in R^D,$$

we obtain that $b \in R^D$ by Lemma 4.4.6 and $a \in R^D$ by Proposition 4.6.5 and Corollary 4.6.6. Note that $(p - q)^2 = ap + b(1 - p)$. We get $(p - q)^2 \in R^D$ by Lemma 4.6.2. Hence, $p - q \in R^D$.

$\square$

Cvetković-Ilić, Deng [30] considered the Drazin invertibility of product and difference of idempotents in a Banach algebra $\mathscr{A}$. Moreover, they proved that if one of $pq$, $1 - p - q$ and $(1 - p)(1 - q)$ belongs to $\mathscr{A}^D$ then they all do. We extend the result in [30] to the ring cases.

**Theorem 4.6.9 ([20])**  *The following statements are equivalent:*

(1) $pq \in R^D$.

(2) $1 - p - q \in R^D$.

(3) $(1 - p)(1 - q) \in R^D$.

*Proof*

(1) $\Leftrightarrow$ (2).    Let $p_1 = 1 - p$ and $q_1 = q$. Then $p_1 - q_1 \in R^D$ if and only if $q_1 - p_1 q_1 \in R^D$ by Proposition 4.6.5 and Theorem 4.6.8. Since $p_1 - q_1 = 1 - p - q$ and $q_1 - p_1 q_1 = pq$, (1) $\Leftrightarrow$ (2) holds.

(1) $\Leftrightarrow$ (3).    Set $p_1 = 1 - p$, $q_1 = q$. Then $p_1 - p_1 q_1 \in R^D$ if and only if $q_1 - p_1 q_1 \in R^D$ by Proposition 4.6.5. Since $p_1 - p_1 q_1 = (1 - p)(1 - q)$ and $q_1 - p_1 q_1 = pq$, the result follows.

$\square$

**Theorem 4.6.10 ([20])**  *The following statements are equivalent:*

(1) $pq - qp \in R^D$.

(2) $pq \in R^D$ and $p - q \in R^D$.

*Proof* Suppose $b = pq(1 - p)$ and $c = (1 - p)qp$. It follows that $b - c = pq - qp$.

(1) $\Rightarrow$ (2).    By hypothesis $b - c \in R^D$, we obtain $pqp - (pqp)^2 = pq(1 - p)qp = bc \in R^D$ by Lemma 4.6.3. It follows that $pqp \in R^D$ by Lemma 4.6.4. Hence, $pq \in R^D$.

Similarly,

$$(1 - p)q - q(1 - p) = -(pq - qp) \in R^D$$

implies $q - pq = (1 - p)q \in R^D$. Therefore, $p - q \in R^D$ by Proposition 4.6.5 and Theorem 4.6.8.

(2) $\Rightarrow$ (1).    By Theorem 4.4.7, both $pqp$ and $(p-q)^2$ are Drazin invertible. Note that $bc = pq(1 - p)qp = pqp(p - q)^2 = (p - q)^2 pqp$. It follows that $bc$ is Drazin invertible by Lemma 4.5.2(2). Hence, according to Lemma 4.6.3, we have $pq - qp = b - c \in R^D$.

□

We can give an interesting result similar to Theorem 4.6.10.

**Theorem 4.6.11 ([20])**  *The following statements are equivalent:*

(1) $pq + qp \in R^D$.
(2) $pq \in R^D$ and $p + q \in R^D$.

*Proof*

(1) $\Rightarrow$ (2).    Since $pq + qp = -(p+q) + (p+q)^2 = (p+q-1) + (p+q-1)^2 \in R^D$, $p + q \in R^D$ and $p + q - 1 \in R^D$ according to Lemma 4.6.4. Therefore, $pq \in R^D$ by Theorem 4.6.8.

(2) $\Rightarrow$ (1).    Since $pq + qp = (p + q)(p + q - 1) = (p + q - 1)(p + q)$, $pq + qp \in R^D$ by Lemma 4.5.2(2) and Theorem 4.6.9.

□

**Remark 4.6.12 ([20])** Let $p$, $q$ be two idempotents in a Banach algebra. Then, $p + q$ is Drazin invertible if and only if $p - q$ is Drazin invertible [75]. Hence, $pq + qp$ is Drazin invertible is equivalent to $pq - qp$ is Drazin invertible in a Banach algebra. However, in general, this need not be true in a ring. For example, let $R = \mathbb{Z}$, $p = q = 1$. Then $p - q = 0 \in R^D$, but $p + q = 2 \notin R^D$.

## 4.7   Drazin Inverses of Matrices Over a Ring

Let $A$ be an $m \times n$ regular matrix over a ring, and $T$ a square matrix with $T^k = PAQ$ for some matrices $P$ and $Q$. In this section, we give necessary and sufficient conditions for the product $PAQ$ with $P'PA = A = AQQ'$ to have a Drazin inverse and give formulae for $(PAQ)^D$ when it exists.

**Theorem 4.7.1 ([16])**  *Let $A$ be an $m \times n$ regular matrix over a ring, and $T$ a square matrix with $T^k = PAQ$ for some positive integer $k$ and matrices $P$ and $Q$. Then the following statements are equivalent:*

(1) *$T$ has a Drazin inverse $T^D$ with $\mathrm{ind}(T) \leq k$ and there exist matrices $P'$ and $Q'$ such that $P'PA = A = AQQ'$.*
(2) *$A^- AQTPA + I_n - A^- A$ is invertible.*

*In this case, $T^D = PA(A^- AQTPA + I_n - A^- A)^{-1}Q$.*

## Proof

(1)$\Rightarrow$(2).   Let $X = T^D$. Then $T^k = T^{k+1}X$, $TX = XT$, $X = TX^2$. Hence,

$$T^k = T^{k+1}X = T^{k+2}X^2 = \cdots = T^{2k}X^k = T^k X^k T^k, \qquad (4.7.1)$$

$$X = TX^2 = T^2X^3 = \cdots = T^k X^{k+1}. \qquad (4.7.2)$$

Thus, by (4.7.1), we have

$$PAQ = PAQX^k PAQ.$$

Multiplying the equality above by $P'$ and $Q'$ on the left hand and right hand sides repectively, we obtain

$$A = AQX^k PA \qquad (4.7.3)$$

by the hypothesis. Therefore we get

$$(A^- AQTPA)(A^- AQX^{k+1}PA)^{k+1}$$

$$= A^- AQTPAA^- AQX^{k+1}PA(A^- AQX^{k+1}PA)^k$$

$$= A^- AQTT^k X^{k+1}PA(A^- AQX^{k+1}PA)^k$$

$$= A^- AQTXPA(A^- AQX^{k+1}PA)(A^- AQX^{k+1}PA)^{k-1} \qquad \text{by (4.7.2)}$$

$$= A^- AQTXT^k X^{k+1}PA(A^- AQX^{k+1}PA)^{k-1}$$

$$= A^- AQXPA(A^- AQX^{k+1}PA)^{k-1} \qquad \text{by (4.7.2)}$$

$$= A^- AQXPA(A^- AQX^{k+1}PA)(A^- AQX^{k+1}PA)^{k-2}$$

$$= A^- AQXT^k X^{k+1}PA(A^- AQX^{k+1}PA)^{k-2}$$

$$= A^- AQX^2 PA(A^- AQ^{k+1}PA)^{k-2} \qquad \text{by (4.7.2)}$$

$$= \cdots \cdots$$

$$= A^- AQX^{k-1}PA(A^- AQX^{k+1}PA)$$

$$= A^- AQX^k PA = A^- A. \qquad \text{by (4.7.3)}$$

Similarly, we have

$$(A^- AQX^{k+1}PA)^{k+1}A^- AQTPA = A^- A.$$

Thus, it follows that

$$(A^- AQTPA + I_n - A^- A)\left((A^- AQX^{k+1}PA)^{k+1} + I_n - A^- A\right) = I_n$$

and

$$\left((A^- AQX^{k+1}PA)^{k+1} + I_n - A^- A\right)(A^- AQTPA + I_n - A^- A) = I_n,$$

that is, $A^- AQTPA + I_n - A^- A$ is invertible.

(2) $\Rightarrow$ (1).    If we write $N = A^- AQTPA + I_n - A^- A$, then $N$ is invertible. Thus there exists $Y$ such that $I_n = YN = NY$. Hence

$$A = AA^- A = AI_n A^- A$$

$$= AYNA^- A$$

$$= AY(A^- AQTPA + I_n - A^- A)A^- A$$

$$= AYA^- AQTPA,$$

and

$$A = AI_n = ANY = A(A^- AQTPA + I_n - A^- A)Y$$

$$= AQTPAY.$$

Put $P' = AYA^- AQT$, $Q' = TPAY$. Then $P'PA = A = AQQ'$.
Let $X = PAN^{-1}Q$. We shall show that $X$ is a Drazin inverse of $T$ with $\mathrm{ind}(T) \leq k$. First of all, it is easy to verify that

$$AN = AQTPA \tag{4.7.4}$$

and

$$A^- AN = NA^- A = A^- AQTPA. \tag{4.7.5}$$

Thus

$$XTX = PAN^{-1}QTPAN^{-1}Q$$

$$= PAA^- AN^{-1}QTPAN^{-1}Q$$

$$= PAN^{-1}A^- AQTPAN^{-1}Q \qquad \text{by (4.7.4)}$$

$$= PAN^{-1}NA^- AN^{-1}Q \qquad \text{by (4.7.4)}$$

$$= PAA^- AN^{-1}Q$$

$$= PAN^{-1}Q = X.$$

Next, we have

$$T^{k+1}X = T^k T X$$

$$= PAQTPAN^{-1}Q$$

$$= P(AN)N^{-1}Q \qquad\qquad \text{by (4.7.5)}$$

$$= PAQ = T^k.$$

Note that $T^k = PAQ$ and we have

$$N^2 = (A^- AQTPA + I_n - A^- A)^2$$

$$= (A^- AQTPA)^2 + I_n - A^- A$$

$$= A^- AQTPAA^- AQTPA + I_n - A^- A$$

$$= A^- AQTT^k TPA + I_n - A^- A$$

$$= A^- AQPAQT^2 PA + I_n - A^- A$$

$$= (A^- AQPA + I_n - A^- A)(A^- AQT^2 PA + I_n - A^- A).$$

Similarly,

$$N^2 = (A^- AQT^2 PA + I_n - A^- A)(A^- AQPA + I_n - A^- A).$$

Set $M = (A^- AQPA + I_n - A^- A)$ and $S = (A^- AQT^2 PA + I_n - A^- A)$. Then $N^2 = MS = SM$. Since $N$ is invertible, so is $M$, and $M^{-1} = N^{-2}S = SN^{-2}$. Thus

$$TX = TPAN^{-1}Q$$

$$= TPANN^{-2}Q$$

$$= TP(AQTPA)N^{-2}Q \qquad\qquad \text{by (4.7.4)}$$

$$= TT^k TPAN^{-2}Q$$

$$= T^k T^2 PAN^{-2}Q$$

$$= PAQT^2 PAN^{-2}Q$$

$$= PA(A^- AQT^2 PA + I_n - A^- A)N^{-2}Q$$

$$= PASN^{-2}Q$$

$$= PAM^{-1}Q$$

and

$$XT = PAN^{-1}QT$$
$$= PAA^-AN^{-1}QT$$
$$= PAN^{-1}A^-AQT \qquad \text{by (4.7.5)}$$
$$= PAN^{-2}NA^-AQT$$
$$= PAN^{-2}(A^-AQTPA)QT \qquad \text{by (4.7.5)}$$
$$= PAN^{-2}A^-AQTT^kT$$
$$= PAN^{-2}A^-AQT^2T^k$$
$$= PAN^{-2}A^-AQT^2PAQ$$
$$= PAN^{-2}A^-A(A^-AQT^2PA + I_n - A^-A)Q$$
$$= PAA^-AN^{-2}SQ \qquad \text{by (4.7.5)}$$
$$= PAM^{-1}Q,$$

This completes the proof of Theorem 4.7.1.

□

Let $k = 1$ in Theorem 4.7.1. Then we have

**Corollary 4.7.2** ([16]) *Let $A$ be an $m \times n$ regular matrix over a ring, and $T$ a square matrix with $T = PAQ$ for some matrices $P$ and $Q$. Then the following statements are equivalent:*

(1) *$T$ has a group inverse $T^{\#}$ and there exist matrices $P'$ and $Q'$ such that $P'PA = A = AQQ'$.*

(2) *$A^-AQPA + I_n - A^-A$ is invertible.*

*In this case,*

$$T^{\#} = PA(A^-AQPA + I_n - A^-A)^{-2}Q.$$

**Proof** Since $T = PAQ$, one have

$$A^-AQTPA + I_n - A^-A$$
$$= A^-AQPAQPA + I_n - A^-A$$
$$= (A^-AQPA + I_n - A^-A)^2.$$

Thus, Corollary 4.7.2 follows from Theorem 4.7.1.                              □

**Corollary 4.7.3 ([16])** *Let E be an n × n idempotent matrix, and T a square matrix such that $T^k = PEQ$ for some positive integer k and matrices P and Q. Then there exist matrices P' and Q' such that $P'PE = E = EQQ'$ and T has a Drazin inverse $T^D$ with ind(T) ≤ k if and only if $EQTPE + I_n - E$ is invertible. In this case,*

$$T^D = PE(EQTPE + I_n - E)^{-1}Q.$$

***Proof*** Since E is idempotent, we have $E^- = E$. Thus Corollary 4.7.3 follows from Theorem 4.7.1. □

Next, we give characterizations for existence of the Drazin inverse of a matrix over an arbitrary ring. Moreover, the Drazin inverse of a product $PAQ$ for which there exist P' and Q' such that $P'PA = A = AQQ'$ can be characterized and computed. This generalizes some results obtained for the group inverse of such products.

**Lemma 4.7.4 ([104])** *Let T be an n × n matrix over R. The following conditions are equivalent:*

(1) *T is Drazin invertible with index k.*
(2) *There exists a matrix L and k is the smallest positive integer such that $T^k = T^{k+1}L = LT^{k+1}$.*
(3) *There exist matrices M and N and k is the smallest positive integer such that $T^k = T^{k+1}N = MT^{k+1}$.*
(4) *k is the smallest positive integer such that $T^k$ has a group inverse $(T^k)^\#$ and, independent of the choice of $(T^k)^-$, also equivalent with.*
(5) *k is the smallest positive integer such that $T^k$ is regular and $T^{2k}(T^k)^- + I_n - T^k(T^k)^-$ is invertible.*
(6) *k is the smallest positive integer such that $T^k$ is regular and $(T^k)^- T^{2k} + I_n - (T^k)^- T^k$ is invertible.*

*In this case, $T^{k+n}$ is regular for all positive integer n with $M^{n(k+1)}T^{(n-1)k}$ belonging to $T^{k+n}\{1\}$ and*

$$T^D = T^k(T^{2k})^- T^{2k-1}(T^{2k})^- T^k$$

$$= T^k(T^{2k+1})^- T^k$$

$$= T^{k-1}(T^k)^\#$$

$$= T^{k-1}(M^{k+1}T^{k+2}N^{k+1})$$

$$= T^{k-1}\left(T^{2k}(T^k)^- + I_n - T^k(T^k)^-\right)^{-1}T^k\left((T^k)^- T^{2k} + I_n - (T^k)^- T^k\right)^{-1}$$

$$= T^{k-1}\left(T^{2k}(T^k)^- + I_n - T^k(T^k)^-\right)^{-2}T^k$$

$$= T^{2k-1}\left((T^k)^- T^{2k} + I_n - (T^k)^- T^k\right)^{-2},$$

which shows that $T^D$ is always equivalent with $T^{2k-1}$. Moreover, $(T^D)^{\#}$ exists and equals $T^2 T^D$.

**Proof**

(1)$\Rightarrow$(2)$\Rightarrow$(3).   It is clear.
(3)$\Rightarrow$(1).   Suppose that $T^k = M T^{k+1} = T^{k+1} N$. Then

$$T^k = T^{k+(k+1)} N^{k+1}$$

$$= T^k \left( T^{k+1} N \right) N^k$$

$$= T^k \left( M T^{k+1} N^k \right)$$

$$= \cdots$$

$$= T^k \left( M^{k+1} T \right) T^k$$

and

$$T^{k+1} = T^{k+1} \left( M^{k+1} \right) T^{k+1},$$

$$T^{k+2} = T^{k+2(k+1)+2} N^{2(k+1)}$$

$$= T^{k+2} \left( T^{2(k+1)} N^{2(k+1)} \right)$$

$$= T^{k+2} \left( M^{2(k+1)} T^{2(k+1)} \right)$$

$$= T^{k+2} \left( M^{2(k+1)} T^k \right) T^{k+2},$$

more generally,

$$T^{k+n} = T^{k+n} \left( M^{n(k+1)} T^{(n-1)k} \right) T^{k+n}, \quad n = 1, 2, \ldots$$

It follows that $\left( T^{2k} \right)^-$ and $\left( T^{2k+1} \right)^-$ exist.
Now, we prove by straightforward computation that

$$T^k \left( T^{2k} \right)^- T^{2k-1} \left( T^{2k} \right)^- T^k = T^k \left( T^{2k+1} \right)^- T^k$$

is the Drazin inverse of $T$ of index $k$.

(I) We claim that

$$T^D = T^k \left( T^{2k} \right)^- T^{2k-1} \left( T^{2k} \right)^- T^k.$$

It follows from (3) that

$$T^k = DT^{2k+1} \quad \text{with } D = M^{k+1}$$
$$= T^{2k+1}C \quad \text{with } C = N^{k+1}.$$

Then

$$T^k \left(T^{2k}\right)^- T^{2k-1} \left(T^{2k}\right)^- T^k = DT^{2k+1} \left(T^{2k}\right)^- T^{2k+1} CT^{k-1} \left(T^{2k}\right)^- T^k$$
$$= DT^{k+1}T^{k-1} \left(T^{2k}\right)^- T^k$$
$$= DT^{2k} \left(T^{2k}\right)^- T^{2k+1}C$$
$$= DT^{2k+1}C$$
$$= T^kC.$$

We verify now the three definition equations:

$$T \left(T^kC\right) = T^{k+1}N^{k+1}$$
$$= M^{k+1}T^{k+1}$$
$$= \left(M^{k+1}T^{2k+1}\right)CT$$
$$= \left(T^kC\right)T,$$

$$T^k \left(T^kC\right)T = T^{k+1} \left(T^kC\right) = T^k,$$

$$\left(T^kC\right)T \left(T^kC\right) = T^{k+1}CT^kC$$
$$= T \left(T^kCT\right)T^{k-1}C$$
$$= T \left(TT^kC\right)T^{k-1}C$$
$$= T^{k+2}CT^{k-1}C$$
$$= T^{2k+1}CC$$
$$= T^kC.$$

(II)  We claim that

$$T^D = T^k \left(T^{2k+1}\right)^- T^k.$$

We verify the three definition equations:

$$TT^k \left(T^{2k+1}\right)^- T^k = TDT^{2k+1} \left(T^{2k+1}\right)^- T^{2k+1}C$$
$$= TDT^{2k+1}C$$
$$= T^{k+1}C,$$

$$T^k \left(T^{2k+1}\right)^- T^kT = DT^{2k+1} \left(T^{2k+1}\right)^- TT^{2k+1}C$$
$$= DT^{2k+1}TC$$
$$= T^{k+1}C,$$

$$T^{k+1}T^k \left(T^{2k+1}\right)^- T^k = T^{2k+1} \left(T^{2k+1}\right)^- T^{2k+1}C$$
$$= T^k,$$

$$\left(T^k \left(T^{2k+1}\right)^- T^k\right)^2 T = T^k \left(T^{2k+1}\right)^- T^k \left(T^k \left(T^{2k+1}\right)^- T^{k+1}\right)$$
$$= T^k \left(T^{2k+1}\right)^- T^kT^{k+1}C$$
$$= T^k \left(T^{2k+1}\right)^- T^k.$$

(3)⇔(4).   It follows from (3) that:

$$T^k = \left(M^{k+1}T\right) T^{2k}$$
$$= T^{2k} \left(TN^{k+1}\right)$$

which implies the existence of $\left(T^k\right)^\#$. Conversely, if $\left(T^k\right)^\#$ exists, for smallest $k$, then $T^{k-1} \left(T^k\right)^\#$ is the Drazin inverse of $T$ since there exists a matrix $G$ such that $T^k = GT^{2k} = T^{2k}G$, from which follows that:

$$T^k(T^{k-1}\left(T^k\right)^{\#})T = GT^{2k-1}\left(T^k\left(T^k\right)^{\#}\right)T$$

$$= GT^{2k-1}T$$

$$= T^k,$$

$$\left(T^{k-1}(T^k)^{\#}\right)T\left(T^{k-1}(T^k)^{\#}\right) = T^{k-1}(T^k)^{\#}$$

and

$$T^k\left(T^{k-1}\left(T^k\right)^{\#}\right) = T^{k-1}\left(T^k\left(T^k\right)^{\#}\right)$$

$$= \left(T^{k-1}\left(T^k\right)^{\#}\right)T^k.$$

Moreover,

$$T^D = T^{k-1}\left(T^k\right)^{\#}$$

implies

$$T^D = T^{k-1}\left(M^{k+1}TT^kTN^{k+1}\right)$$

$$= T^{k-1}\left(M^{k+1}T^{k+2}N^{k+1}\right).$$

(4)$\Leftrightarrow$(5)$\Leftrightarrow$(6).   It follows from Theorem 3.3.1. Moreover,

$$T^D = T^{k-1}\left(T^k\right)^{\#} = T^{2k-1}\left((T^k)^- T^{2k} + I_n - (T^k)^- T^k\right)^{-2}$$

i.e., $T^D$ is always equivalent with $T^{2k-1}$.
Finally, it follows from (2) that:

$$T^k = T^{2k}L^k = L^kT^{2k}$$

and therefore that $\left(T^D\right)^{\#}$ always exists, if $T^D$ exists. Since

$$T^DTT^D = T^D = T\left(T^D\right)^2 = \left(T^D\right)^2 T,$$

we have

$$\left(T^D\right)^{\#} = T^2 T^D.$$

□

**Theorem 4.7.5 ([104])** *Let A be an n × n matrix over an arbitrary ring R and P, Q matrices over R for which there exist matrices P′, Q′ such that P′PA = A = AQQ′. Let A₁ := A and for all positive integer i > 1, let*

$$A_i := AQ(PAQ)^{i-2}PA.$$

*Then the following statements are equivalent:*

(1) $T = PAQ$ *is Drazin invertible with index k.*
(2) *k is the smallest positive integer such that $A_k$ is regular and $A^k Q P A_k A_k^- + I_n - A_k A_k^-$ is invertible.*
(3) *k is the smallest positive integer such that $A_k$ is regular and $A_k^- A_k Q P A_k + I_n - A_k^- A_k$ is invertible and independent of the choice of $A_k^-$.*

*In that case, with $U_k := A_k Q P A_k A_k^- + 1_n - A_k A_k^-$ and $V_k := A_k^- A_k Q P A_k + I_n - A_k^- A_k$, we have*

$$T^D = T^{k-1}(PU_k^{-1}A_k V_k^{-1}Q)$$
$$= T^{k-1}(PU_k^{-2}A_k Q)$$
$$= T^{k-1}(PA_k V_k^{-2}Q).$$

**Proof** *T* has Drazin index *k* if and only if *k* is the smallest positive integer such that $T^k$ has Drazin index 1. But,

$$T^k = P\left(AQT^{k-2}PA\right)Q$$
$$= PA_k Q$$

and $AQT^{k-2}PA$ is regular if and only if $T^k$ is regular, because there exist $P′$ and $Q′$ such that $P′PA = A = AQQ′$. Indeed,

$$AQT^{k-2}PA\left(Q\left(T^k\right)^- P\right)AQT^{k-2}PA = AQT^{k-2}PA$$

is equivalent to

$$T^k \left(T^k\right)^- T^k = T^k.$$

Moreover, since $P'PA = A = AQQ'$, we have

$$P'\left(P(AQT^{k-2}PA)\right) = AQT^{k-2}PA = \left((AQT^{k-2}PA)Q\right)Q'$$

which means that $P'PA_k = A_k = A_kQQ'$.

We therefore can apply Corollary 3.3.3 with $p = P$, $a = AQT^{k-2}PA$ and $q = Q$, which implies $T^k$ has Drazin index 1 if and only if $T^k$ is regular and $U_k := A_kQPA_kA_k^- + I_n - A_kA_k^-$ is invertible if and only if $T^k$ is regular and $V_k := A_k^-A_kQPA_k + I_n - A_k^-A_k$ is invertible.                          □

Indeed, apply the theorem to the factorization $T = TT^{(1,2)}T$ with $A = T^{(1,2)}$, $P = Q = T$ and $P' = Q' = T^{(1,2)}$.

**Corollary 4.7.6 ([104])** *If $T$ is regular, then the following are equivalent for any reflexive inverse $T^{(1,2)}$ of $T$:*

(1) *$T$ is Drazin invertible with index $k$.*
(2) *$k$ is the smallest positive integer such that $T_k := T^{(1,2)}T^kT^{(1,2)}$ is regular and $T_kT^2T_kT_k^{(1,2)} + I_n - T_kT_k^{(1,2)}$ is invertible.*
(3) *$k$ is the smallest positive integer such that $T_k := T^{(1,2)}T^kT^{(1,2)}$ is regular and $T_k^{(1,2)}T_kT^2T_k + I_n - T_k^{(1,2)}T_k$ is invertible.*

*In that case, and with*

$$U_k := T_kT^2T_kT_k^{(1,2)} + I_n - T_kT_k^{(1,2)},$$
$$V_k := T_k^{(1,2)}T_kT^2T_k + I_n - T_k^{(1,2)}T_k,$$

*we obtain*

$$T^{D_k} = T^kU_k^{-1}T_kV_k^{-1}T$$
$$= T^kU_k^{-2}T^{(1,2)}T^k$$
$$= [T^kT^{(1,2)}]^2V_k^{-2}T.$$

## 4.8   The Drazin Inverse of a Sum of Morphisms

Let $C$ be an additive category. Suppose that $\varphi$, $\eta\colon X \longrightarrow X$ are two morphisms of $C$ with Drazin inverse $\varphi^D$ and $\eta^D$. In this section, we mainly investigate the Drazin inverse of a sum of morphisms.

In 1958, Drazin [39] proved that if two ring elements $a$, $b$ are Drazin invertible and $ab = ba = 0$, then the sum $a + b$ is also Drazin invertible with $(a + b)^D = a^D + b^D$. The fact is also true for morphisms in an additive category. Moreover, the following result shows that the condition is $ab = 0$ is enough for $a + b$ to be Drazin invertible.

**Theorem 4.8.1 ([21])** *Let $C$ be an additive category. Suppose that $\varphi$, $\eta\colon X \longrightarrow X$ are two morphisms of $C$ with Drazin inverse $\varphi^D$ and $\eta^D$ such that $\varphi\eta = 0$. Then $\varphi + \eta$ has the Drazin inverse with* $\operatorname{ind}(\varphi + \eta) \leq m$, *and*

$$(\varphi + \eta)^D = (1_X - \eta\eta^D)\sum_{i=0}^{k_2-1}\eta^i(\varphi^D)^i\varphi^D + \eta^D\sum_{j=0}^{k_1-1}(\eta^D)^j\varphi^j(1_X - \varphi\varphi^D),$$

*where $k_1 = \operatorname{ind}(\varphi)$, $k_2 = \operatorname{ind}(\eta)$, $m = k_1 + k_2 - 1$.*

*Proof* Let

$$\omega_1 = \left(1_X - \eta\eta^D\right)\sum_{i=0}^{k_2-1}\eta^i\left(\varphi^D\right)^i\varphi^D, \qquad \omega_2 = \eta^D\sum_{j=0}^{k_1-1}\left(\eta^D\right)^j\varphi^j\left(1_X - \varphi\varphi^D\right).$$

Set $\omega = \omega_1 + \omega_2$. Since $\varphi\eta = 0$, we have $\varphi\eta^D = \varphi\eta(\eta^D)^2 = 0$ and $\varphi^D\eta = (\varphi^D)^2\varphi\eta = 0$. Then $\varphi\omega_2 = 0$,

$$\varphi\omega_1 = \varphi\left(1_X - \eta\eta^D\right)\sum_{i=0}^{k_2-1}\eta^i\left(\varphi^D\right)^i\varphi^D$$

$$= \varphi\sum_{i=0}^{k_2-1}\eta^i\left(\varphi^D\right)^i\varphi^D = \varphi\varphi^D,$$

$$\eta\omega_1 = \eta\left(1_X - \eta\eta^D\right)\sum_{i=0}^{k_2-1}\eta^i\left(\varphi^D\right)^i\varphi^D$$

$$= \left(1_X - \eta\eta^D\right)\sum_{i=0}^{k_2-1}\eta^{i+1}\left(\varphi^D\right)^{i+1}$$

$$= \left(1_X - \eta\eta^D\right)\sum_{i=1}^{k_2}\eta^i\left(\varphi^D\right)^i$$

$$= \left(1_X - \eta\eta^D\right)\sum_{i=1}^{k_2-1}\eta^i\left(\varphi^D\right)^i + \left(1_X - \eta\eta^D\right)\eta^{k_2}\left(\varphi^D\right)^{k_2}$$

$$= \left(1_X - \eta\eta^D\right)\sum_{i=1}^{k_2-1}\eta^i\left(\varphi^D\right)^i,$$

$$\eta\omega_2 = \eta\eta^D \sum_{j=0}^{k_1-1} \left(\eta^D\right)^j \varphi^j \left(1_X - \varphi\varphi^D\right)$$

$$= \eta\eta^D \left(1_X - \varphi\varphi^D\right) + \eta\eta^D \sum_{j=1}^{k_1-1} \left(\eta^D\right)^j \varphi^j \left(1_X - \varphi\varphi^D\right)$$

$$= \eta\eta^D \left(1_X - \varphi\varphi^D\right) + \sum_{j=1}^{k_1-1} \left(\eta^D\right)^j \varphi^j \left(1_X - \varphi\varphi^D\right).$$

Hence

$$(\varphi + \eta)\omega = \varphi\omega_1 + \varphi\omega_2 + \eta\omega_1 + \eta\omega_2$$

$$= \varphi\varphi^D + \left(1_X - \eta\eta^D\right) \sum_{i=1}^{k_2-1} \eta^i \left(\varphi^D\right)^i + \eta\eta^D \left(1_X - \varphi\varphi^D\right)$$

$$+ \sum_{j=1}^{k_1-1} \left(\eta^D\right)^j \varphi^j \left(1_X - \varphi\varphi^D\right)$$

$$= \varphi\varphi^D + \eta\eta^D - \eta\eta^D\varphi\varphi^D + \left(1_X - \eta\eta^D\right) \sum_{i=1}^{k_2-1} \eta^i \left(\varphi^D\right)^i$$

$$+ \sum_{j=1}^{k_1-1} \left(\eta^D\right)^j \varphi^j \left(1_X - \varphi\varphi^D\right).$$

On the other hand,

$$\omega_1\varphi = \left(1_X - \eta\eta^D\right) \sum_{i=0}^{k_2-1} \eta^i \left(\varphi^D\right)^i \varphi^D\varphi$$

$$= \left(1_X - \eta\eta^D\right) \varphi^D\varphi + \left(1_X - \eta\eta^D\right) \sum_{i=1}^{k_2-1} \eta^i \left(\varphi^D\right)^i \varphi^D\varphi$$

$$= \left(1_X - \eta\eta^D\right) \varphi^D\varphi + \left(1_X - \eta\eta^D\right) \sum_{i=1}^{k_2-1} \eta^i \left(\varphi^D\right)^i,$$

$$\omega_2\varphi = \eta^D \sum_{j=0}^{k_1-1} \left(\eta^D\right)^j \varphi^j \left(1_X - \varphi\varphi^D\right) \varphi$$

$$= \sum_{j=0}^{k_1-1} \left(\eta^D\right)^{j+1} \varphi^{j+1} \left(1_X - \varphi\varphi^D\right)$$

$$= \sum_{j=1}^{k_1} \left(\eta^D\right)^{j} \varphi^{j} \left(1_X - \varphi\varphi^D\right)$$

$$= \sum_{j=1}^{k_1-1} \left(\eta^D\right)^{j} \varphi^{j} \left(1_X - \varphi\varphi^D\right) + \left(\eta^D\right)^{k_1} \varphi^{k_1} \left(1_X - \varphi\varphi^D\right)$$

$$= \sum_{j=1}^{k_1-1} \left(\eta^D\right)^{j} \varphi^{j} \left(1_X - \varphi\varphi^D\right),$$

and

$$\omega_1 \eta = \left(1_X - \eta\eta^D\right) \sum_{i=0}^{k_2-1} \eta^i \left(\varphi^D\right)^{i} \varphi^D \eta = 0,$$

$$\omega_2 \eta = \eta^D \sum_{j=0}^{k_1-1} \left(\eta^D\right)^{j} \varphi^{j} \left(1_X - \varphi\varphi^D\right) \eta = \eta^D \sum_{j=0}^{k_1-1} \left(\eta^D\right)^{j} \varphi^{j} \eta = \eta^D \eta.$$

Therefore

$$\omega(\varphi + \eta) = \omega_1 \varphi + \omega_2 \varphi + \omega_1 \eta + \omega_2 \eta$$

$$= \left(1_X - \eta\eta^D\right) \varphi^D \varphi + \left(1_X - \eta\eta^D\right) \sum_{i=1}^{k_2-1} \eta^i \left(\varphi^D\right)^{i}$$

$$+ \sum_{j=1}^{k_1-1} \left(\eta^D\right)^{j} \varphi^{j} \left(1_X - \varphi\varphi^D\right) + \eta^D \eta$$

$$= (\varphi + \eta)\omega.$$

Next we prove that $\omega(\varphi + \eta)\omega = \omega$. Let

$$\tilde{\omega}_1 = \left(1_X - \eta\eta^D\right) \sum_{i=1}^{k_2-1} \eta^i \left(\varphi^D\right)^{i}, \quad \tilde{\omega}_2 = \sum_{j=1}^{k_1-1} \left(\eta^D\right)^{j} \varphi^{j} \left(1_X - \varphi\varphi^D\right).$$

Then we have

$$\omega_1 \varphi^D \varphi = \omega_1,$$

$$\omega_1 \eta \eta^D \left(1_X - \varphi \varphi^D\right) = \left(1_X - \eta \eta^D\right) \sum_{i=0}^{k_2-1} \eta^i \left(\varphi^D\right)^i \varphi^D \eta \eta^D \left(1_X - \varphi \varphi^D\right) = 0,$$

$$\varphi^D \tilde{\omega}_1 = \varphi^D \left(1_X - \eta \eta^D\right) \sum_{i=1}^{k_2-1} \eta^i \left(\varphi^D\right)^i = \varphi^D \sum_{i=1}^{k_2-1} \eta^i \left(\varphi^D\right)^i = 0.$$

Hence $\omega_1 \tilde{\omega}_1 = 0$. From $\varphi^D \eta^D = (\varphi^D)^2 \varphi \eta^D = 0$, we have $\varphi^D \tilde{\omega}_2 = 0$. Then $\omega_1 \tilde{\omega}_2 = 0$, $\omega_2 \varphi^D \varphi = 0$ and we have

$$\omega_2 \eta = \eta^D \sum_{j=0}^{k_1-1} \left(\eta^D\right)^j \varphi^j \left(1_X - \varphi \varphi^D\right) \eta = \eta^D \sum_{j=0}^{k_1-1} \left(\eta^D\right)^j \varphi^j \eta = \eta^D \eta,$$

or

$$\omega_2 \eta \eta^D \left(1_X - \varphi \varphi^D\right) = \eta^D \eta \eta^D \left(1_X - \varphi \varphi^D\right) = \eta^D \left(1_X - \varphi \varphi^D\right),$$

and

$$\sum_{j=1}^{k_1-1} \left(\eta^D\right)^j \varphi^j \left(1_X - \varphi \varphi^D\right) \tilde{\omega}_1$$

$$= \sum_{j=1}^{k_1-1} \left(\eta^D\right)^j \varphi^j \left(1_X - \varphi \varphi^D\right) \left(1_X - \eta \eta^D\right) \sum_{i=1}^{k_2-1} \eta^i \left(\varphi^D\right)^i$$

$$= \sum_{j=1}^{k_1-1} \left(\eta^D\right)^j \varphi^{j-1} \left(1_X - \varphi \varphi^D\right) \varphi \eta \left(1_X - \eta \eta^D\right) \sum_{i=1}^{k_2-1} \eta^{i-1} \left(\varphi^D\right)^i$$

$$= 0,$$

$$\omega_2 \tilde{\omega}_1 = \left(\eta^D \left(1_X - \varphi \varphi^D\right) + \eta^D \sum_{j=1}^{k_1-1} \left(\eta^D\right)^j \varphi^j \left(1_X - \varphi \varphi^D\right)\right) \tilde{\omega}_1$$

$$= \eta^D \left(1_X - \varphi \varphi^D\right) \tilde{\omega}_1$$

$$= \eta^D \left(1_X - \varphi\varphi^D\right)\left(1_X - \eta\eta^D\right) \sum_{i=1}^{k_2-1} \eta^i \left(\varphi^D\right)^i$$

$$= \eta^D \left(1_X - \varphi\varphi^D\right) \eta \left(1_X - \eta\eta^D\right) \sum_{i=1}^{k_2-1} \eta^{i-1} \left(\varphi^D\right)^i$$

$$= \eta^D \eta \left(1_X - \eta\eta^D\right) \sum_{i=1}^{k_2-1} \eta^{i-1} \left(\varphi^D\right)^i$$

$$= 0,$$

and

$$\omega_2\tilde{\omega}_2 = \left(\eta^D \left(1_X - \varphi\varphi^D\right) + \eta^D \sum_{j=1}^{k_1-1} \left(\eta^D\right)^j \varphi^j \left(1_X - \varphi\varphi^D\right)\right) \tilde{\omega}_2$$

$$= \eta^D \left(1_X - \varphi\varphi^D\right) \sum_{j=1}^{k_1-1} \left(\eta^D\right)^j \varphi^j \left(1_X - \varphi\varphi^D\right)$$

$$+ \eta^D \sum_{j=1}^{k_1-1} \left(\eta^D\right)^j \varphi^j \left(1_X - \varphi\varphi^D\right) \sum_{j=1}^{k_1-1} \left(\eta^D\right)^j \varphi^j \left(1_X - \varphi\varphi^D\right)$$

$$= \eta^D \sum_{j=1}^{k_1-1} \left(\eta^D\right)^j \varphi^j \left(1_X - \varphi\varphi^D\right)$$

$$+ \eta^D \sum_{j=1}^{k_1-1} \left(\eta^D\right)^j \varphi^{j-1}\varphi \left(1_X - \varphi\varphi^D\right) \eta^D \sum_{j=1}^{k_1-1} \left(\eta^D\right)^{j-1} \varphi^j \left(1_X - \varphi\varphi^D\right)$$

$$= \eta^D \sum_{j=1}^{k_1-1} \left(\eta^D\right)^j \varphi^j \left(1_X - \varphi\varphi^D\right),$$

$$\omega(\varphi + \eta)\omega = (\omega_1 + \omega_2)\left((\varphi + \eta)\omega\right)$$

$$= (\omega_1 + \omega_2)\left(\varphi\varphi^D + \eta\eta^D(1_X - \varphi\varphi^D) + \tilde{\omega}_1 + \tilde{\omega}_2\right)$$

$$= \omega_1\left(\varphi\varphi^D + \eta\eta^D(1_X - \varphi\varphi^D)\right) + \omega_1\tilde{\omega}_1 + \omega_1\tilde{\omega}_2$$

$$+ \omega_2\left(\varphi\varphi^D + \eta\eta^D(1_X - \varphi\varphi^D)\right) + \omega_2\tilde{\omega}_1 + \omega_2\tilde{\omega}_2$$

$$= \omega_1 + \eta^D \left(1_X - \varphi\varphi^D\right) + \eta^D \sum_{j=1}^{k_1-1} \left(\eta^D\right)^j \varphi^j (1_X - \varphi\varphi^D)$$

$$= \omega_1 + \eta^D \sum_{j=0}^{k_1-1} \left(\eta^D\right)^j \varphi^j \left(1_X - \varphi\varphi^D\right)$$

$$= \omega_1 + \omega_2 = \omega.$$

Finally, we prove that $(\varphi + \eta)^m \omega (\varphi + \eta) = (\varphi + \eta)^m$, where $m = k_1 + k_2 - 1$. From $\varphi\eta = 0$, we have

$$(\varphi + \eta)^m = \sum_{s=0}^{m} \eta^s \varphi^{m-s},$$

$$(\varphi + \eta)^m \eta\eta^D \left(1_X - \varphi\varphi^D\right) = \sum_{s=0}^{m} \eta^s \varphi^{m-s} \eta\eta^D \left(1_X - \varphi\varphi^D\right)$$

$$= \eta^m \eta\eta^D \left(1_X - \varphi\varphi^D\right) = \eta^m \left(1_X - \varphi\varphi^D\right),$$

$$(\varphi + \eta)^m \tilde{\omega}_1 = (\varphi + \eta)^m \left(1_X - \eta\eta^D\right) \sum_{i=1}^{k_2-1} \eta^i \left(\varphi^D\right)^i$$

$$= (\varphi + \eta)^m \sum_{i=1}^{k_2-1} \eta^i \left(\varphi^D\right)^i - (\varphi + \eta)^m \eta\eta^D \sum_{i=1}^{k_2-1} \eta^i \left(\varphi^D\right)^i$$

$$= \sum_{s=0}^{m} \eta^s \varphi^{m-s} \sum_{i=1}^{k_2-1} \eta^i \left(\varphi^D\right)^i - \sum_{s=0}^{m} \eta^s \varphi^{m-s} \eta\eta^D \sum_{i=1}^{k_2-1} \eta^i \left(\varphi^D\right)^i$$

$$= \eta^m \sum_{i=1}^{k_2-1} \eta^i \left(\varphi^D\right)^i - \eta^m \eta\eta^D \sum_{i=1}^{k_2-1} \eta^i \left(\varphi^D\right)^i$$

$$= \eta^m \sum_{i=1}^{k_2-1} \eta^i \left(\varphi^D\right)^i - \eta^m \sum_{i=1}^{k_2-1} \eta^i \left(\varphi^D\right)^i$$

$$= 0.$$

Now for $j = 1, 2, \cdots, k_1 - 1$, we have

$$\eta^m \left(\eta^D\right)^j = \eta^{m-1}\eta \left(\eta^D\right)^2 \left(\eta^D\right)^{j-2} = \eta^{m-1} \left(\eta^D\right)^{j-1} = \cdots\cdots$$

$$= \eta^{m-(j-1)} \left(\eta^D\right)^{j-(j-1)} = \eta^{m-(j-1)}\eta^D$$

$$= \eta^{m-k_2-j}\eta^{k_2+1}\eta^D = \eta^{m-k_2-j}\eta^{k_2} = \eta^{m-j}.$$

Hence,

$$(\varphi + \eta)^m \tilde{\omega}_2 = \sum_{s=0}^{m} \eta^s \varphi^{m-s} \sum_{j=1}^{k_1-1} \left(\eta^D\right)^j \varphi^j \left(1_X - \varphi\varphi^D\right)$$

$$= \eta^m \sum_{j=1}^{k_1-1} \left(\eta^D\right)^j \varphi^j \left(1_X - \varphi\varphi^D\right)$$

$$= \sum_{j=1}^{k_1-1} \eta^m \left(\eta^D\right)^j \varphi^j \left(1_X - \varphi\varphi^D\right) = \sum_{j=1}^{k_1-1} \eta^{m-j} \varphi^j \left(1_X - \varphi\varphi^D\right),$$

$$(\varphi + \eta)^m \varphi\varphi^D = \sum_{s=0}^{m} \eta^s \varphi^{m-s} \varphi\varphi^D = \sum_{s=0}^{m} \eta^{m-s} \varphi^s \varphi\varphi^D$$

$$= \eta^m \varphi\varphi^D + \sum_{s=1}^{k_1-1} \eta^{m-s} \varphi^s \varphi\varphi^D + \sum_{s=k_1}^{m} \eta^{m-s} \varphi^s \varphi\varphi^D$$

$$= \eta^m \varphi\varphi^D + \sum_{s=1}^{k_1-1} \eta^{m-s} \varphi^s \varphi\varphi^D + \sum_{s=k_1}^{m} \eta^{m-s} \varphi^s,$$

$$(\varphi + \eta)^m \omega(\varphi + \eta)$$

$$= (\varphi + \eta)^m \left(\varphi\varphi^D + \eta\eta^D \left(1_X - \varphi\varphi^D\right) + \tilde{\omega}_1 + \tilde{\omega}_2\right)$$

$$= (\varphi + \eta)^m \varphi\varphi^D + (\varphi + \eta)^m \eta\eta^D \left(1_X - \varphi\varphi^D\right) + (\varphi + \eta)^m \tilde{\omega}_1 + (\varphi + \eta)^m \tilde{\omega}_2$$

$$= \eta^m \varphi\varphi^D + \sum_{s=1}^{k_1-1} \eta^{m-s} \varphi^s \varphi\varphi^D + \sum_{s=k_1}^{m} \eta^{m-s} \varphi^s + \eta^m \left(1_X - \varphi\varphi^D\right)$$

$$+ \sum_{j=1}^{k_1-1} \eta^{m-j} \varphi^j \left(1_X - \varphi\varphi^D\right)$$

$$= \eta^m + \sum_{s=1}^{k_1-1} \eta^{m-s} \varphi^s + \sum_{s=k_1}^{m} \eta^{m-s} \varphi^s$$

$$= \eta^m + \sum_{s=1}^{m} \eta^{m-s} \varphi^s = (\varphi + \eta)^m.$$

Therefore, $\omega$ is the Drazin inverse of $\varphi + \eta$, i.e.,

$$(\varphi + \eta)^D = (1_X - \eta\eta^D) \sum_{i=0}^{k_2-1} \eta^i (\varphi^D)^i \varphi^D + \eta^D \sum_{j=0}^{k_1-1} (\eta^D)^j \varphi^j (1_X - \varphi\varphi^D).$$

$\square$

**Corollary 4.8.2 ([21])** *Let $C$ be an additive category. Suppose that $\varphi$, $\eta: X \longrightarrow X$ are two morphisms of $C$ with group inverse $\varphi^{\#}$ and $\eta^{\#}$ such that $\varphi\eta = 0$. Then $\varphi + \eta$ has the group inverse and $(\varphi + \eta)^{\#} = (1_X - \eta\eta^{\#})\varphi^{\#} + \eta^{\#}(1_X - \varphi\varphi^{\#})$.*

For any positive integer $j$, define

$$U_j = \{(p_1, q_1, p_2, q_2, \cdots, p_j, q_j) : \sum_{i=1}^{j} p_i + \sum_{i=1}^{j} q_i = j - 1,$$

$$p_i, q_i \in \{0, 1, \cdots, j - 1\}, i = 1, \cdots, j\}.$$

**Theorem 4.8.3 ([29])** *Let $C$ be an additive category. Suppose that $\varphi$, $\eta$ are two morphisms of $C$ with Drazin inverse $\varphi^D$ and $\eta^D$. Let $\mathrm{ind}(\varphi) = r$, $\mathrm{ind}(\eta) = s$ and $k \in \mathbb{N}$. If*

$$\varphi\eta \prod_{i=1}^{k} \left(\varphi^{p_i} \eta^{q_i}\right) = 0, \tag{4.8.1}$$

*for every $(p_1, q_1, p_2, q_2, \ldots, p_k, q_k) \in U_k$, then $\varphi + \eta$ is Drazin invertible and*

$$(\varphi + \eta)^D = y_1 + y_2 + \sum_{i=1}^{k-1} \left(y_1 \left(\varphi^D\right)^{i+1} + \left(\eta^D\right)^{i+1} y_2\right)$$

$$- \sum_{j=1}^{i+1} \left(\eta^D\right)^j \left(\varphi^D\right)^{i+2-j}\right) \varphi\eta(\varphi + \eta)^{i-1}, \tag{4.8.2}$$

*where*

$$y_1 = \eta^{\pi} \left( \sum_{i=0}^{s-1} \eta^i \left( \varphi^D \right)^i \right) \varphi^D, \quad y_2 = \eta^D \left( \sum_{i=0}^{r-1} \left( \eta^D \right)^i \varphi^i \right) \varphi^{\pi}. \tag{4.8.3}$$

**Proof** Let us denote the element on the right hand side of (4.8.2) by $\chi$. We will prove that $(\varphi + \eta)\chi = \chi(\varphi + \eta)$, $\chi^2(\varphi + \eta) = \chi$ and $(\varphi + \eta)^{l+1}\chi = (\varphi + \eta)^l$, for some $l \in \mathbb{N}$. Denote by

$$y_3 = \sum_{i=1}^{k-1} \left( y_1 \left( \varphi^D \right)^{i+1} + \left( \eta^D \right)^{i+1} y_2 - \sum_{j=1}^{i+1} \left( \eta^D \right)^j \left( \varphi^D \right)^{i+2-j} \right) \varphi \eta (\varphi + \eta)^{i-1}.$$

Hence, $\chi = y_1 + y_2 + y_3$. From (4.8.1) it follows that $\varphi \eta \varphi^D = 0$ and $\varphi \eta^D = 0$. Also, we have that

$$y_1 \varphi = \eta y_1 + \eta^{\pi} \varphi \varphi^D,$$
$$\eta y_2 = \eta \eta^D \varphi^{\pi} + y_2 \varphi,$$
$$\varphi y_1 = \varphi \varphi^D,$$
$$\varphi y_2 = 0. \tag{4.8.4}$$

Therefore, by (4.8.4) we get

$$(\varphi + \eta)\chi = \varphi \varphi^D + \eta y_1 + \eta \eta^D \varphi^{\pi} + y_2 \varphi + (\varphi + \eta)y_3$$

and

$$\chi(\varphi + \eta) = \eta y_1 + \eta^{\pi} \varphi \varphi^D + y_1 \eta + y_2(\varphi + \eta) + y_3(\varphi + \eta).$$

By computation using that $\varphi \eta^D = 0$,

$$(\varphi + \eta)y_3 = \varphi y_3 + \eta y_3 = \sum_{i=1}^{k-1} S_i \varphi \eta (\varphi + \eta)^{i-1}. \tag{4.8.5}$$

where $S_i = \varphi y_1 \left( \varphi^D \right)^{i+1} + \eta y_1 \left( \varphi^D \right)^{i+1} + \left( \eta^D \right)^i y_2 - \sum_{j=1}^{i+1} \eta \left( \eta^D \right)^j \left( \varphi^D \right)^{i+2-j}$. Using (4.8.4), we have

$$S_i = \left( \varphi^D \right)^{i+1} + \left( y_1 \varphi - \eta^{\pi} \varphi \varphi^D \right) \left( \varphi^D \right)^{i+1} + \left( \eta^D \right)^i y_2 - \sum_{j=1}^{i+1} \eta \left( \eta^D \right)^j \left( \varphi^D \right)^{i+2-j}$$

$$= \left(\varphi^D\right)^{i+1} + y_1\left(\varphi^D\right)^i - \eta^\pi\left(\varphi^D\right)^{i+1} + \left(\eta^D\right)^i y_2 - \sum_{j=1}^{i+1}\eta\left(\eta^D\right)^j\left(\varphi^D\right)^{i+2-j}$$

$$= \eta\eta^D\left(\varphi^D\right)^{i+1} + y_1\left(\varphi^D\right)^i + \left(\eta^D\right)^i y_2 - \eta\eta^D\left(\varphi^D\right)^{i+1}$$

$$- \sum_{j=2}^{i+1}\eta\left(\eta^D\right)^j\left(\varphi^D\right)^{i+2-j}$$

$$= y_1\left(\varphi^D\right)^i + \left(\eta^D\right)^i y_2 - \sum_{j=1}^{i}\left(\eta^D\right)^j\left(\varphi^D\right)^{i+1-j}. \tag{4.8.6}$$

On the other hand,

$$y_3(\varphi+\eta) = \sum_{i=1}^{k-1}\left(y_1\left(\varphi^D\right)^{i+1} + \left(\eta^D\right)^{i+1}y_2\right.$$

$$\left. - \sum_{j=1}^{i+1}\left(\eta^D\right)^j\left(\varphi^D\right)^{i+2-j}\right)\varphi\eta(\varphi+\eta)^i$$

$$= \sum_{i=2}^{k}\left(y_1\left(\varphi^D\right)^i + \left(\eta^D\right)^i y_2 - \sum_{j=1}^{i}\left(\eta^D\right)^j\left(\varphi^D\right)^{i+1-j}\right)\varphi\eta(\varphi+\eta)^{i-1}.$$

Hence,

$$y_3(\varphi+\eta) = \sum_{i=2}^{k} S_i\varphi\eta(\varphi+\eta)^{i-1}. \tag{4.8.7}$$

Since $\varphi\eta(\varphi+\eta)^{k-1} = 0$, we have that $y_3(\varphi+\eta) = \sum_{i=2}^{k-1} S_i\varphi\eta(\varphi+\eta)^{i-1}$, so by (4.8.5) and (4.8.7),

$$(\varphi+\eta)y_3 = y_3(\varphi+\eta) + S_1\varphi\eta = y_3(\varphi+\eta) + y_1\varphi^D\varphi\eta + \eta^D y_2\varphi\eta - \eta^D\varphi^D\varphi\eta.$$

Now, by (4.8.4) and using that $y_1\varphi^D\varphi = y_1$ and $\eta^D y_2\varphi = y_2 - \eta^D\varphi^\pi$, we get that $(\varphi+\eta)\chi = \chi(\varphi+\eta)$.

In order to prove that $\chi^2(\varphi+\eta) = \chi$, remark that $y_1 y_2 = 0$, $y_3 y_1 = 0$, $y_3 y_2 = 0$ and $y_3^2 = 0$. Thus we have

$$\chi^2(\varphi+\eta) = \left(y_1^2 + y_2^2 + y_2 y_1 + y_1 y_3 + y_2 y_3\right)(\varphi+\eta). \tag{4.8.8}$$

We can easy verify the following

$$y_1^2 = y_1\varphi^D, \quad y_2^2 = \eta^D y_2, \quad y_2 y_1 = -\eta^D \varphi^D,$$

$$y_1 y_3 = \sum_{i=1}^{k-1} y_1 \left(\varphi^D\right)^{i+2} \varphi\eta(\varphi+\eta)^{i-1}.$$

Now, we will compute $y_2 y_3$. Since $\varphi y_1 - \varphi\eta^D$, we get that $\varphi^\pi \varphi y_1 = 0$, which together with $\varphi\eta^D = 0$ implies that $\varphi^\pi \varphi y_3 = 0$. Now,

$$y_2 y_3 = \eta^D \left(\sum_{i=0}^{r-1} \left(\eta^D\right)^i \varphi^i\right)\varphi^\pi y_3$$

$$= \eta^D \varphi^\pi y_3 + \eta^D \left(\sum_{i=1}^{r-1} \left(\eta^D\right)^i \varphi^i\right)\varphi^\pi y_3$$

$$= \eta^D \varphi^\pi y_3 = \eta^D y_3 - \sum_{i=1}^{k-1} \eta^D \left(\varphi^D\right)^{i+2} \varphi\eta(\varphi+\eta)^{i-1}$$

$$= \sum_{i=1}^{k-1} \left(\left(\eta^D\right)^{i+2} y_2 - \sum_{j=1}^{i+2} \left(\eta^D\right)^j \left(\varphi^D\right)^{i+3-j}\right)\varphi\eta(\varphi+\eta)^{i-1}. \qquad (4.8.9)$$

Upon substituting these identities into (4.8.8) , we have the following

$$\chi^2(\varphi+\eta) = y_1\varphi^D(\varphi+\eta) + \eta^D y_2(\varphi+\eta) - \eta^D \varphi^D(\varphi+\eta)$$

$$+ \sum_{i=1}^{k-1} \left(y_1 \left(\varphi^D\right)^{i+2} + \left(\eta^D\right)^{i+2} y_2 - \sum_{j=1}^{i+2} \left(\eta^D\right)^j \left(\varphi^D\right)^{i+3-j}\right)\varphi\eta(\varphi+\eta)^i$$

$$= y_1\varphi^D(\varphi+\eta) + \eta^D y_2(\varphi+\eta) - \eta^D \varphi^D(\varphi+\eta)$$

$$+ \sum_{i=2}^{k-1} \left(y_1 \left(\varphi^D\right)^{i+1} + \left(\eta^D\right)^{i+1} y_2 - \sum_{j=1}^{i+1} \left(\eta^D\right)^j \left(\varphi^D\right)^{i+2-j}\right)\varphi\eta(\varphi+\eta)^{i-1}.$$

Now, since $\varphi\eta(\varphi+\eta)^{k-1} = 0$, $y_1\varphi^D\varphi = y_1$ and $\eta^D y_2\varphi = y_2 - \eta^D\varphi^\pi$, we get

$$\chi^2(\varphi+\eta) = y_1\varphi^D(\varphi+\eta) + \eta^D y_2(\varphi+\eta) - \eta^D \varphi^D(\varphi+\eta)$$

$$+ y_3 - y_1\varphi^D\eta - \left(\eta^D\right)^2 y_2\varphi\eta + \eta^D\varphi^D\eta + \left(\eta^D\right)^2 \varphi^D\varphi\eta$$

$$= y_1 + y_2 - \eta^D\varphi^\pi + \eta^D y_2\eta - \eta^D\varphi^D\varphi$$

$$+ y_3 - \left(\eta^D\right)\left(y_2 - \eta^D\varphi^\pi\right)\eta + \left(\eta^D\right)^2\varphi^D\varphi\eta$$

$$= \chi. \tag{4.8.10}$$

Let $l$, $i \in \mathbb{N}$ be such that $i \geqslant r + s - 1$ and $l \geqslant k + i - 1$. We will prove that

$$\chi(\varphi + \eta)^{l+1} = (\varphi + \eta)^l. \tag{4.8.11}$$

Since $y_3(\varphi + \eta)^l = 0$, to prove (4.8.11), it is sufficient to prove that $(y_1 + y_2)(\varphi + \eta)^{l+1} = (\varphi + \eta)^l$. By $\varphi\eta(\varphi + \eta)^l = 0$, we get that $y_1\eta(\varphi + \eta)^l = 0$. Also, we have that $y_2\eta(\varphi + \eta)^l = \eta^D\eta(\varphi + \eta)^l$, $y_2\varphi(\varphi + \eta)^l = 0$ and $y_1\varphi(\varphi + \eta)^l = y_1\varphi^{i+1}(\varphi + \eta)^{l-i}$. Now

$$(y_1 + y_2)(\varphi + \eta)^{l+1} = \eta^D\eta(\varphi + \eta)^l + y_1\varphi^{i+1}(\varphi + \eta)^{l-i}, \tag{4.8.12}$$

so to prove (4.8.11), it is sufficient to prove that

$$y_1\varphi^{i+1}(\varphi + \eta)^{l-i} = \eta^\pi(\varphi + \eta)^l. \tag{4.8.13}$$

Since $i \geqslant r + s - 2$,

$$y_1\varphi^{i+1}(\varphi + \eta)^{l-i} = \eta^\pi\left(\sum_{j=0}^{s-1}\eta^j\left(\varphi^D\right)^{j+1}\right)\varphi^{i+1}(\varphi + \eta)^{l-i}$$

$$= \eta^\pi\left(\sum_{j=0}^{s-1}\eta^j\varphi^D\varphi^{i+1-j}\right)(\varphi + \eta)^{l-i}$$

$$= \eta^\pi\left(\sum_{j=0}^{s-1}\eta^j\varphi^{i-j}\right)(\varphi + \eta)^{l-i}$$

and

$$\eta^\pi(\varphi + \eta)^l = \eta^\pi\left(\varphi^2 + \eta\varphi + \eta^2\right)(\varphi + \eta)^{l-2}$$

$$= \eta^\pi\left(\varphi^3 + \eta\varphi^2 + \eta^2\varphi + \eta^3\right)(\varphi + \eta)^{l-3}$$

$$= \cdots$$

$$= \eta^\pi\left(\sum_{j=0}^{s-1}\eta^j\varphi^{i-j}\right)(\varphi + \eta)^{l-i},$$

we get that (4.8.13) holds. Hence, $\chi = (\varphi + \eta)^D$.                       $\square$

**Remark 4.8.4** Theorem 4.8.3 gives representations of $(\varphi + \eta)^D$ in the following particular cases:

(1) $\varphi\eta = 0$.
(2) $\varphi\eta\varphi = 0$, $\varphi\eta^2 = 0$.
(3) $\varphi\eta\varphi\eta = 0$, $\varphi\eta\varphi^2 = 0$, $\varphi\eta^2\varphi = 0$, $\varphi\eta^3 = 0$.

**Theorem 4.8.5** ([21]) Let $C$ be an additive category. Suppose that $\varphi: X \longrightarrow X$ is a morphism of $C$ with Drazin inverse $\varphi^D$ and $\mathrm{ind}(\varphi) = k_1$ and that $\eta : X \longrightarrow X$ is a morphism of $C$ such that $\varphi\varphi^D\eta = \eta$. If $\gamma = (\varphi + \eta)\varphi\varphi^D$ has the Drazin inverse $\gamma^D$ with $\mathrm{ind}(\gamma) = k_2$, then $\varphi + \eta$ has the Drazin inverse with $\mathrm{ind}(\varphi + \eta) \leq k_1 + k_2$ and

$$(\varphi + \eta)^D = \gamma^D + \sum_{i=0}^{k_1-1} \left(\gamma^D\right)^{i+2} \eta\varphi^i \left(1_X - \varphi\varphi^D\right).$$

**Proof** Let $q = \varphi^2\varphi^D + \eta$, $s = \eta\left(1_X - \varphi\varphi^D\right)$, $\gamma = (\varphi + \eta)\varphi\varphi^D$. Then $q = s + \gamma$ and

$$s\gamma = \eta\left(1_X - \varphi\varphi^D\right)(\varphi + \eta)\varphi\varphi^D$$

$$= \eta\left(1_X - \varphi\varphi^D\right)\varphi\varphi\varphi^D + \eta\left(1_X - \varphi\varphi^D\right)\eta\varphi\varphi^D = 0,$$

$$s^2 = \eta\left(1_X - \varphi\varphi^D\right)\eta\left(1_X - \varphi\varphi^D\right) = 0.$$

Hence $s$ has the Drazin inverse with $\mathrm{ind}(s) = 2$ and $s^D = 0$. From Theorem 4.8.1, $q$ has the Drazin inverse with $\mathrm{ind}(q) \leq k_2 + 1$, and

$$q^D = \gamma^D\left(1_X + \gamma^D s\right) = \gamma^D + \left(\gamma^D\right)^2 s.$$

Let $p = \varphi\left(1_X - \varphi\varphi^D\right)$. Then $\varphi + \eta = p + q$,

$$pq = \varphi\left(1_X - \varphi\varphi^D\right)\left(\varphi^2\varphi^D + \eta\right) = \varphi\left(1_X - \varphi\varphi^D\right)\varphi^2\varphi^D + \varphi\left(1_X - \varphi\varphi^D\right)\eta = 0,$$

$$p^{k_1} = \left(\varphi\left(1_X - \varphi\varphi^D\right)\right)^{k_1} = \varphi^{k_1}\left(1_X - \varphi\varphi^D\right) = 0.$$

Hence, $p$ has the Drazin inverse with $\mathrm{ind}(p) = k_1$ and $p^D = 0$. From Theorem 4.8.1, $\varphi + \eta = p + q$ has the Drazin inverse with $\mathrm{ind}(\varphi + \eta) \leq m$ (where

$m = k_1 + k_2)$ and

$$(\varphi + \eta)^D = (p + q)^D = q^D \sum_{j=0}^{k_1-1} \left(q^D\right)^j p^j \left(1_X - pp^D\right)$$

$$= q^D \sum_{j=0}^{k_1-1} \left(q^D\right)^j p^j = q^D + \sum_{j=1}^{k_1-1} \left(q^D\right)^{j+1} p^j.$$

From $s\gamma = 0$, we have $s\gamma^D = 0$. Then

$$\left(q^D\right)^2 = \left(\gamma^D + \left(\gamma^D\right)^2 s\right)\left(\gamma^D + \left(\gamma^D\right)^2 s\right) = \left(\gamma^D\right)^2 + \left(\gamma^D\right)^3 s.$$

Furthermore, $\left(q^D\right)^i = \left(\gamma^D\right)^i + \left(\gamma^D\right)^{i+1} s$ for arbitrary $i \geq 1$. Note that

$$sp = \eta\left(1_X - \varphi\varphi^D\right)\varphi\left(1_X - \varphi\varphi^D\right) = \eta\varphi\left(1_X - \varphi\varphi^D\right) = \eta p,$$

$$\gamma p = (\varphi + \eta)\varphi\varphi^D\varphi\left(1_X - \varphi\varphi^D\right) = 0, \quad \gamma^D p = 0,$$

$$\left(q^D\right)^{i+1} p = \left(\left(\gamma^D\right)^{i+1} + \left(\gamma^D\right)^{i+2} s\right) p = \left(\gamma^D\right)^{i+1} p + \left(\gamma^D\right)^{i+2} \eta p$$

$$= \left(\gamma^D\right)^{i+2} \eta p,$$

$$\left(q^D\right)^{i+1} p^i = \left(\gamma^D\right)^{i+2} \eta p^i = \left(\gamma^D\right)^{i+2} \eta \left(\varphi\left(1_X - \varphi\varphi^D\right)\right)^i$$

$$= \left(\gamma^D\right)^{i+2} \eta\varphi^i\left(1_X - \varphi\varphi^D\right).$$

Therefore,

$$(\varphi + \eta)^D = q^D + \sum_{i=1}^{k_1-1} \left(q^D\right)^{i+1} p^i = \gamma^D + \left(\gamma^D\right)^2 s$$

$$+ \sum_{i=1}^{k_1-1} \left(\gamma^D\right)^{i+2} \eta\varphi^i\left(1_X - \varphi\varphi^D\right)$$

$$= \gamma^D + \sum_{i=0}^{k_1-1} \left(\gamma^D\right)^{i+2} \eta\varphi^i\left(1_X - \varphi\varphi^D\right).$$

$\square$

**Corollary 4.8.6 ([127])** *Let $C$ be an additive category. Suppose that $\varphi: X \longrightarrow X$ is a morphism of $C$ with Drazin inverse $\varphi^D$ and $\mathrm{ind}(\varphi) = k_1$ and that $\eta : X \longrightarrow X$ is a morphism of $C$ such that $\varphi\varphi^D\eta = \eta$ and $1_X + \varphi^D\eta$ is invertible, then $\varphi + \eta$ has the Drazin inverse with $\mathrm{ind}(\varphi + \eta) \leq k_1 + 1$ and*

$$(\varphi + \eta)^D = \xi + \sum_{i=0}^{k_1-1} \xi^{i+2}\eta\varphi^i \left(1_X - \varphi\varphi^D\right),$$

*where $\xi = (1_X + \varphi^D\eta)^{-1}\varphi^D$.*

**Proof** Let $\gamma = (\varphi + \eta)\varphi\varphi^D$. Next we prove $\gamma^\# = \xi$.

$$\gamma\xi = (\varphi + \eta)\varphi\varphi^D \left(1_X + \varphi^D\eta\right)^{-1}\varphi^D = (\varphi + \eta)\varphi\varphi^D\varphi^D \left(1_X + \eta\varphi^D\right)^{-1}$$

$$= (\varphi + \eta)\varphi^D \left(1_X + \eta\varphi^D\right)^{-1} = \left(\varphi\varphi^D + \eta\varphi^D\right)\left(1_X + \eta\varphi^D\right)^{-1}$$

$$= \left(\varphi\varphi^D + \varphi\varphi^D\eta\varphi^D\right)\left(1_X + \eta\varphi^D\right)^{-1} = \varphi\varphi^D,$$

$$\xi\gamma = \left(1_X + \varphi^D\eta\right)^{-1}\varphi^D(\varphi + \eta)\varphi\varphi^D$$

$$= \left(1_X + \varphi^D\eta\right)^{-1}\varphi^D\varphi + \left(1_X + \varphi^D\eta\right)^{-1}\varphi^D\eta\varphi\varphi^D$$

$$= \left(1_X + \varphi^D\eta\right)^{-1}\left(1_X + \varphi^D\eta\right)\varphi\varphi^D = \varphi\varphi^D.$$

Hence

$$\gamma\xi = \xi\gamma,$$

$$\gamma\xi\gamma = \varphi\varphi^D\gamma = \varphi\varphi^D(\varphi + \eta)\varphi\varphi^D = \varphi^2\varphi^D + \eta\varphi\varphi^D = (\varphi + \eta)\varphi\varphi^D = \gamma,$$

$$\xi\gamma\xi = \left(1_X + \varphi^D\eta\right)^{-1}\varphi^D\varphi\varphi^D = \left(1_X + \varphi^D\eta\right)^{-1}\varphi^D = \xi.$$

Then we have $\gamma^\# = \xi$. From Theorem 4.8.5, $\varphi + \eta$ has the Drazin inverse with $\mathrm{ind}(\varphi + \eta) \leq k_1 + 1$ and

$$(\varphi+\eta)^D = \gamma^D + \sum_{i=0}^{k_1-1} \left(\gamma^D\right)^{i+2}\eta\varphi^i \left(1_X - \varphi\varphi^D\right) = \xi + \sum_{i=0}^{k_1-1} \xi^{i+2}\eta\varphi^i \left(1_X - \varphi\varphi^D\right).$$

$\square$

**Theorem 4.8.7 ([21])** *Let $C$ be an additive category. Suppose that $\varphi : X \longrightarrow X$ is a morphism of $C$ with Drazin inverse $\varphi^D$ and $\mathrm{ind}(\varphi) = k_1$ and that $\eta : X \longrightarrow X$ is a* *morphism of* $C$ *such that* $\varphi^D \eta = 0$. *If* $\gamma = (\varphi + \eta)(1_X - \varphi\varphi^D)$ *has the Drazin inverse* $\gamma^D$ *with* $\text{ind}(\varphi) = k_2$, *then* $\varphi + \eta$ *has the Drazin inverse with* $\text{ind}(\varphi + \eta) \le k_2 + 1$ *and*

$$(\varphi + \eta)^D = \gamma^D + \left(1_X - \gamma^D\eta\right)\varphi^D + \left(1_X - \gamma\gamma^D\right)\sum_{i=0}^{k_2-1}(\varphi + \eta)^i\eta\left(\varphi^D\right)^{i+2}.$$

**Proof** Let $p = \varphi^2\varphi^D$. Then $p$ has $p^\# = \varphi^D$. Let $q = \varphi\left(1_X - \varphi\varphi^D\right) + \eta$. Then we have

$$\varphi + \eta = p + q,$$

$$pq = \varphi^2\varphi^D\varphi\left(1_X - \varphi\varphi^D\right) + \varphi^2\varphi^D\eta = 0.$$

Let $s = \eta\varphi\varphi^D$. Then $q = s + \gamma$ and $s^2 = \eta\varphi\varphi^D\eta\varphi\varphi^D = 0$. Hence $s$ has the Drazin inverse with $\text{ind}(s) = 2$ and $s^D = 0$,

$$s\gamma = \eta\varphi\varphi^D\left((\varphi + \eta)\left(1_X - \varphi\varphi^D\right)\right) = \eta\varphi\varphi^D\varphi\left(1_X - \varphi\varphi^D\right) = 0.$$

Then we have $s\gamma^D = 0$. From Theorem 4.8.1, $q$ has the Drazin inverse with $\text{ind}(q) = k$, where $k \le k_2 + 1$.

$$q^D = (s + \gamma)^D = \gamma^D + \left(\gamma^D\right)^2 s,$$

$$qq^D = (s + \gamma)\left(\gamma^D + \left(\gamma^D\right)^2 s\right) = \gamma\left(\gamma^D + \left(\gamma^D\right)^2 s\right) = \gamma\gamma^D + \gamma^D s.$$

From Theorem 4.8.1, $\varphi + \eta$ has the Drazin inverse with $\text{ind}(\varphi + \eta) \le k$, and

$$(\varphi + \eta)^D = (p + q)^D = \left(1_X - qq^D\right)\sum_{i=0}^{k-1}q^i\left(p^D\right)^i p^D + q^D\left(1_X - pp^D\right)$$

$$= \left(1_X - qq^D\right)\sum_{i=0}^{k-1}q^i\left(p^\#\right)^i p^\# + q^D\left(1_X - pp^\#\right)$$

$$= \left(1_X - qq^D\right)\sum_{i=0}^{k-1}q^i\left(\varphi^D\right)^i \varphi^D + q^D\left(1_X - \varphi\varphi^D\right).$$

Hence $q\varphi^D = \left(\varphi\left(1_X - \varphi\varphi^D\right) + \eta\right)\varphi^D = \eta\varphi^D$, $q^2\varphi^D = \left(\varphi + \eta - \varphi^2\varphi^D\right)\eta\varphi^D = (\varphi + \eta)\eta\varphi^D$. Note that $\varphi^D(\varphi + \eta) = \varphi^D\varphi$, $\varphi^D(\varphi + \eta)^2 = \varphi^D\varphi(\varphi + \eta) = \varphi^2\varphi^D$. Since $\varphi^D(\varphi + \eta)^i = \varphi^i\varphi^D$, we have $\varphi^D(\varphi + \eta)^i\eta = \varphi^i\varphi^D\eta = 0$. By induction, we know $q^i \varphi^D = (\varphi + \eta)^{i-1} \eta \varphi^D$ for arbitrary $i \geq 1$. Hence

$$q^i \left(\varphi^D\right)^{i+1} = (\varphi + \eta)^{i-1} \eta \left(\varphi^D\right)^{i+1},$$

and

$$s(\varphi + \eta)^{i-1} \eta = \eta \varphi \varphi^D (\varphi + \eta)^{i-1} \eta = 0,$$

$$s \left(1_X - \varphi \varphi^D\right) = \eta \varphi \varphi^D \left(1_X - \varphi \varphi^D\right) = 0,$$

$$s \varphi^D = \eta \varphi \varphi^D \varphi^D = \eta \varphi^D,$$

$$\gamma \varphi^D = (\varphi + \eta) \left(1_X - \varphi \varphi^D\right) \varphi^D = 0.$$

Then

$$\gamma^D \varphi^D = 0,$$

$$q^D \left(1_X - \varphi \varphi^D\right) = \left(\gamma^D + \left(\gamma^D\right)^2 s\right) \left(1_X - \varphi \varphi^D\right) = \gamma^D \left(1_X - \varphi \varphi^D\right) = \gamma^D,$$

$$\left(1_X - q q^D\right) \varphi^D = \left(1_X - \gamma \gamma^D - \gamma^D s\right) \varphi^D = \varphi^D - \gamma \gamma^D \varphi^D - \gamma^D s \varphi^D$$

$$= \varphi^D - \gamma^D \eta \varphi^D,$$

and

$$\left(1_X - q q^D\right) q^i \left(\varphi^D\right)^{i+1} = \left(1_X - \gamma \gamma^D - \gamma^D s\right) q^i \left(\varphi^D\right)^{i+1}$$

$$= \left(1_X - \gamma \gamma^D\right) q^i \left(\varphi^D\right)^{i+1} - \gamma^D s q^i \left(\varphi^D\right)^{i+1}$$

$$= \left(1_X - \gamma \gamma^D\right) (\varphi + \eta)^{i-1} \eta \left(\varphi^D\right)^{i+1}$$

$$- \gamma^D s (\varphi + \eta)^{i-1} \eta \left(\varphi^D\right)^{i+1}$$

$$= \left(1_X - \gamma \gamma^D\right) (\varphi + \eta)^{i-1} \eta \left(\varphi^D\right)^{i+1}.$$

Therefore, we have

$$(\varphi + \eta)^D = \left(1_X - q q^D\right) \sum_{i=0}^{k-1} q^i \left(\varphi^D\right)^{i+1} + q^D \left(1_X - \varphi \varphi^D\right)$$

$$= \left(1_X - qq^D\right) \varphi^D + q^D \left(1_X - \varphi\varphi^D\right) + \sum_{i=1}^{k-1} \left(1_X - qq^D\right) q^i \left(\varphi^D\right)^{i+1}$$

$$= \varphi^D - \gamma^D \eta \varphi^D + \gamma^D + \sum_{i=1}^{k-1} \left(1_X - \gamma\gamma^D\right) (\varphi + \eta)^{i-1} \eta \left(\varphi^D\right)^{i+1}$$

$$= \gamma^D + \left(1_X - \gamma^D \eta\right) \varphi^D + \sum_{i=0}^{k-2} \left(1_X - \gamma\gamma^D\right) (\varphi + \eta)^i \eta \left(\varphi^D\right)^{i+2}$$

$$= \gamma^D + \left(1_X - \gamma^D \eta\right) \varphi^D + \left(1_X - \gamma\gamma^D\right) \sum_{i=0}^{k_2-1} (\varphi + \eta)^i \eta \left(\varphi^D\right)^{i+2}.$$

□

Let $B = A + E$ be a complex matrix and $I + A^D E$ invertible. Wei and Wang [120], Wei [119] investigated the perturbation problem of group inverse and gave the necessary and sufficient condition of $B^\# = (I + A^D E)^{-1} A^D$. Castro-González et al. [11, 12] investigated the necessary and sufficient condition of $B^D = (I + A^D E)^{-1} A^D$. Chen et al. [21] generalized these results to morphisms of a category, the following are the results.

**Theorem 4.8.8 ([21])** *Let $C$ be an additive category. Suppose that $\varphi\colon X \longrightarrow X$ is a morphism of $C$ with Drazin inverse $\varphi^D$ and $\mathrm{ind}(\varphi) = k$ and that $\eta : X \longrightarrow X$ is a morphism of $C$ such that $1_X + \varphi^D \eta$ is invertible. Let*

$$\gamma = \alpha \left(1_X - \varphi^D \varphi\right) \eta \varphi^D \beta,$$

$$\delta = \alpha \varphi^D \eta \left(1_X - \varphi\varphi^D\right) \beta,$$

$$\varepsilon = \left(1_X - \varphi\varphi^D\right) \eta \alpha \left(1_X - \varphi^D \varphi\right),$$

*where $\alpha = \left(1_X + \varphi^D \eta\right)^{-1}$, $\beta = \left(1_X + \eta \varphi^D\right)^{-1}$. If $1_X - \gamma$ and $1_X - \delta$ are invertible and satisfying*

$$\eta \left(\varphi^D \varphi - 1_X\right) \varphi = 0, \quad \varphi \left(\varphi\varphi^D - 1_X\right) \eta = 0,$$

*then $f = \varphi + \eta - \varepsilon$ has the Drazin inverse with $\mathrm{ind}(f) \le k$.*

**Proof** Let $f_0^{(2)} = \alpha \varphi^D$. It is easy to have $\alpha \varphi^D = \varphi^D \beta$, $\varphi^D \varepsilon = \varepsilon \varphi^D = 0$. Then

$$f_0^{(2)} f f_0^{(2)} = \alpha \varphi^D (\varphi + \eta - \varepsilon) \alpha \varphi^D = \alpha \varphi^D \varphi \alpha \varphi^D + \alpha \varphi^D \eta \alpha \varphi^D$$

$$= \alpha \varphi^D \varphi \varphi^D \beta + \alpha \varphi^D \eta \alpha \varphi^D = \alpha \varphi^D \beta + \alpha \varphi^D \eta \alpha \varphi^D$$

$$= \alpha\varphi^D\left(\beta + \eta\alpha\varphi^D\right) = \alpha\varphi^D\left(\beta + \beta\eta\varphi^D\right)$$

$$= \alpha\varphi^D\beta\left(1_X + \eta\varphi^D\right) = \alpha\varphi^D = f_0^{(2)},$$

$$ff_0^{(2)}f = f\alpha\varphi^Df = (\varphi + \eta - \varepsilon)\alpha\varphi^D(\varphi + \eta - \varepsilon) = (\varphi + \eta)\alpha\varphi^D(\varphi + \eta)$$

$$= \varphi\alpha\varphi^D(\varphi + \eta) + \eta\alpha\varphi^D(\varphi + \eta)$$

$$= \varphi\left(1_X - \alpha\varphi^D\eta\right)\varphi^D(\varphi + \eta) + \eta\alpha\varphi^D(\varphi + \eta)$$

$$= \varphi\varphi^D(\varphi + \eta) - \varphi\alpha\varphi^D\eta\varphi^D(\varphi + \eta) + \eta\alpha\varphi^D(\varphi + \eta)$$

$$= \varphi\varphi^D\varphi + \eta - \eta + \varphi\varphi^D\eta - \varphi\varphi^D\eta\alpha\varphi^D(\varphi + \eta) + \eta\alpha\varphi^D(\varphi + \eta)$$

$$= \varphi\varphi^D\varphi + \eta - \left(1_X - \varphi\varphi^D\right)\eta + \left(1_X - \varphi\varphi^D\right)\eta\alpha\varphi^D(\varphi + \eta)$$

$$= \varphi\varphi^D\varphi + \eta - \left(1_X - \varphi\varphi^D\right)\eta\left(1_X - \alpha\varphi^D(\varphi + \eta)\right)$$

$$= \varphi\varphi^D\varphi + \eta - \left(1_X - \varphi\varphi^D\right)\eta\left(1_X - \alpha\varphi^D\varphi - \alpha\varphi^D\eta\right)$$

$$= \varphi\varphi^D\varphi + \eta - \left(1_X - \varphi\varphi^D\right)\eta\alpha\left(1_X - \varphi\varphi^D\right)$$

$$= \varphi\varphi^D\varphi + \eta - \varepsilon = \left(\varphi\varphi^D\varphi - \varphi\right) + f.$$

And we have

$$\varepsilon\varphi = \left(1_X - \varphi\varphi^D\right)\eta\alpha\left(1_X - \varphi\varphi^D\right)\varphi = \left(1_X - \varphi\varphi^D\right)\beta\eta\left(1_X - \varphi\varphi^D\right)\varphi = 0.$$

Similarly, $\varphi\varepsilon = 0$.

$$f^2 f_0^{(2)}f = f\left(f + \varphi\varphi^D\varphi - \varphi\right) = f^2 + (\varphi + \eta - \varepsilon)\left(\varphi^D\varphi - 1_X\right)\varphi$$

$$= f^2 + \left(\varphi^D\varphi - 1_X\right)\varphi^2 + \eta\left(\varphi^D\varphi - 1_X\right)\varphi - \varepsilon\varphi\left(\varphi^D\varphi - 1_X\right)$$

$$= f^2 + \left(\varphi^D\varphi - 1_X\right)\varphi^2.$$

By induction we have $f^s f_0^{(2)}f = f^s + \left(\varphi^D\varphi - 1_X\right)\varphi^s$, for arbitrary $s \geq 1$. In particular,

$$f^k f_0^{(2)}f = f^k + \left(\varphi^D\varphi - 1_X\right)\varphi^k = f^k.$$

Similarly, since $\varphi \left( \varphi \varphi^D - 1_X \right) \eta = 0$, we have $f^k = f f_0^{(2)} f^k$ and

$$\varphi^D f = \varphi^D (\varphi + \eta - \varepsilon) = \varphi^D \varphi \left( 1_X + \varphi^D \eta \right) = \varphi^D \varphi \alpha^{-1},$$

$$f_0^{(2)} f = \alpha \varphi^D f = \alpha \varphi^D \varphi \alpha^{-1}.$$

Similarly we have

$$f \varphi^D = \beta^{-1} \varphi \varphi^D, \quad f f_0^{(2)} = \beta^{-1} \varphi \varphi^D \beta,$$

$$1_X - f f_0^{(2)} = \beta^{-1} \left( 1_X - \varphi \varphi^D \right) \beta = \left( 1_X + \eta \varphi^D \right) \left( 1_X - \varphi \varphi^D \right) \beta = \left( 1_X - \varphi \varphi^D \right) \beta,$$

$$1_X - f_0^{(2)} f = \alpha \left( 1_X - \varphi \varphi^D \right) \alpha^{-1} = \alpha \left( 1_X - \varphi \varphi^D \right) \left( 1_X + \varphi^D \eta \right) = \alpha \left( 1_X - \varphi \varphi^D \right),$$

$$\begin{aligned}
1_X - f_0^{(2)} f \left( 1_X - f f_0^{(2)} \right) &= 1_X - \alpha \varphi^D \varphi \alpha^{-1} \left( 1_X - \varphi \varphi^D \right) \beta \\
&= 1_X - \alpha \varphi^D \varphi \left( 1_X + \varphi^D \eta \right) \left( 1_X - \varphi \varphi^D \right) \beta \\
&= 1_X - \alpha \left( \varphi^D \varphi + \varphi^D \eta \right) \left( 1_X - \varphi \varphi^D \right) \beta \\
&= 1_X - \alpha \varphi^D \eta \left( 1_X - \varphi \varphi^D \right) \beta = 1_X - \delta.
\end{aligned}$$

Similarly we have

$$1_X - \left( 1_X - f_0^{(2)} f \right) f f_0^{(2)} = 1_X - \gamma.$$

$$\begin{aligned}
f^{k+1} f_0^{(2)} = f^k f f_0^{(2)} = f^k - f^k \left( 1_X - f f_0^{(2)} \right) &= f^k - f^k f_0^{(2)} f \left( 1_X - f f_0^{(2)} \right) \\
&= f^k \left( 1_X - f_0^{(2)} f \left( 1_X - f f_0^{(2)} \right) \right) = f^k (1_X - \delta).
\end{aligned}$$

$$f_0^{(2)} f^{k+1} = (1_X - \gamma) f^k.$$

Since $1_X - \gamma$ and $1_X - \delta$ are invertible, we have

$$f^k = f^{k+1} f_0^{(2)} (1_X - \delta)^{-1} = f^{k+1} x, \quad x = f_0^{(2)} (1_X - \delta)^{-1},$$

$$f^k = (1_X - \gamma)^{-1} f_0^{(2)} f^{k+1} = y f^{k+1}, \quad y = (1_X - \gamma)^{-1} f_0^{(2)}.$$

Hence, $f$ has the Drazin inverse with $\mathrm{ind}(f) \leq k$, and

$$f^D = y^k f^k x = \left( (1_X - \gamma)^{-1} f_0^{(2)} \right)^k f^k f_0^{(2)} (1_X - \delta)^{-1}$$

$$= (1_X - \gamma)^{-1} \left( f_0^{(2)} (1_X - \gamma)^{-1} \right)^{k-1} f_0^{(2)} f^k f_0^{(2)} (1_X - \delta)^{-1}.$$

□

**Corollary 4.8.9 ([126])** *Let $C$ be an additive category. Suppose that $\varphi : X \longrightarrow X$ is a morphism of $C$ with group inverse $\varphi^{\#}$ and that $\eta : X \longrightarrow X$ is a morphism of $C$ such that $1_X + \varphi^{\#}\eta$ is invertible. Let*

$$\gamma = \alpha \left( 1_X - \varphi^{\#}\varphi \right) \eta \varphi^{\#} \beta,$$

$$\delta = \alpha \varphi^{\#} \eta \left( 1_X - \varphi \varphi^{\#} \right) \beta,$$

$$\varepsilon = \left( 1_X - \varphi \varphi^{\#} \right) \eta \alpha \left( 1_X - \varphi^{\#}\varphi \right),$$

*where $\alpha = \left( 1_X + \varphi^{\#}\eta \right)^{-1}$, $\beta = \left( 1_X + \eta \varphi^{\#} \right)^{-1}$. If $1_X - \gamma$ and $1_X - \delta$ are invertible, then $f = \varphi + \eta - \varepsilon$ has the group inverse and $f^{\#} = (1_X - \gamma)^{-1} \alpha \varphi^{\#} (1_X - \delta)^{-1}$.*

**Corollary 4.8.10 ([21])** *Let $R$ be a unital ring and $J(R)$ its Jacobson radical. If $a \in R$ with $a^D$ exists and $j \in J(R)$. If $j \left( a^D a - 1 \right) a = 0$, $a \left( a^D a - 1 \right) j = 0$, then $a + j - \varepsilon$ has the Drazin inverse, where $\varepsilon = \left( 1 - aa^D \right) j \left( 1 + a^D j \right)^{-1} \left( 1 - a^D a \right)$.*

**Proof** Since $j \in J(R)$, we have that $1 + a^D j$, $1 - \gamma$, $1 - \delta$ are invertible. From Theorem 4.8.8, the conclusion holds.                                                                 □

Theorem 4.8.11 generalizes the conclusion in [12] to morphisms in additive category.

**Theorem 4.8.11** *Let $C$ be an additive category. Suppose that $\varphi : X \longrightarrow X$ is a morphism of $C$ with Drazin inverse $\varphi^D$ and $\mathrm{ind}(\varphi) = k$ and that $\eta : X \longrightarrow X$ is a morphism of $C$ such that $1_X + \varphi^D \eta$ is invertible and*

$$\varphi \varphi^D \eta = \eta \varphi \varphi^D, \quad \varphi \left( 1_X - \varphi \varphi^D \right) \eta = \eta \varphi \left( 1_X - \varphi \varphi^D \right).$$

*Let $f = \varphi + \eta$. Then the following statements are equivalent:*

(1) *$f$ has Drazin inverse, and $f^D = \left( 1_X + \varphi^D \eta \right)^{-1} \varphi^D$.*
(2) *$\varphi \varphi^D \eta^m = \eta^m$, where $m$ is a positive integer.*
(3) *$\eta^m \varphi \varphi^D = \eta^m$, where $m$ is a positive integer.*

*In this case, $ff^D = \varphi \varphi^D$.*

***Proof*** Let $f_0^{(2)} = \left(1_X + \varphi^D \eta\right)^{-1} \varphi^D$, the proof is similar to Theorem 4.8.8,

$$1_X - f_0^{(2)} f = \alpha \left(1_X - \varphi^D \varphi\right), \quad 1_X - f f_0^{(2)} = \left(1_X - \varphi^D \varphi\right) \beta,$$

from $\varphi \varphi^D \eta = \eta \varphi^D \varphi$, we have

$$\left(1_X - \varphi \varphi^D\right) \eta = \eta \left(1_X - \varphi^D \varphi\right). \tag{4.8.14}$$

Then

$$\alpha^{-1} \left(1_X - \varphi^D \varphi\right) = \left(1_X + \varphi^D \eta\right) \left(1_X - \varphi^D \varphi\right) = 1_X - \varphi^D \varphi + \varphi^D \eta \left(1_X - \varphi^D \varphi\right)$$

$$= 1_X - \varphi^D \varphi + \varphi^D \left(1_X - \varphi^D \varphi\right) \eta = 1_X - \varphi^D \varphi.$$

Similarly $\left(1_X - \varphi^D \varphi\right) \beta^{-1} = 1_X - \varphi^D \varphi$. Hence

$$\alpha \left(1_X - \varphi^D \varphi\right) = \left(1_X - \varphi^D \varphi\right) \beta = 1_X - \varphi^D \varphi.$$

Since $1_X - f_0^{(2)} f = 1_X - f f_0^{(2)} = 1_X - \varphi^D \varphi$, we have

$$f_0^{(2)} f = f f_0^{(2)} = \varphi^D \varphi. \tag{4.8.15}$$

Similarly to the proof of Theorem 4.8.8, we have

$$f f_0^{(2)} f = f + \left(\varphi \varphi^D \varphi - \varphi - \varepsilon\right),$$

where

$$\varepsilon = \left(1_X - \varphi \varphi^D\right) \eta \alpha \left(1_X - \varphi \varphi^D\right) = \eta \left(1_X - \varphi \varphi^D\right) \alpha \left(1_X - \varphi \varphi^D\right)$$

$$= \eta \left(1_X - \varphi \varphi^D\right) \left(1_X - \varphi \varphi^D\right) = \eta \left(1_X - \varphi \varphi^D\right).$$

From (4.8.14) and hypothesis, we have $\varphi \eta \left(1_X - \varphi \varphi^D\right) = \eta \varphi \left(1_X - \varphi \varphi^D\right)$. By induction we have

$$\varphi^s \eta \left(1_X - \varphi^D \varphi\right) = \eta \varphi^s \left(1_X - \varphi^D \varphi\right), \text{ for arbitrary } s \geq 0. \tag{4.8.16}$$

Next we prove

$$f^s \left(1_X - \varphi \varphi^D\right) = \sum_{i=0}^{s} C_s^i \varphi^{s-i} \eta^i \left(1_X - \varphi^D \varphi\right).$$   (4.8.17)

In fact, by (4.8.14)

$$\left(1_X - \varphi \varphi^n\right) \eta^i = \eta^j \left(1_X - \varphi \varphi^D\right), \text{ for arbitrary } j \geq 1,$$   (4.8.18)

$$\varphi^{s-j} \eta^{j+1} \left(1_X - \varphi^D \varphi\right) = \varphi^{s-j} \eta \eta^j \left(1_X - \varphi^D \varphi\right)$$

$$= \varphi^{s-j} \eta \left(1_X - \varphi \varphi^D\right) \eta^j \qquad \text{by (4.8.18)}$$

$$= \eta \varphi^{s-j} \left(1_X - \varphi \varphi^D\right) \eta^j \qquad \text{by (4.8.16)}$$

$$= \eta \varphi^{s-j} \eta^j \left(1_X - \varphi \varphi^D\right). \qquad \text{by (4.8.18)} \qquad (4.8.19)$$

It is easy to check when $s = 1$, (4.8.17) is true. Suppose that the conclusion holds at $s$. Then

$$f^{s+1} \left(1_X - \varphi^D \varphi\right) = f f^s \left(1_X - \varphi^D \varphi\right)$$

$$= (\varphi + \eta) \sum_{i=0}^{s} C_s^i \varphi^{s-i} \eta^i \left(1_X - \varphi^D \varphi\right)$$

$$= \sum_{i=0}^{s} C_s^i \varphi^{s-i+1} \eta^i \left(1_X - \varphi^D \varphi\right) + \sum_{i=0}^{s} C_s^i \eta \varphi^{s-i} \eta^i \left(1_X - \varphi^D \varphi\right)$$

$$= \varphi^{s+1} \left(1_X - \varphi^D \varphi\right) + \sum_{i=1}^{s} C_s^i \varphi^{s-i+1} \eta^i \left(1_X - \varphi^D \varphi\right)$$

$$+ \sum_{i=0}^{s-1} C_s^i \eta \varphi^{s-i} \eta^i \left(1_X - \varphi^D \varphi\right) + \eta^{s+1} \left(1_X - \varphi^D \varphi\right)$$

$$= \varphi^{s+1} \left(1_X - \varphi^D \varphi\right) + \sum_{i=0}^{s-1} C_s^{i+1} \varphi^{s-i} \eta^{i+1} \left(1_X - \varphi^D \varphi\right)$$

$$+ \sum_{i=0}^{s-1} C_s^i \varphi^{s-i} \eta^{i+1} \left(1_X - \varphi^D \varphi\right) + \eta^{s+1} \left(1_X - \varphi^D \varphi\right) \text{ by (4.8.19)}$$

$$= \varphi^{s+1}\left(1_X - \varphi^D\varphi\right) + \sum_{i=0}^{s-1} C_{s+1}^{i+1}\varphi^{s-i}\eta^{i+1}\left(1_X - \varphi^D\varphi\right) + \eta^{s+1}\left(1_X - \varphi^D\varphi\right)$$

$$= \sum_{i=0}^{s+1} C_{s+1}^{i}\varphi^{(s+1)-i}\eta^{i}\left(1_X - \varphi^D\varphi\right).$$

Hence for arbitrary $s$, (4.8.17) holds. And we have

$$f f_0^{(2)} f = f + \varphi\varphi^D\varphi - \varphi - \varepsilon = f - \varphi\left(1_X - \varphi^D\varphi\right) - \eta\left(1_X - \varphi^D\varphi\right)$$

$$= f - (\varphi + \eta)\left(1_X - \varphi^D\varphi\right) = f - f\left(1_X - \varphi^D\varphi\right).$$

Therefore

$$f^s f_0^{(2)} f = f^s - f^s\left(1_X - \varphi^D\varphi\right). \qquad (4.8.20)$$

$(2) \Leftrightarrow (3)$.   It is easy to obtain by (4.8.18).

$(2) \Rightarrow (1)$.   If there exists $m \geq 1$ such that $\varphi\varphi^D\eta^m = \eta^m$, then

$$\left(1_X - \varphi^D\varphi\right)\eta^m = 0. \qquad (4.8.21)$$

Since $\varphi^D$ is the Drazin inverse of $\varphi$ and $\text{ind}(\varphi) = k$. Let $s = m + k - 1$. Then we have

$$f^s\left(1_X - \varphi^D\varphi\right)$$

$$= \varphi^s\left(1_X - \varphi^D\varphi\right) + C_s^1\varphi^{s-1}\left(1_X - \varphi^D\varphi\right)\eta$$

$$+ \cdots + C_s^{s-k}\varphi^k\left(1_X - \varphi^D\varphi\right)\eta^{s-k}$$

$$+ C_s^{s-k+1}\varphi^{k-1}\left(1_X - \varphi^D\varphi\right)\eta^{s-k+1} + \cdots + \left(1_X - \varphi^D\varphi\right)\eta^s$$

$$= 0 \quad \text{by (4.8.21)}.$$

Hence by (4.8.20), we have

$$f^s = f^s f_0^{(2)} f, \quad f_0^{(2)} f = f f_0^{(2)}, \quad f_0^{(2)} f f_0^{(2)} = f_0^{(2)}.$$

Therefore $f$ has the Drazin inverse with $\text{ind}(f) \leq s$ and $f^D = f_0^{(2)} = \left(1_X + \varphi^D\eta\right)^{-1}\varphi^D$. By (4.8.15), we have $f f^D = \varphi\varphi^D$.

$(1) \Rightarrow (2)$.  Suppose $f^D = \left(1_X + \varphi^D \eta\right)^{-1} \varphi^D$ is the Drazin inverse of $f$ and $\mathrm{ind}(f) = s$, $\varphi^D$ is the Drazin inverse of $\varphi$ and $\mathrm{ind}(\varphi) = k$. Let

$$u = f - f^2 f^D, \quad v = \varphi - \varphi^2 \varphi^D.$$

Then $u^s = 0$, $v^k = 0$, and by (4.8.15) we have $ff^D = \varphi\varphi^D$,

$$u = f \left(1_X \quad ff^D\right) = f \left(1_X - \varphi\varphi^D\right),$$

$$v = \varphi \left(1_X - \varphi\varphi^D\right).$$

Then

$$uv = f \left(1_X - \varphi\varphi^D\right) \varphi \left(1_X - \varphi\varphi^D\right) = (\varphi + \eta)\varphi \left(1_X - \varphi\varphi^D\right)$$

$$= \varphi^2 \left(1_X - \varphi\varphi^D\right) + \eta\varphi \left(1_X - \varphi\varphi^D\right)$$

$$= \varphi^2 \left(1_X - \varphi\varphi^D\right) + \varphi \left(1_X - \varphi\varphi^D\right) \eta,$$

$$vu = \varphi \left(1_X - \varphi\varphi^D\right) f \left(1_X - \varphi\varphi^D\right)$$

$$= \varphi \left(1_X - \varphi\varphi^D\right) (\varphi + \eta) \left(1_X - \varphi\varphi^D\right)$$

$$= \varphi^2 \left(1_X - \varphi\varphi^D\right) + \varphi \left(1_X - \varphi\varphi^D\right) \eta \left(1_X - \varphi\varphi^D\right)$$

$$= \varphi^2 \left(1_X - \varphi\varphi^D\right) + \varphi \left(1_X - \varphi\varphi^D\right) \left(1_X - \varphi\varphi^D\right) \eta \quad \text{by (4.8.14)}$$

$$= \varphi^2 \left(1_X - \varphi\varphi^D\right) + \varphi \left(1_X - \varphi\varphi^D\right) \eta,$$

and we have $uv = vu$. Let $m = s + k - 1$. Then $(u - v)^m = 0$. But

$$u - v = \eta \left(1_X - \varphi\varphi^D\right).$$

Note that (4.8.14), we have $\eta \left(1_X - \varphi\varphi^D\right) = \left(1_X - \varphi\varphi^D\right) \eta$. Hence

$$0 = (u - v)^m = \left(\eta \left(1_X - \varphi\varphi^D\right)\right)^m = \eta^m \left(1_X - \varphi\varphi^D\right)^m = \eta^m \left(1_X - \varphi\varphi^D\right).$$

That is, there exists $m \geq 1$ such that $\eta^m \left(1_X - \varphi\varphi^D\right) = 0$ and $\eta^m = \eta^m \varphi\varphi^D$.

$\square$

Theorem 4.8.12 generalizes the conclusion in [119] to morphisms in addition category.

**Theorem 4.8.12** *Let $C$ be an additive category. Suppose that $\varphi : X \longrightarrow X$ is a morphism of $C$ with Drazin inverse $\varphi^D$ and $\mathrm{ind}(\varphi) = k$ and that $\eta : X \longrightarrow X$ is a morphism of $C$. Then the following statements are equivalent:*

(1) $\varphi\varphi^D\eta = \eta\varphi\varphi^D = \varphi - \varphi^2\varphi^D + \eta$.

(2) $f = \varphi + \eta$ has the group inverse and $f^\# = \left(1_X + \varphi^D\eta\right)^{-1}\varphi^D$.

*Proof*

$(1) \Rightarrow (2)$.   Let $f_0^{(2)} = \left(1_X + \varphi^D\eta\right)^{-1}\varphi^D$. By the condition (1), we have

$$f = \varphi + \eta = \varphi + \left(\varphi\varphi^D\eta + \varphi^2\varphi^D - \varphi\right) = \varphi\left(1_X + \varphi^D\eta\right) + \varphi\left(\varphi\varphi^D - 1_X\right),$$

$$ff_0^{(2)} = \varphi\left(1_X + \varphi^D\eta\right)f_0^{(2)} + \varphi\left(\varphi\varphi^D - 1_X\right)f_0^{(2)}$$

$$= \varphi\left(1_X + \varphi^D\eta\right)\left(1_X + \varphi^D\eta\right)^{-1}\varphi^D$$

$$+ \varphi\left(\varphi\varphi^D - 1_X\right)\varphi^D\left(1_X + \eta\varphi^D\right)^{-1}$$

$$= \varphi\varphi^D.$$

Similarly, since $\eta = \eta\varphi^D\varphi + \varphi^2\varphi^D - \varphi$, we have $f_0^{(2)}f = \varphi^D\varphi$. Hence

$$ff_0^{(2)} = f_0^{(2)}f,$$

$$f - f^2f_0^{(2)} = f\left(1_X - ff_0^{(2)}\right) = f\left(1_X - \varphi\varphi^D\right)$$

$$= \varphi\left(1_X - \varphi\varphi^D\right) + \eta\left(1_X - \varphi\varphi^D\right)$$

$$= \varphi\left(1_X - \varphi\varphi^D\right) + \varphi\left(\varphi^D\varphi - 1_X\right) \quad \text{by the condition (1)}$$

$$= 0.$$

Hence, $f = ff_0^{(2)}f$, $f_0^{(2)}ff_0^{(2)} = f_0^{(2)}$. So $f_0^{(2)}$ is the group inverse of $f$,

$$f^\# = f_0^{(2)} = \left(1_X + \varphi^D\eta\right)^{-1}\varphi^D,$$

and $f^\# f = \varphi^D\varphi$.

$(2) \Rightarrow (1)$.   It is similar to the proof of Theorem 4.8.8, we have

$$1_X - ff_0^{(2)} = \left(1_X - \varphi^D\varphi\right)\beta, \quad 1_X - f_0^{(2)}f = \alpha\left(1_X - \varphi^D\varphi\right).$$

Since $f_0^{(2)} = f^\#$, we have $\left(1_X - \varphi^D\varphi\right)\beta = \alpha\left(1_X - \varphi^D\varphi\right)$. Hence

$$\alpha^{-1}\left(1_X - \varphi^D\varphi\right) = \left(1_X - \varphi^D\varphi\right)\beta^{-1}. \tag{4.8.22}$$

Equation (4.8.22) postmultiplication $\varphi^D$, we have

$$0 = \left(1_X - \varphi^D\varphi\right)\beta^{-1}\varphi^D = \left(1_Y - \varphi^D\varphi\right)\left(1_X + \eta\varphi^D\right)\varphi^D$$
$$= \left(1_X - \varphi^D\varphi\right)\eta\left(\varphi^D\right)^2,$$

that is

$$\left(1_X - \varphi^D\varphi\right)\eta\varphi^D = 0 \tag{4.8.23}$$

Similarly (4.8.22) premultiplication $\varphi^D$, we have

$$\varphi^D\eta\left(1_X - \varphi^D\varphi\right) = 0.$$

Then we obtain $\eta\varphi^D = \varphi^D\varphi\eta\varphi^D$ and $\varphi^D\eta = \varphi^D\eta\varphi\varphi^D$. Therefore

$$\varphi\varphi^D\eta = \varphi\varphi^D\eta\varphi\varphi^D = \varphi^D\varphi\eta\varphi^D\varphi = \eta\varphi^D\varphi = \eta\varphi\varphi^D. \tag{4.8.24}$$

Next we calculate $ff_0^{(2)}f$. By the proof of Theorem 4.8.8, it is known

$$ff_0^{(2)}f = \varphi\varphi^D\varphi + \eta - \left(1_X - \varphi\varphi^D\right)\eta + \left(1_X - \varphi\varphi^D\right)\eta\alpha\varphi^D(\varphi + \eta)$$
$$= \varphi\varphi^D\varphi + \eta - \left(1_X - \varphi\varphi^D\right)\eta + \left(1_X - \varphi\varphi^D\right)\eta\varphi^D\beta(\varphi + \eta)$$
$$= \varphi\varphi^D\varphi + \eta - \left(1_X - \varphi\varphi^D\right)\eta, \quad \text{by (4.8.23)}$$

but $f_0^{(2)} = f^\#$, $ff_0^{(2)}f = f$, $\varphi\varphi^D\varphi + \eta - \left(1_X - \varphi\varphi^D\right)\eta = \varphi + \eta$, that is $\varphi - \varphi^2\varphi^D + \eta = \varphi\varphi^D\eta$. By (4.8.24), we have $\varphi\varphi^D\eta = \eta\varphi\varphi^D = \varphi - \varphi^2\varphi^D + \eta$.

$\square$

# Chapter 5
# Core Inverses

In the previous chapters, we introduced Moore-Penrose inverses, group inverses and Drazin inverses, which are the most well-known generalized inverses. Although these generalized inverses coincide in some special cases, they behave rather differently in general. In this chapter, we shall introduce a "mixed" generalized inverse, which is obtained by combining the Moore-Penrose inverse and the group inverse and therefore has many properties of both these two generalized inverses.

Actually, it was first noticed by Randall E. Cline [26] in 1968 that every square complex matrix $A$ with index 1 has a unique $\{1, 2, 3\}$-inverse $(A)_R$ such that $(A)_R A = A^{\#} A$, and dually a unique $\{1, 2, 4\}$-inverse $(A)_L$ such that $A(A)_L = AA^{\#}$. Such two generalized inverses were also discussed by C. Radhakrishna Rao and Sujit Kumar Mitra [109] for their typical properties of range spaces and null spaces (where two other symbols $A_{\rho^* \chi}^-$ and $A_{\rho \chi^*}^-$ were used to denote $(A)_R$ and $(A)_L$ respectively).

Since Cline's generalized inverses $(A)_R$ and $(A)_L$ are of index at most 1 in which case they are also called core matrices (or group matrices), and they share many properties with the group inverse $A^{\#}$, following Oskar Maria Baksalary and Götz Trenkler [2] the generalized inverse $(A)_R$ (resp., $(A)_L$) is now commonly called the core inverse (resp., dual core inverse) of $A$ and denoted by $A^{\circledR}$ (resp., $A_{\circledR}$). In 2014, Dragan S. Rakić, Nebojša Č. Dinčić and Dragan S. Djordjević [108] extended the notion of core inverses from complex matrices to elements in an arbitrary ∗-ring. Since then many achievements have been made on the core inverse.

In this chapter we introduce the main results for the core inverse, most of which were developed in the recent decade.

Throughout this chapter, unless otherwise stated, $R$ is a ∗-ring with identity.

© The Author(s), under exclusive license to Springer Nature Singapore Pte Ltd. 2024
J. Chen, X. Zhang, *Algebraic Theory of Generalized Inverses*,
https://doi.org/10.1007/978-981-99-8285-1_5

## 5.1  Core Inverses of Complex Matrices

In this section, we consider basic properties and representations for core inverses of complex matrices.

**Definition 5.1.1 ([2])**  Let $A, X \in \mathbb{C}^{n \times n}$. If $X$ satisfies

$$AX = P_A \quad \text{and} \quad \mathcal{R}(X) \subseteq \mathcal{R}(A),$$

where $P_A$ is the orthogonal projector onto $\mathcal{R}(A)$, then $X$ is called a core inverse of $A$ and denoted by $A^{\circledcirc}$. Dually, if $X$ satisfies

$$XA = P_{A^*} \quad \text{and} \quad \mathcal{R}(X) \subseteq \mathcal{R}(A^*),$$

then $X$ is called a dual core inverse of $A$ and denoted by $A_{\circledcirc}$.

Making use of the Hartwig-Spindelböck decomposition (as in Theorem 1.1.2), the next result gives the necessary and sufficient condition for the existence of $A^{\circledcirc}$ and shows its uniqueness.

**Theorem 5.1.2 ([2])**  *Let  $A \in \mathbb{C}^{n \times n}$  with  rank$(A) = r$  and let  $A = U \begin{bmatrix} \Sigma K & \Sigma L \\ 0 & 0 \end{bmatrix} U^*$  be the Hartwig-Spindelböck decomposition of  A. Then the necessary and sufficient condition for the existence of  $A^{\circledcirc}$  is that  K  is nonsingular. Furthermore, when A is core invertible,*

$$A^{\circledcirc} = U \begin{bmatrix} (\Sigma K)^{-1} & 0 \\ 0 & 0 \end{bmatrix} U^*.$$

***Proof***  First, observe that

$$P_A = U \begin{bmatrix} I_r & 0 \\ 0 & 0 \end{bmatrix} U^*.$$

Suppose now that $B \in \mathbb{C}^{n \times n}$ is partitioned as

$$B = U \begin{bmatrix} W & X \\ Y & Z \end{bmatrix} U^*$$

where $W \in \mathbb{C}^{r \times r}$ and $Z \in \mathbb{C}^{(n-r) \times (n-r)}$. Direct calculations show that $AB = P_A$ if and only if $\Sigma K W + \Sigma L Y = I_r$, $KX + LZ = 0$. Since $\mathcal{R}(A^{\circledcirc}) \subseteq \mathcal{R}(A)$ can be equivalently expressed as $P_A A^{\circledcirc} = A^{\circledcirc}$, it follows that $\mathcal{R}(B) \subseteq \mathcal{R}(A)$ if and only if $Y = 0$ and $Z = 0$. Hence, we obtain $\Sigma K W = I_r$, $KX = 0$. Therefore, $A$ has a core inverse $B$ if and only if $K$ is nonsingular, in which case

$$A^{\circledast} = B = U \begin{bmatrix} (\Sigma K)^{-1} & 0 \\ 0 & 0 \end{bmatrix} U^*.$$

$\square$

Since $K$ being nonsingular is also equivalent to the requirement that $A$ is of index one, we obtain the following.

**Theorem 5.1.3 ([2])** Let $A \in \mathbb{C}^{n \times n}$. Then $A^{\circledast}$ exists if and only if $\mathrm{ind}(A) = 1$. In this case $A^{\circledast} = A^{\#}AA^{\dagger}$.

**Proof** Let $X = A^{\#}AA^{\dagger}$. Then it is straightforward to check that $AX = AA^{\dagger} = P_A$ and $\mathcal{R}(X) = \mathcal{R}(A^{\#}) = \mathcal{R}(A)$, which implies that $X = A^{\#}AA^{\dagger}$ is exactly the core inverse of $A$. $\square$

Recall from Theorems 2.1.2 and 3.1.3 that if $A \in \mathbb{C}^{n \times n}$ is of index 1 and has a full-rank decomposition $A = GH$ then

$$A^{\#} = G(HG)^{-2}H,$$

$$A^{\dagger} = H^* \left(HH^*\right)^{-1} \left(G^*G\right)^{-1} G^*.$$

Now, for the core inverse, we have

**Theorem 5.1.4 ([115, Theorem 2.4])** Let $A \in \mathbb{C}^{n \times n}$ with $\mathrm{rank}(A) = r$. If $A = GH$ is a full-rank decomposition, then $A^{\circledast}$ exists if and only if $HG$ is nonsingular. In this case,

$$A^{\circledast} = G(HG)^{-1} \left(G^*G\right)^{-1} G^*.$$

**Proof** It follows by Theorems 5.1.3 and 1.1.7 directly. $\square$

## 5.2 Core Inverses of Elements in Rings with Involution

In this section, we consider the core and dual core inverses in a ring with involution.

**Definition 5.2.1 ([108])** An element $a \in R$ is said to be core invertible if there exists $a^{\circledast} \in R$ such that

$$aa^{\circledast}a = a, \; a^{\circledast}R = aR, \; Ra^{\circledast} = Ra^*,$$

in which case such $a^{\circledR}$ is called a core inverse of $a$. Dually, $a$ is said to be dual core invertible if there exists $a_{\circledR} \in R$ such that

$$aa_{\circledR}a = a, \quad a_{\circledR}R = a^*R, \quad Ra_{\circledR} = Ra,$$

in which case such $a_{\circledR}$ is called a dual core inverse of $a$.

In what follows, the symbol $R^{\circledR}$ (resp., $R_{\circledR}$) will denote set of all core invertible elements (resp., dual core invertible elements) of a ring $R$.

### 5.2.1   Equivalent Definitions and Characterizations of Core Inverses

**Theorem 5.2.2 ([108])** *The following statements are equivalent for any $a \in R$:*

(1) *$a$ is core invertible.*
(2) *There exists $x \in R$ such that $axa = a$, $^{\circ}x = {}^{\circ}a$ and $x^{\circ} = (a^*)^{\circ}$.*
(3) *There exists $x \in R$ such that*

$$(i) \quad axa = a, \quad (ii) \quad xax = x, \quad (iii) \quad (ax)^* = ax,$$

$$(iv) \quad xa^2 = a, \quad (v) \quad ax^2 = x.$$

(4) *There exist a projection $p \in R$ and an idempotent $q \in R$ such that $pR = aR$, $qR = aR$ and $Rq = Ra$.*
(5) *$a$ is regular and there exist a projection $p \in R$ and an idempotent $q \in R$ such that $^{\circ}a = {}^{\circ}p$, $^{\circ}a = {}^{\circ}q$ and $a^{\circ} = q^{\circ}$.*

*If the previous assertions are valid then $x = a^{\circledR}$, $a^{\circledR}$ is unique and the assertions (4) and (5) deal with the same pair of unique idempotents $p$ and $q$. Moreover, $qa^{(1)}p$ is invariant under the choice of $a^{(1)} \in a\{1\}$ and*

$$a = \begin{bmatrix} a & 0 \\ 0 & 0 \end{bmatrix}_{p \times q}, \quad a^{\circledR} = \begin{bmatrix} qa^{(1)}p & 0 \\ 0 & 0 \end{bmatrix}_{q \times p}. \tag{5.2.1}$$

**Proof** $(1) \Rightarrow (2)$. Suppose that $a$ is core invertible and let $x = a^{\circledR}$. By definition, $axa = a$, $xR = aR$ and $Rx = Ra^*$. By Lemmas 1.2.21, it follows that $^{\circ}x = {}^{\circ}a$ and $x^{\circ} = (a^*)^{\circ}$.

$(2) \Rightarrow (3)$. Suppose that there exists $x \in R$ such that $axa = a$, $^{\circ}x = {}^{\circ}a$ and $x^{\circ} = (a^*)^{\circ}$. We can obtain that

$$x = ax^2, \quad ax = (ax)^* \quad \text{and} \quad xax = x.$$

From $xa - 1 \in {}^{\circ}x \subseteq {}^{\circ}a$, we have $a = xa^2$.

$(3)\Rightarrow(4)$. Set $p = ax$ and $q = xa$. From $axa = a$ it follows that $p$ and $q$ are idempotents such that $pR = aR$ and $Rq = Ra$. Equation (iii) shows that $p$ is Hermitian. From $a = xa^2 = qa$ and $q = xa = ax^2a$ we conclude that $qR = aR$.

$(4)\Rightarrow(5)$. The proof is similar to the proof of Theorem 3.2.6 (2) $\Rightarrow$ (3).

$(5)\Rightarrow(1)$. Suppose that $a \in R^{(1)}$ and suppose that there exist a projection $p \in R$ and idempotent $q \in R$ such that $^\circ a = {}^\circ p, {}^\circ a = {}^\circ q$ and $a^\circ = q^\circ$. Fix $a^{(1)} \in a\{1\}$. In the proof of Theorem 3.2.6 we showed that $a = qa = aq$ and $q = qa^{(1)}a = aa^{(1)}q$. In the proof of Theorem 2.2.13 we showed that $a = pa$ and $p = aa^{(1)}p$. Let $a^- \in a\{1\}$ be arbitrary. Then $qa^-p = qa^{(1)}aa^-aa^{(1)}p = qa^{(1)}aa^{(1)}p = qa^{(1)}p$, so $qa^-p$ is invariant under the choice of $a^- \in a\{1\}$. Set $x = qa^{(1)}p$. We have $axa = aqa^{(1)}pa = aa^{(1)}a = a$. Also, $x = qa^{(1)}p = aa^{(1)}qa^{(1)}p$ and $xa^2 = qa^{(1)}pa^2 = qa^{(1)}aa = qa = a$, so $xR = aR$. Moreover,

$$x = qa^{(1)}p^* = qa^{(1)}\left(aa^{(1)}p\right)^* = qa^{(1)}p\left(a^{(1)}\right)^* a^* \quad \text{and}$$

$$a^*ax = a^*aqa^{(1)}p = a^*aa^{(1)}p = a^*p = (pa)^* = a^*,$$

so $Rx = Ra^*$. It follows that $x = a^{\circledast}$, i.e., $a$ is core invertible.

The uniqueness of $p$ and $q$ follows by Lemma 1.2.16. If $x$ is core inverse of $a$ then we showed that $x$ has properties given in (2) and (3). Suppose that there exist two elements $x$ and $y$ satisfying equations in (3). By the proof of (3) $\Rightarrow$ (4) and the uniqueness of $p$ and $q$ we conclude that $p = ax = ay$ and $q = xa = ya$. Therefore, $x = xax = yay = y$. We also proved that if there exists some $x$ satisfying equations in (3) then $a$ is core invertible but its core inverse must satisfies equations in (3) which uniquely determine $x$. It follows that $x$ appearing in (2) and $x$ appearing in (3) are both equal to $a^{\circledast}$ and that core inverse of $a$ is unique. Representations (5.2.1) follow by $a = paq$ and $a^{\circledast} = x = qa^{(1)}p$.

$\square$

**Theorem 5.2.3 ([108])** *Let $a \in R$. The following assertions are equivalent:*

(1) *$a$ is dual core invertible.*
(2) *There exists $x \in R$ such that $axa = a$, $^\circ x = {}^\circ(a^*)$ and $x^\circ = a^\circ$.*
(3) *There exists $x \in R$ such that (i) $axa = a$, (ii) $xax = x$, (iii) $(xa)^* = xa$, (iv) $a^2x = a$, (v) $x^2a = x$.*
(4) *There exist a projection $r \in R$ and an idempotent $q \in R$ such that $Rr = Ra$, $qR = aR$ and $Rq = Ra$.*
(5) *$a$ is regular and there exist a projection $r \in R$ and an idempotent $q \in R$ such that $a^\circ = r^\circ, {}^\circ a = {}^\circ q$ and $a^\circ = q^\circ$.*

*If the previous assertions are valid then $x = a_{\circledast}$, $a_{\circledast}$ is unique and the assertions (4) and (5) deal with the same pair of unique idempotents $r$ and $q$. Moreover, $ra^{(1)}q$ is invariant under the choice of $a^{(1)} \in a\{1\}$ and*

$$a = \begin{bmatrix} a & 0 \\ 0 & 0 \end{bmatrix}_{q\times r}, \quad a_{\circledast} = \begin{bmatrix} ra^{(1)}q & 0 \\ 0 & 0 \end{bmatrix}_{r\times q}.$$

### 5.2.2   Relationship Between Group Invertibility and Core Invertibility

**Lemma 5.2.4 ([108])** *Let $a \in R$. Then:*

(1) *If $a$ is core invertible, then it is group invertible and $a^\# = (a^\circledR)^2 a$.*
(2) *$a$ is group invertible if and only if there exists an idempotent $q \in R$ such that $qR = uR$ and $Rq = Ra$.*

From Theorem 5.2.2, it can be seen that $a \in R^\circledR$ if and only if there exist a projection $p$ and an idempotent $q$ such that $aR = pR$ and $qR = aR$ and $Rq = Ra$. Thus by Lemmas 5.2.4 and 2.2.5, we have the following theorem. For the convenience of the reader, we give another method to prove this result.

**Theorem 5.2.5 ([124])** *Let $a \in R$. Then $a \in R^\circledR$ if and only if $a \in R^\# \cap R^{\{1,3\}}$. In this case, $a^\circledR = a^\# a a^{(1,3)}$.*

**Proof** Suppose $a \in R^\circledR$, then $a^\# = (a^\circledR)^2 a$ by Lemma 5.2.4 and $a \in R^{\{1,3\}}$. Conversely, suppose $a \in R^\# \cap R^{\{1,3\}}$, then $aa^{(1,3)}a = a$ and $(aa^{(1,3)})^* = aa^{(1,3)}$, we have

$$a = aa^{(1,3)}a = (aa^{(1,3)})^* a = (a^{(1,3)})^* a^* a.$$

Let $y = a^\# a a^{(1,3)}$, then $aya = aa^\# a a^{(1,3)}a = aa^\# a = a$. Since $y = a^\# a a^{(1,3)} = a(a^\#)^2 a a^{(1,3)}$ and $a = a^\# a^2 = a^\# (a^{(1,3)})^* a^* a^2 = ya^2$, we get $yR = aR$. We also have

$$a^* = a^* a a^{(1,3)} = a^* a a^\# a a^{(1,3)} = a^* ay.$$

So $Ry = Ra^*$. Thus $a \in R^\circledR$ by the definition of the core inverse.    $\square$

**Corollary 5.2.6** *Let $a \in R$. Then the following conditions are equivalent:*

(1) *$a \in R^\circledR$.*
(2) *$a \in R^\#$ and there exists $x \in R$ such that $(ax)^* = ax$ and $xa^2 = a$.*
(3) *$a \in R^\#$ and there exists $x \in R$ such that $(ax)^* = ax$ and $xa = a^\# a$.*

**Proof** (1)$\Rightarrow$(2). It is clear by Theorem 5.2.2 and Lemma 5.2.4.

(2)$\Rightarrow$(3). It is sufficient to prove that $xa^2 = a$ implies $xa = a^\# a$. It is easy to see that by $xa = xa^2 a^\# = aa^\# = a^\# a$.

(3)$\Rightarrow$(1). Suppose $a \in R^\#$ and there exists $x \in R$ such that $(ax)^* = ax$ and $xa = a^\# a$. Then $axa = aa^\# a = a$, that is $a \in R^{\{1,3\}}$. Thus by the hypothesis $a \in R^\#$ and Theorem 5.2.5, we have $a \in R^\circledR$.    $\square$

**Proposition 5.2.7** *The following conditions are equivalent for any $a \in R$:*

(1) *$a$ is core invertible.*
(2) *$R = aR \oplus (a^*)^\circ = aR \oplus a^\circ$.*
(3) *$R = aR + (a^*)^\circ = aR \oplus a^\circ$.*

(4) $R = Ra^* \oplus {}^{\circ}a = aR \oplus a^{\circ}$.
(5) $R = Ra^* + {}^{\circ}a = aR \oplus a^{\circ}$.
(6) $R = aR \oplus (a^*)^{\circ} = Ra \oplus {}^{\circ}a$.
(7) $R = aR + (a^*)^{\circ} = Ra \oplus {}^{\circ}a$.
(8) $R = Ra^* \oplus {}^{\circ}a = Ra \oplus {}^{\circ}a$.
(9) $R = Ra^* + {}^{\circ}a = Ra \oplus {}^{\circ}a$.

*In this case,*

$$a^{\oplus} = ay_1^2 ax_1 = ay_1^2 ax_2^* = y_2 ax_1 = y_2 ax_2^*,$$

*where* $1 = ax_1 + u_1 = x_2 a^* + u_2 = ay_1 + v_1 = y_2 a + v_2$ *for some* $x_1, x_2, y_1, y_2 \in R$, $u_1 \in (a^*)^{\circ}$, $v_1 \in a^{\circ}$ *and* $u_2, v_2 \in {}^{\circ}a$.

**Proof** By Theorem 5.2.5, we have $a \in R^{\oplus}$ if and only if $a \in R^{\#} \cap R^{\{1,3\}}$ and

$$a^{\oplus} = a^{\#} a a^{(1,3)}. \tag{5.2.2}$$

Thus it is easy to see (1)–(9) are equivalent by Theorem 2.2.11 and Proposition 3.2.3. Suppose $1 = ax_1 + u_1 = x_2 a^* + u_2 = ay_1 + v_1 = y_2 a + v_2$, for some $x_1, x_2, y_1, y_2 \in R$, $u_1 \in (a^*)^{\circ}$, $v_1 \in a^{\circ}$ and $u_2, v_2 \in {}^{\circ}a$. Then we have

$$a^{(1,3)} = x_1 = x_2^* \quad and \quad a^{\#} = ay_1^2 = y_2^2 a. \tag{5.2.3}$$

$$a = y_2 a^2 + v_2 a = y_2 a^2. \tag{5.2.4}$$

Thus by (5.2.2) and (5.2.3), we have

$$a^{\oplus} = ay_1^2 ax_1 = ay_1^2 ax_2^* = y_2^2 a^2 x_1 = y_2^2 a^2 x_2^*.$$

Hence $a^{\oplus} = ay_1^2 ax_1 = ay_1^2 ax_2^* = y_2 ax_1 = y_2 ax_2^*$ by (5.2.4). $\qquad\square$

**Proposition 5.2.8** *The following conditions are equivalent for any $a \in R$:*

(1) *a is dual core invertible.*
(2) $R = a^* R \oplus a^{\circ} = aR \oplus a^{\circ}$.
(3) $R = a^* R + a^{\circ} = aR \oplus a^{\circ}$.
(4) $R = Ra \oplus {}^{\circ}(a^*) = aR \oplus a^{\circ}$.
(5) $R = Ra + {}^{\circ}(a^*) = aR \oplus a^{\circ}$.
(6) $R = a^* R \oplus a^{\circ} = Ra \oplus {}^{\circ}a$.
(7) $R = a^* R + a^{\circ} = Ra \oplus {}^{\circ}a$.
(8) $R = Ra \oplus {}^{\circ}(a^*) = Ra \oplus {}^{\circ}a$.
(9) $R = Ra + {}^{\circ}(a^*) = Ra \oplus {}^{\circ}a$.

*In this case,*

$$a_\circledcirc = x_1^* a y_1 = x_1^* a y_2^2 a = x_2 a y_1 = x_2 a y_2^2 a,$$

*where* $1 = a^* x_1 + u_1 = x_2 a + u_2 = a y_1 + v_1 = y_2 a + v_2$ *for some* $x_1, x_2, y_1, y_2 \in R$, $u_2 \in {}^\circ(a^*)$, $v_2 \in {}^\circ a$ *and* $u_1, v_1 \in a^\circ$.

### 5.2.3 Characterizations of Core Inverses by Algebraic Equations

**Theorem 5.2.9 ([124])** *Let* $a, x \in R$, *then* $a \in R^\circledcirc$ *with* $a^\circledcirc = x$ *if and only if* $(ax)^* = ax$, $xa^2 = a$ *and* $ax^2 = x$.

**Proof** Suppose $a \in R^\circledcirc$, then we have $(ax)^* = ax$, $xa^2 = a$ and $ax^2 = x$ by Theorem 5.2.2. Conversely, if $(ax)^* = ax$, $xa^2 = a$ and $ax^2 = x$, then we have

$$x = ax^2 = xa^2 x^2 = xa(ax^2) = xax. \tag{5.2.5}$$

$$a = xa^2 = ax^2 a^2 = ax(xa^2) = axa. \tag{5.2.6}$$

Thus by Theorem 5.2.2, we have $a \in R^\circledcirc$ and $a^\circledcirc = x$.                                  $\square$

For complex matrices, Wang and Liu [115] proved that if $A \in \mathbb{C}^{n \times n}$ is core invertible, then the core inverse of $A$ is the unique matrix $X \in \mathbb{C}^{n \times n}$ satisfying the following three equations:

$$AXA = A, \quad AX^2 = X, \quad (AX)^* = AX.$$

The next result extends this fact to the ring case.

**Theorem 5.2.10 ([124])** *Let* $a, b \in R$. *If* $Ra = Ra^2$, *then* $b = a^\circledcirc$ *if and only if* $aba = a$, $(ab)^* = ab$ *and* $ab^2 = b$.

**Proof** The necessity follows by Theorem 5.2.2. For the sufficiency, suppose that $aba = a$, $ab^2 = b$ and $(ab)^* = ab$. Then we have $a = aba = a^2 b^2 a \in a^2 R$, and thus $a$ is group invertible by the hypothesis $Ra = Ra^2$. So $a + 1 - ab$ is invertible by Corollary 3.2.8. Since

$$(a + 1 - ab)a^\# ab = aa^\# ab = ab, \tag{5.2.7}$$

it follows that (5.2.7) is equivalent to

$$a^\# ab = (a + 1 - ab)^{-1} ab. \tag{5.2.8}$$

By (5.2.8) and Theorem 5.2.5 we have

$$a^{\oplus} = (a + 1 - ab)^{-1}ab. \tag{5.2.9}$$

Since $ab^2 = b$, it follows that $0 = b - ab^2 = (a + 1 - ab)b - ab$, thus

$$b = (a + 1 - ab)^{-1}ab. \tag{5.2.10}$$

Hence $b = a^{\oplus}$ by (5.2.9) and (5.2.10). $\qquad\square$

An infinite matrix $M$ is said to be bi-finite if it is both row-finite and column-finite.

**Example 5.2.11** The condition $Ra = Ra^2$ in Theorem 5.2.10 cannot be dropped. Let $R$ be the ring of all bi-finite matrices over the real field $\mathbb{R}$ with transpose as involution. Consider the following elements $A$, $B$ of $R$. $A = \sum\limits_{i=1}^{\infty} e_{i,i+1}$ and $B = A^*$, where $e_{i,j}$ denotes the matrix with $(i, j)$-th is 1 and other entries are zero. Then $BA = \sum\limits_{i=2}^{\infty} e_{i,i}$ and $AB = I$, thus $ABA = A$, $(AB)^* = AB$, $AB^2 = B$. Since $A$ is not group invertible, $A$ is not core invertible.

**Theorem 5.2.12 ([124])** *Let $a, b \in R$. If $aR = a^2R$, then $b = a^{\oplus}$ if and only if $bab = b$, $(ab)^* = ab$ and $ba^2 = a$.*

**Proof** The necessity follows by Theorem 5.2.2. For the sufficiency, suppose that $bab = b$, $(ab)^* = ab$ and $ba^2 = a$. Then we have $a = ba^2 \in Ra^2$, and thus $a \in R^{\#}$ by the hypothesis $aR = a^2R$. Post-multiplication by $a^{\#}$ on equation $ba^2 = a$ yields $ba = aa^{\#}$, thus

$$aba = a^2a^{\#} = a.$$

Hence $b \in a\{1, 3\}$ by $(ab)^* = ab$. By Theorem 5.2.5, we have $a^{\oplus} = a^{\#}aa^{(1,3)}$. Therefore

$$a^{\oplus} = a^{\#}aa^{(1,3)} = aa^{\#}b = bab = b.$$

$\qquad\square$

**Example 5.2.13** The condition $aR = a^2R$ in Theorem 5.2.12 cannot be dropped. Let $R$ be the ring of all bi-finite matrices with entries over field $\mathbb{R}$ with transpose as involution. Consider the following matrices $A$, $B$ over $R$. $A = \sum\limits_{i=1}^{\infty} e_{i+1,i}$ and $B = A^*$, then $AB = \sum\limits_{i=2}^{\infty} e_{i,i}$, $BA = I$, thus $BAB = B$, $(AB)^* = AB$ and $BA^2 = A$. Since $A$ is not group invertible, whence $A$ is not core invertible.

By Theorems 5.2.10 and 5.2.12, we have the following corollary.

**Corollary 5.2.14** *Let* $a, b \in R$. *If* $a \in R^{\#}$, *then the following conditions are equivalent:*

(1) $b = a^{\oplus}$.
(2) $aba = a$, $ab^2 = b$, $(ab)^* = ab$.
(3) $bab = b$, $ba^2 = a$, $(ab)^* = ab$.

**Theorem 5.2.15 ([124])** *A ring* $R$ *is Dedekind-finite if and only if, for any elements* $a, b \in R$, $aba = a$, $(ab)^* = ab$ *and* $ab^2 = b$ *imply that* $a$ *is core invertible with* $a^{\oplus} = b$.

**Proof** Suppose $R$ is a Dedekind-finite ring and $aba = a$, $(ab)^* = ab$, $ab^2 = b$. Then $b \in a\{1, 3\}$. Since

$$a = aba = a(ab^2)a = a^2b^2a \in a^2R,$$

it follows by Theorem 3.3.1 that $a + 1 - ab$ is right invertible. Thus $a + 1 - ab$ is invertible by Dedekind-finiteness. By Corollary 3.2.8, we have $a \in R^{\#}$, therefore $a^{\oplus} = b$ by Corollary 5.2.14.

Conversely, for arbitrary $m, n \in R$, if $mn = 1$, then $mnm = m$, $(mn)^* = mn$, $mn^2 = n$. Thus by the hypothetical condition, we have $m \in R^{\oplus}$ with $m^{\oplus} = n$, then $m \in R^{\#}$ by Lemma 5.2.4. Post-multiplication by $n$ on equation $m = mm^{\#}m$ yields

$$1 = mn = mm^{\#}mn = mm^{\#},$$

thus, $mm^{\#} = m^{\#}m = 1$ which imply $m$ is invertible. Then $nm = 1$, so $R$ is a Dedekind-finite ring.                                                                                $\square$

**Theorem 5.2.16 ([124])** *Consider the following conditions:*

(1) *$R$ is a Dedekind-finite ring.*
(2) *Let* $a, b \in R$, *if* $bab = b$, $(ab)^* = ab$ *and* $ba^2 = a$, *then* $a \in R^{\oplus}$ *with* $a^{\oplus} = b$.
(3) *Let* $a, b \in R$, *if* $aba = a$, $(ab)^* = ab$ *and* $ba^2 = a$, *then* $a \in R^{\oplus}$ *with* $a^{\oplus} = bab$.
(4) *Let* $a \in R$, *if* $a^*a = 1$, *then* $aa^* = 1$.

*Then, we have* (1) $\Rightarrow$ (2) $\Rightarrow$ (3) $\Rightarrow$ (4).

**Proof** (1)$\Rightarrow$(2). Suppose $R$ is a Dedekind-finite ring, by $bab = b$ and $ba^2 = a$, we have

$$b = bab = b(ba^2)b = b^2a^2b \in b^2R.$$

Since $b = bab$ gives $a \in b\{1\}$, by Theorem 3.3.1, $b + 1 - ba$ is right invertible. As $R$ is a Dedekind-finite ring, hence $b + 1 - ba$ is invertible, which gives $b \in R^{\#}$ by

Corollary 3.2.8. Pre-multiplication of $b = b^2 a^2 b$ by $b^\#$ now yields $b^\# b = ab$, thus

$$ab^2 = b^\# b^2 = b.$$

Whence by Theorem 5.2.9, we have $a \in R^\circledcirc$ with $a^\circledcirc = b$.

(2) $\Rightarrow$ (3). Let $c = bab$, by $bab = b$, $(ab)^* = ab$ and $ba^2 = a$, we have $cac = c$, $(ac)^* = ac$ and $ca^2 = a$, thus by (2), we have $a \in R^\circledcirc$ with $a^\circledcirc = c = bab$.

(3) $\Rightarrow$ (4). Suppose $a^* a = 1$, let $b = a^*$, then $aba = a$, $(ab)^* = ab$ and $ba^2 = a$. By (3), we have $a \in R^\circledcirc$, thus $1 = a^* a = a^* aa^\circledcirc a = a^\circledcirc a = a(a^\circledcirc)^2 a$, and so $a$ is invertible, which gives $aa^* = 1$. $\qquad\square$

**Proposition 5.2.17 ([124])** *Let* $a, b \in R$. *If* $aR = a^2 R$, *then the following conditions are equivalent:*

(1) $a \in R^\circledcirc$ *with* $a^\circledcirc = b$.
(2) $ba^2 = a$, $(ab)^* = ab$, $bR \subseteq aR$.
(3) $a \in R^{\{1,3\}}$ *and satisfies* $ba^2 = a$, $b = ba\hat{a}$ *for some* $\hat{a} \in a\{1, 3\}$.

**Proof** (1) $\Rightarrow$ (2) Suppose $a \in R^\circledcirc$ with $a^\circledcirc = b$, then $ba^2 = a$, $(ab)^* = ab$ and $b = ab^2$ by Theorem 5.2.2, thus $bR \subseteq aR$.

(1) $\Rightarrow$ (3) Suppose $a \in R^\circledcirc$ with $a^\circledcirc = b$, then $ba^2 = a$ by Theorem 5.2.2. By Theorem 5.2.5, we have

$$a^\circledcirc = a^\# aa^{(1,3)} = (a^\# aa^{(1,3)})aa^{(1,3)} = a^\circledcirc aa^{(1,3)} = baa^{(1,3)}.$$

(2) $\Rightarrow$ (1) Suppose $ba^2 = a$, $(ab)^* = ab$, $bR \subseteq aR$, then $b = ax$, for some $x \in R$. By $ba^2 = a$, we have $b = ax = ba^2 x = bab$. Therefore, $a \in R^\circledcirc$ with $a^\circledcirc = b$ by Theorem 5.2.12.

(3) $\Rightarrow$ (1) Suppose $ba^2 = a$, $b = ba\hat{a}$, for some $\hat{a} \in a\{1, 3\}$ and $aR = a^2 R$, thus $a \in R^\#$ by $a \in a^2 R \cap Ra^2$. Hence

$$b = ba\hat{a} = ba^2 a^\# \hat{a} = aa^\# \hat{a} = a^\circledcirc$$

by Theorem 5.2.5. $\qquad\square$

Since an element $a \in R$ is group invertible if and only if $a \in Ra^2 \cap a^2 R$, we obtain that $a \in R^\circledcirc$ if and only if $a \in Ra^* a \cap Ra^2 \cap a^2 R$ by Lemmas 2.2.3 and 5.2.5. We aim at characterizing the core invertibility by the intersection of two left principal ideals.

**Proposition 5.2.18 ([81])** *Let* $a \in R$ *and* $n \geq 1$. *We have the following results:*

(I) *the following conditions are equivalent:*

(1) $a \in R(a^*)^n a$.
(2) $a \in Ra^* a \cap a^n R$.
(3) $R = {}^\circ a \oplus R(a^*)^n$.
(4) $R = {}^\circ a + R(a^*)^n$.

(II) *the following conditions are equivalent:*

(1) $a \in a(a^*)^n R$.
(2) $a \in aa^* R \cap Ra^n$.
(3) $R = a^\circ \oplus (a^*)^n R$.
(4) $R = a^\circ + (a^*)^n R$.

**Proof** (I): (1)$\Rightarrow$(2). It is clear to see that $a \in Ra^*a$ follows from $a \in R(a^*)^n a$, and there exists $r \in R$ such that

$$a = r(a^*)^n a = r(a^*)^{n-1} a^* a,$$

so $a \in R^{\{1,3\}}$ and $(r(a^*)^{n-1})^* = a^{n-1} r^* \in a\{1, 3\}$ by Lemma 2.2.3. Moreover,

$$a = aa^{(1,3)}a = a(a^{n-1}r^*)a = a^n r^* a \in a^n R.$$

Thus $a \in Ra^* a \cap a^n R$.

(2) $\Rightarrow$ (1). Suppose $a \in Ra^* a \cap a^n R$, there exist $s, t \in R$ such that $a = sa^* a = a^n t$. Thus we get

$$a = sa^* a = s(a^n t)^* a = st^*(a^*)^n a \in R(a^*)^n a.$$

(1) $\Rightarrow$ (3). Assume that $a = s(a^*)^n a$ for some $s \in R$, which gives that $1 - s(a^*)^n \in {}^\circ a$. Since we can write $r$ as $r = r(1 - s(a^*)^n) + rs(a^*)^n$ for any $r \in R$, where $r(1 - s(a^*)^n) \in {}^\circ a$ and $rs(a^*)^n \in R(a^*)^n$, thus $R = {}^\circ a + R(a^*)^n$.
If $x \in R(a^*)^n \cap {}^\circ a$, then $xa = 0$ and $x = t(a^*)^n$ for some $t \in R$. Moreover,

$$x = t(a^*)^n = t(a^*)^{n-1} a^* = t(a^*)^{n-1}(s(a^*)^n a)^*$$
$$= t(a^*)^n a^n s^* = xa^n s^* = 0.$$

Hence $R = {}^\circ a \oplus R(a^*)^n$.
(3) $\Rightarrow$ (4) is trivial.
(4) $\Rightarrow$ (1). Since $a \in Ra = ({}^\circ a + R(a^*)^n)a \subseteq R(a^*)^n a$, which gives the condition (1).
(II): The proof is similar to the proof of (I).                        $\square$

Using Proposition 5.2.18, we obtain the following new characterizations of core and dual core inverses which will be useful in the upcoming results.

**Theorem 5.2.19 ([81])** *Let* $a \in R$, $n \geqslant 2$. *We have the following results:*

(1) $a \in R^\oplus$ *if and only if* $a \in R(a^*)^n a \cap Ra^n$. *In this case,* $a^\oplus = a^{n-1} s^*$ *for some* $s \in R$ *such that* $a = s(a^*)^n a$.
(2) $a \in R_\oplus$ *if and only if* $a \in a(a^*)^n R \cap a^n R$. *In this case,* $a_\oplus = t^* a^{n-1}$ *for some* $t \in R$ *such that* $a = a(a^*)^n t$.

***Proof*** (1) Since $a \in R^{\#}$ if and only if $a \in a^2 R \cap Ra^2$ by Theorem 3.2.2, thus $a = a^2 x = ya^2$ for some $x, y \in R$. Further, we have

$$a = a^2 x = a(a^2 x)x = a^3 x = \cdots = a^n x \in a^n R$$

and

$$a = ya^2 = y(ya^2)a = ya^3 = \cdots = ya^n \in Ra^n,$$

where $n \geqslant 2$. Hence it is easy to deduce that $a \in R^{\#}$ if and only if $a \in a^n R \cap Ra^n$ for $n \geqslant 2$.

Applying Proposition 5.2.18, $a \in R(a^*)^n a \cap Ra^n$ if and only if $a \in Ra^* a \cap a^n R \cap Ra^n$, which shows that $a \in R(a^*)^n a \cap Ra^n$ if and only if $a \in Ra^* a \cap R^{\#}$. Since $a \in Ra^* a$ is equivalent to $a \in R^{\{1,3\}}$ by Lemma 2.2.3, thus $a \in R(a^*)^n a \cap Ra^n$ if and only if $a \in R^{\oplus}$ by Theorem 5.2.5.

Next, we give the representation of $a^{\oplus}$. Since $a \in R(a^*)^n a \cap Ra^n$, there exists $s \in R$ such that $a = s(a^*)^n a$. By Lemma 2.2.3, we have

$$(s(a^*)^{n-1})^* = a^{n-1} s^* \in a\{1, 3\}.$$

Using Theorem 5.2.5, we obtain

$$a^{\oplus} = a^{\#} a a^{(1,3)} = a^{\#} a(a^{n-1} s^*) = a^{n-1} s^*.$$

(2) Similarly as (1).                                                                                       □

It is easy to see that Theorem 5.2.19 is also true in a semigroup by the above proof. From Theorem 5.2.5 and dual theorem, we can obtain that $a \in R^{\oplus} \cap R_{\oplus}$ if and only if $a \in R^{\dagger} \cap R^{\#}$. Therefore, we have the following result by applying Proposition 5.2.18 and Theorem 5.2.19.

**Theorem 5.2.20 ([81])** *Let $a \in R$, $n \geqslant 2$. The following conditions are equivalent:*

(1) $a \in R^{\dagger} \cap R^{\#}$.
(2) $a \in R^{\oplus} \cap R_{\oplus}$.
(3) $a \in a(a^*)^n R \cap R(a^*)^n a$.
(4) $R = {}^{\circ}a \oplus R(a^*)^n$, $R = a^{\circ} \oplus (a^*)^n R$.
(5) $R = {}^{\circ}a + R(a^*)^n$, $R = a^{\circ} + (a^*)^n R$.
(6) $R = {}^{\circ}a \oplus R(a^*)^n$, $R = a^{\circ} + (a^*)^n R$.
(7) $R = {}^{\circ}a + R(a^*)^n$, $R = a^{\circ} \oplus (a^*)^n R$.

*In this case,*

$$a^{®} = a^{n-1}s^*,$$

$$a_{®} = t^*a^{n-1},$$

$$a^\dagger = t^*a^{2n-1}s^*,$$

$$a^{\#} = (a^{n-1}s^*)^2a = a(t^*a^{n-1})^2,$$

*where $a = s(a^*)^n a = a(a^*)^n t$ for some $s, t \in R$.*

**Proof** The equivalences of seven conditions above and the representations of $a^{®}$ and $a_{®}$ can be easily obtained by Proposition 5.2.18 and Theorem 5.2.19. We will give the representations of $a^\dagger$ and $a^{\#}$ in the following.

Suppose $a = s(a^*)^n a = a(a^*)^n t$ for some $s, t \in R$, so $a^{n-1}s^* \in a\{1, 3\}$ and $t^*a^{n-1} \in a\{1, 4\}$ follow from Lemma 2.2.3. Applying Theorem 2.2.6 and Lemma 5.2.4, we get

$$a^\dagger = a^{(1,4)}aa^{(1,3)} = (t^*a^{n-1})a(a^{n-1}s^*) = t^*a^{2n-1}s^*,$$

$$a^{\#} = (a^{®})^2a = (a^{n-1}s^*)^2a = a(a_{®})^2 = a(t^*a^{n-1})^2.$$

$\square$

## 5.3   Characterizations of Core Invertibility by Special Elements

In this section, we will characterize core inverses by special elements, such as Hermitian, projections, regular, invertible and one-sided invertible elements.

### 5.3.1   Characterizations of Core Invertibility by Hermitian Elements or Projections in a Ring

In the following, we will characterize core inverses by Hermitian elements or projections.

**Theorem 5.3.1 ([81])** *Let $a \in R$ and $n \geq 2$. The following conditions are equivalent:*

(1) $a \in R^{®}$.
(2) *There exists a unique projection $p$ such that $pa = 0$, $u = a^n + p \in R^{-1}$.*
(3) *There exists an Hermitian element $p$ such that $pa = 0$, $u = a^n + p \in R^{-1}$.*

*In this case,*

$$a^{\circledR} = a^{n-1}u^{-1}.$$

**Proof** $(1) \Rightarrow (2)$. Let $p = 1 - aa^{\circledR}$, we observe first that $p$ is a projection satisfying $pa = 0$. It is necessary for us to show that $^{\circ}(a^n) = {^{\circ}}(1 - p)$. If $xa^n = 0$, then

$$0 = xa^n = x(1 - p)a^n = x(1 - p)(a^n + p).$$

By $a^n + p \in R^{-1}$, we have $x(1 - p) = 0$. Conversely, if $y(1 - p) = 0$, then $ya^n = y(1-p)a^n = 0$. Thus $^{\circ}(a^n) = {^{\circ}}(1-p)$. Assume that $p, q$ are both projections which satisfy the condition (2), then $^{\circ}(1 - p) = {^{\circ}}(a^n) = {^{\circ}}(1-q)$. By $p \in {^{\circ}}(1-p) = {^{\circ}}(1 - q)$, we obtain $p = pq$. Similarly, we can get $q = qp$ from $q \in {^{\circ}}(1 - q) = {^{\circ}}(1 - p)$. Thus

$$p = p^* = (pq)^* = q^*p^* = qp = q.$$

Next, we prove the invertibility of $u$ by induction on $n$.

When $n = 2$, it is easy to verify that

$$(a + 1 - aa^{\circledR})(a^{\circledR} + 1 - a^{\circledR}a) = 1 = (a^{\circledR} + 1 - a^{\circledR}a)(a + 1 - aa^{\circledR}),$$

thus $a + 1 - aa^{\circledR} = 1 + aa^{\circledR}(a - 1)$ is invertible. Moreover, $1 + (a - 1)aa^{\circledR} = a^2a^{\circledR} + 1 - aa^{\circledR}$ is invertible by Jacobson's lemma. Therefore, $a^2 + 1 - aa^{\circledR} = (a^2a^{\circledR} + 1 - aa^{\circledR})(a + 1 - aa^{\circledR})$ is invertible.

We assume that $n > 2$ and the result is true for the case $n - 1$. By assumption, $a \in R^{\circledR}$ implies that $a^{n-1} + 1 - aa^{\circledR}$ is invertible. Hence, $a^n + 1 - aa^{\circledR} = (a^2a^{\circledR} + 1 - aa^{\circledR})(a^{n-1} + 1 - aa^{\circledR})$ is invertible.

$(2) \Rightarrow (3)$. It is trivial.

$(3) \Rightarrow (1)$. Assume that $u = a^n + p \in R^{-1}$, where $p = p^*, n \geqslant 2$, and then $u^* = (a^*)^n + p$ is also invertible.

Since $ua = a^{n+1}$ and $u^*a = (a^*)^na$, we obtain

$$a = u^{-1}a^{n+1} \in Ra^n$$

and

$$a = (u^*)^{-1}(a^*)^na \in R(a^*)^na.$$

Thus $a \in R(a^*)^na \cap Ra^n$, and then we have $a \in R^{\circledR}$ and $a^{\circledR} = a^{n-1}u^{-1}$ by Theorem 5.2.19. $\square$

There is a corresponding result for the dual core inverse of $a \in R_{\circledR}$. The following theorem shows that Theorem 5.3.1 is true when taking $n = 1$, but its proof is different from that of Theorem 5.3.1, and so is the expression of the core inverse of $a$.

**Theorem 5.3.2** ([81]) *Let $a \in R$. The following conditions are equivalent:*

(1) $a \in R^{\oplus}$.
(2) *There exists a unique projection $p$ such that $pa = 0$, $u = a + p \in R^{-1}$.*
(3) *There exists an Hermitian element $p$ such that $pa = 0$, $u = a + p \in R^{-1}$.*

*In this case,*

$$a^{\oplus} = u^{-1}au^{-1} = (u^*u)^{-1}a^*.$$

**Proof** (1) $\Rightarrow$ (2). Let $p = 1 - aa^{\oplus}$, $p$ is a projection satisfying $pa = 0$, and the proof of the uniqueness of $p$ is similar to Theorem 5.3.1. It is easy to verify

$$(a + 1 - aa^{\oplus})(a^{\oplus} + 1 - a^{\oplus}a) = 1 = (a^{\oplus} + 1 - a^{\oplus}a)(a + 1 - aa^{\oplus}).$$

Thus $a + 1 - aa^{\oplus}$ is invertible.

(2) $\Rightarrow$ (3). Obviously.

(3) $\Rightarrow$ (1). Assume that $u = a + p \in R^{-1}$, where $p = p^*$, and then $u^* = a^* + p$ is also invertible.

Since $ua = a^2$ and $u^*a = a^*a$, we obtain

$$a = u^{-1}a^2 \in Ra^2 \tag{5.3.1}$$

and

$$a = (u^*)^{-1}a^*a \in Ra^*a. \tag{5.3.2}$$

By Lemma 2.2.3, it is easily seen that $a \in R^{\{1,3\}}$ with $u^{-1} \in a\{1, 3\}$ from Eq. (5.3.2), so we have

$$a = aa^{(1,3)}a = au^{-1}a. \tag{5.3.3}$$

Moreover, $pu = p(a + p) = p^2$ implies $p = p^2u^{-1}$, thus we have

$$p^2u^{-1} = p = p^* = (u^*)^{-1}p^2, \tag{5.3.4}$$

direct calculations with the use of (5.3.4) show that

$$
\begin{aligned}
pu^{-2} &= ((u^*)^{-1}p^2)u^{-2} = (u^*)^{-1}(p^2u^{-1})u^{-1} \\
&= (u^*)^{-1}((u^*)^{-1}p^2)u^{-1} = (u^*)^{-2}p^2u^{-1} \\
&= (u^*)^{-2}p.
\end{aligned} \tag{5.3.5}
$$

Therefore

$$u^{-1} = uu^{-2} = (a + p)u^{-2} = au^{-2} + pu^{-2}$$

$$\stackrel{(5.3.5)}{=} au^{-2} + (u^*)^{-2}p.$$

(5.3.6)

Thus we obtain

$$a \stackrel{(5.3.3)}{=} au^{-1}a \stackrel{(5.3.6)}{=} a(au^{-2} + (u^*)^{-2}p)a$$

$$= a^2u^{-2}a \in a^2R.$$

(5.3.7)

Equations (5.3.1), (5.3.2), and (5.3.7) lead to $a \in R^{\circledR}$ by Theorem 5.2.5.

Applying Theorem 3.2.2, we obtain $a^{\#} = u^{-2}a$, again by Theorem 5.2.5, we have

$$a^{\circledR} = a^{\#}aa^{(1,3)} = (u^{-2}a)au^{-1} = u^{-1}(u^{-1}a^2)u^{-1} \stackrel{(5.3.1)}{=} u^{-1}au^{-1}.$$

Further, since $u^{-1} \in a\{1, 3\}$, thus

$$a^{\circledR} = u^{-1}au^{-1} = u^{-1}(au^{-1})^* = u^{-1}(u^*)^{-1}a^* = (u^*u)^{-1}a^*.$$

$\square$

The analogous results for the dual core inverse of $a \in R_{\circledR}$ are valid.

**Remark 5.3.3 ([81])** Theorems 5.3.1 and 5.3.2 show that $a \in R^{\circledR}$ if and only if there exists $p = p^*(= p^2)$ such that $pa = 0$, $u = a^n + p \in R^{-1}$ for all choices $n \geqslant 1$. Dually, $a \in R_{\circledR}$ if and only if there exists $q = q^*(= q^2)$ such that $aq = 0$, $u = a^n + q \in R^{-1}$ for all choices $n \geqslant 1$.

Under the condition (2) of Theorem 5.3.2, since $u^*u = a^*a + p$, the expression of the core inverse of $a$ can be showed as $a^{\circledR} = (a^*a + p)^{-1}a^*$. Through this expression, we naturally want to know whether $a$ is core invertible when there is a unique projection $p$ such that $pa = 0$, $a^*a + p \in R^{-1}$. The next example gives a negative answer.

**Example 5.3.4 ([81])** Let $R$ be an infinite matrix ring over complex filed whose rows and columns are both finite, let conjugate transpose be the involution and $a = \sum_{i=1}^{\infty} e_{i+1,i}$. Then $a^*a = 1$, $aa^* = \sum_{i=2}^{\infty} e_{i,i}$. Set $p = 0$, then it is a projection satisfying $pa = 0$ and $a^*a + p = 1 \in R^{-1}$. But $a$ is not group invertible, thus $a$ is not core invertible.

The above counterexample shows that even if there is a unique projection $p$ such that $pa = 0$ and $a^*a + p \in R^{-1}$, $a$ is not necessary to be core invertible in general rings. However, it is true when we take $R$ as a Dedekind-finite ring.

**Theorem 5.3.5 ([81])** *Let R be a Dedekind-finite ring and a ∈ R. The following conditions are equivalent:*

(1) $a \in R^{\circledcirc}$.
(2) *There exists a unique projection p such that $pa = 0$, $a^*a + p \in R^{-1}$.*
(3) *There exists a unique projection p such that $pa = 0$, $a^*a + p$ is right invertible.*
(4) *there exists a unique projection p such that $pa = 0$, $a^*a + p$ is left invertible.*

*In this case,*

$$a^{\circledcirc} = (a^*a + p)^{-1}a^*.$$

**Proof** Since $a^*a + p$ is Hermitian, thus $a^*a + p$ is one-sided invertible if and only if it is invertible, hence the conditions (2), (3) and (4) are equivalent. Next, we mainly show the equivalence between the conditions (1) and (2).

(1) $\Rightarrow$ (2). Assume that $a \in R^{\circledcirc}$ and let $p = 1 - aa^{\circledcirc}$, we have $a + p \in R^{-1}$ by Theorem 5.3.2, thus $(a+p)^* = a^*+p \in R^{-1}$. Therefore $a^*a+p = (a^*+p)(a+p)$ is invertible.

(2) $\Rightarrow$ (1). Let $u = a + p$, $u^*u = a^*a + p \in R^{-1}$. As $R$ is a Dedekind-finite ring, thus $u \in R^{-1}$, which guarantees $a \in R^{\circledcirc}$ and $a^{\circledcirc} = (a^*a + p)^{-1}a^*$ by Theorem 5.3.2.                                                                              □

As mentioned before, $a \in R^{\circledcirc} \cap R_{\circledcirc}$ if and only if $a \in R^{\dagger} \cap R^{\#}$. Recall that an element $a \in R$ is called EP if $a \in R^{\dagger} \cap R^{\#}$ with $a^{\dagger} = a^{\#}$. It is obvious that $a$ is EP if and only if $a \in R^{\circledcirc} \cap R_{\circledcirc}$ and $a^{\circledcirc} = a_{\circledcirc}$. By Remark 5.3.3, we obtain the following theorem.

**Theorem 5.3.6 ([81])** *Let $a \in R$, $n \geqslant 1$. The following conditions are equivalent:*

(1) *a is EP.*
(2) *There exists a unique projection p such that $pa = ap = 0$, $a^n + p \in R^{-1}$.*
(3) *There exists an Hermitian element p such that $pa = ap = 0$, $a^n + p \in R^{-1}$.*

**Proof** (1) $\Rightarrow$ (2). Since $a$ is EP, we have $a \in R^{\dagger} \cap R^{\#}$ and $a^{\dagger} = a^{\#}$. Let $p = 1 - a^{\#}a = 1 - a^{\dagger}a$. Then it is easily seen that $p$ is a projection satisfying $pa = ap = 0$. Since

$$(a^n + 1 - a^{\#}a)((a^{\#})^n + 1 - a^{\#}a) = 1 = ((a^{\#})^n + 1 - a^{\#}a)(a^n + 1 - a^{\#}a),$$

thus $a^n + 1 - a^{\#}a$ is invertible.

(2) $\Rightarrow$ (3). Obviously.

(3) $\Rightarrow$ (1). We can see that $a \in R^{\circledcirc} \cap R_{\circledcirc}$ follows from Remark 5.3.3, so we only need to show that $a^{\circledcirc} = a_{\circledcirc}$. Write $u = a^n + p$, when taking $n = 1$, we have $a^{\circledcirc} = u^{-1}au^{-1} = a_{\circledcirc}$ by Theorem 5.3.2. When $n \geqslant 2$, according to Theorem 5.3.1, $a^{\circledcirc} = a^{n-1}u^{-1}$, $a_{\circledcirc} = u^{-1}a^{n-1}$. Since $pa = ap = 0$ implies $au = ua$, which obtain

that $a$ commutes with $u^{-1}$, thus $a^{\circledcirc} = a^{n-1}u^{-1} = u^{-1}a^{n-1} = a_{\circledcirc}$. Therefore, $a$ is EP. $\qquad\square$

### 5.3.2 Characterizations of Core Invertibility for a Regular Element by Units in a Ring

**Theorem 5.3.7 ([81])** *Let $a \in R$, $n \geqslant 2$. If $a \in R$ is regular with $a^- \in a\{1\}$, then the following conditions are equivalent:*

(1) $a \in R^\dagger \cap R^\#$.
(2) $a \in R^{\circledcirc} \cap R_{\circledcirc}$.
(3) $u = (a^*)^n a + 1 - a^- a$ is invertible.
(4) $v = a(a^*)^n + 1 - aa^-$ is invertible.
(5) $s = a^- a(a^*)^n a + 1 - a^- a$ is invertible.
(6) $t = a(a^*)^n aa^- + 1 - aa^-$ is invertible.

*In this case,*

$$a^{\circledcirc} = a^{n-1}(v^{-1}a)^*,$$

$$a_{\circledcirc} = (au^{-1})^* a^{n-1},$$

$$a^\dagger = (au^{-1})^* a^{2n-1}(v^{-1}a)^*,$$

$$a^\# = (a^{n-1}(v^{-1}a)^*)^2 a.$$

**Proof** It is obvious that the conditions (1) and (2) are equivalent, and equivalences of the conditions (3), (4), (5) and (6) can be deduced from Jacobson's lemma.

(2)$\Rightarrow$(3). Since

$$u[a^- aa^{\circledcirc}(a^n_{\circledcirc})^* + 1 - a_{\circledcirc}a]$$

$$= [(a^*)^n a + 1 - a^- a][a^- aa^{\circledcirc}(a^n_{\circledcirc})^* + 1 - a_{\circledcirc}a]$$

$$= (a^*)^n aa^- aa^{\circledcirc}(a^n_{\circledcirc})^* + (1 - a^- a)(1 - a_{\circledcirc}a)$$

$$= (a^*)^n aa^{\circledcirc}(a^n_{\circledcirc})^* + 1 - a_{\circledcirc}a$$

$$= (a^*)^n (aa^{\circledcirc})^* (a^n_{\circledcirc})^* + 1 - a_{\circledcirc}a$$

$$= (aa^{\circledcirc}a^n)^* (a^n_{\circledcirc})^* + 1 - a_{\circledcirc}a$$

$$= (a^n)^* (a^n_{\circledcirc})^* + 1 - a_{\circledcirc}a$$

$$= (a^n_{\circledcirc}a^n)^* + 1 - a_{\circledcirc}a$$

$$= (a_{\circledcirc}a)^* + 1 - a_{\circledcirc}a$$

$$= a_{\circledcirc}a + 1 - a_{\circledcirc}a$$

$$= 1,$$

where the eighth equation follows from

$$a_{\circledcirc}^n a^n = a_{\circledcirc}^{n-2}(a_{\circledcirc}^2 a)a^{n-1} = a_{\circledcirc}^{n-2} a_{\circledcirc} a^{n-1} = a_{\circledcirc}^{n-1} a^{n-1}$$

$$- a_{\circledcirc}^{n-3}(a_{\circledcirc}^2 a)a^{n-2} = a_{\circledcirc}^{n-3} a_{\circledcirc} a^{n-2} = a_{\circledcirc}^{n-2} a^{n-2}$$

$$= \cdots = a_{\circledcirc}^2 a^2 = (a_{\circledcirc}^2 a)a = a_{\circledcirc}a,$$

thus $u$ is right invertible with right inverse $a^- aa^{\circledcirc}(a_{\circledcirc}^n)^* + 1 - a_{\circledcirc}a$. Similarly, we have

$$[((a^{\circledcirc})^n)^* a_{\circledcirc} aa^- + 1 - aa^{\circledcirc}]v = 1,$$

which implies that $v$ is left invertible, and thus $u$ is left invertible by Jacobson's lemma. Further, $u = (a^*)^n a + 1 - a^- a$ is invertible.

(3)$\Rightarrow$(2). Since $u$ is invertible if and only if $v$ is invertible follows from Jacobson's lemma, $au = a(a^*)^n a = va$, so we have

$$a = a(a^*)^n au^{-1} \in a(a^*)^n R,$$

$$a = v^{-1}a(a^*)^n a \in R(a^*)^n a.$$

By Theorem 5.2.20, we obtain $a \in R^{\circledcirc} \cap R_{\circledcirc}$ and the following representations

$$a^{\circledcirc} = a^{n-1}(v^{-1}a)^*,$$

$$a_{\circledcirc} = (au^{-1})^* a^{n-1},$$

$$a^{\dagger} = (au^{-1})^* a^{2n-1}(v^{-1}a)^*,$$

$$a^{\#} = [a^{n-1}(v^{-1}a)^*]^2 a.$$

□

**Lemma 5.3.8 ([90, 135])**  *Let $d \in R$ be regular with $d^- \in d\{1\}$. Then the following conditions are equivalent:*

(1)  $d \in Rdad$ *if and only if* $da + 1 - dd^-$ *(resp., $ad + 1 - d^- d$) is left invertible.*
(2)  $d \in dadR$ *if and only if* $da + 1 - dd^-$ *(resp., $ad + 1 - d^- d$) is right invertible.*
(3)  $d \in Rdad \cap dadR$ *if and only if* $da + 1 - dd^-$ *(resp., $ad + 1 - d^- d$) is invertible.*

Comparing Theorem 5.3.7 and Lemma 5.3.8, we have the following corollary.

**Corollary 5.3.9 ([81])**  *Let $a \in R$ and $n \geq 2$. Then the following conditions are equivalent:*

(1)  $a \in R^{\dagger} \cap R^{\#}$.
(2)  $a \in R^{\oplus} \cap R_{\oplus}$.
(3)  $a \in a(a^*)^n a R \cap R a(a^*)^n a$.

*In this case,*

$$a^{\oplus} = a^{n-1} a^* y^*,$$

$$a_{\oplus} = x^* a^* a^{n-1},$$

$$a^{\dagger} = x^* a^* a^{2n-1} a^* y^*,$$

$$a^{\#} = [a^{n-1} a^* y^*]^2 a = a[x^* a^* a^{n-1}]^2,$$

*where $a = a(a^*)^n a x = y a(a^*)^n a$ for some $x, y \in R$.*

Theorem 5.2.20 and Corollary 5.3.9 show that $a \in R^{\oplus} \cap R_{\oplus}$ if and only if $a \in a(a^*)^n R \cap R(a^*)^n a$ if and only if $a \in a(a^*)^n a R \cap R a(a^*)^n a$. And it is easily seen that $a \in Ra(a^*)^n a \subseteq R(a^*)^n a$, but $a \in R(a^*)^n a$ does not lead to $a \in Ra(a^*)^n a$. The following is a counterexample when taking $n = 2$.

**Example 5.3.10 ([81])**  Let $R = M_2(\mathbb{C})$ be the ring of all $2 \times 2$ matrices over the complex field $\mathbb{C}$. Taking transposition as involution, considering the matrix $a = \begin{bmatrix} 1 & i \\ 0 & 0 \end{bmatrix}$, we have $a^2 = a$, $a a^* = 0$, $a^* a = \begin{bmatrix} 1 & i \\ i & -1 \end{bmatrix} \neq 0$. Thus $a = \begin{bmatrix} 1 & 0 \\ 0 & 0 \end{bmatrix} (a^*)^2 a \in R(a^*)^2 a$, but $a \notin Ra(a^*)^2 a$.

In view of Theorem 2.2.6 an element $a \in Raa^*a$ if and only if $a \in aa^*aR$. Reproducing this work, we get the following result.

**Proposition 5.3.11 ([81])**  *For $a \in R$, $n \geq 1$, we have the following results:*

(1)  *If $a \in Ra(a^*)^n a$, then $a \in a^n a^* a^n R$.*
(2)  *If $a \in a(a^*)^n a R$, then $a \in Ra^n a^* a^n$.*

**Proof** (1) Suppose $a \in Ra(a^*)^n a$, there exists $x \in R$ such that

$$a = xa(a^*)^n a. \tag{5.3.8}$$

Taking involution on (5.3.8), we get

$$a^* = a^* a^n a^* x^*. \tag{5.3.9}$$

Again by (5.3.8), we obtain

$$[xa(a^*)^n]^* = a^n(xa)^* = aa^{n-1}(xa)^*$$

$$\overset{(5.3.8)}{=} [xa(a^*)^n a]a^{n-1}(xa)^*$$

$$= xa(a^*)^n a^n(xa)^*.$$

Hence $[xa(a^*)^n]^*$ is symmetric, that is to say,

$$xa(a^*)^n = [xa(a^*)^n]^*.  \qquad (5.3.10)$$

The equalities (5.3.8), (5.3.9) and (5.3.10) give

$$a \overset{(5.3.8)}{=} xa(a^*)^n a \overset{(5.3.10)}{=} [xa(a^*)^n]^* a$$

$$= a^n a^* x^* a \overset{(5.3.9)}{=} a^n(a^* a^n a^* x^*)x^* a$$

$$= a^n a^* a^n(a^* x^* x^* a) \in a^n a^* a^n R.$$

(2) Dually as (1).                                                                                          □

As we all know, $a \in Raa^*a$ if and only if $a \in aa^*aR$, but $a \in Ra(a^*)^n a$ is not equivalent to $a \in a(a^*)^n aR$ when $n \geqslant 2$. Taking the condition $n = 2$ as an example.

**Example 5.3.12 ([81])** Let $R$ be as Example 5.3.4, and let $a = \sum_{i=1}^{\infty} e_{i,i+1}$. Then $aa^* = 1$, $a^*a = \sum_{i=2}^{\infty} e_{i,i}$. Moreover, $a(a^*)^2 a = a^*a$, $a = aa^*a \in Ra(a^*)^2 a$, but $a \notin a(a^*)^2 aR$. However, when taking $R$ as a Dedekind-finite ring, we will get some unexpected results.

**Theorem 5.3.13 ([81])** *For $a \in R$, $n \geqslant 2$, consider the following conditions:*

(1)  *$R$ is a Dedekind-finite ring.*
(2)  *$a \in a(a^*)^n aR$ if and only if $a \in Ra(a^*)^n a$.*
(3)  *$aa^* = 1$ implies $a^*a = 1$ for any $a \in R$.*

*Then we have (1) $\Rightarrow$ (2) $\Rightarrow$ (3).*

**Proof** (1) $\Rightarrow$ (2). Since $R$ is a Dedekind-finite ring, which guarantees that $(a^*)^n a + 1 - a^- a$ is right invertible if and only if $(a^*)^n a + 1 - a^- a$ is left invertible. Hence, $a \in a(a^*)^n aR$ if and only if $a \in Ra(a^*)^n a$ by Lemma 5.3.8.

(2) $\Rightarrow$ (3). Suppose that $aa^* = 1$, and then $a^k(a^*)^k = 1$ for any $k \geqslant 1$. Moreover, $a = a^{n-1}(a^*)^{n-1}a \in R(a^*)^{n-1}a = Ra(a^*)^n a$, which implies $a \in a(a^*)^n aR = (a^*)^{n-1}aR$ by the given condition (2). Thus there exists $t \in R$ such

that $a = (a^*)^{n-1}at$. Furthermore,

$$
\begin{aligned}
(a^*)^{n-1}a^n &= (a^*)^{n-1}a^{n-1}((a^*)^{n-1}at) \\
&= (a^*)^{n-1}(a^{n-1}(a^*)^{n-1})at \\
&= (a^*)^{n-1}at = a.
\end{aligned}
$$

Hence we obtain

$$
\begin{aligned}
a^*a &= (a^{n-2}(a^*)^{n-2})a^*a(a^{n-1}(a^*)^{n-1}) \\
&= a^{n-2}((a^*)^{n-1}a^n)(a^*)^{n-1} = a^{n-2}a(a^*)^{n-1} \\
&= a^{n-1}(a^*)^{n-1} = 1.
\end{aligned}
$$

$\square$

## 5.4   The Core Inverse of the Sum of Two Core Invertible Elements

In this section, we investigate some sufficient conditions which guarantee core invertibility of the sum of two core invertible elements, and some sufficient conditions that guarantee core invertibility of the difference of two core invertible elements.

**Theorem 5.4.1 ([124])** *Let $a, b \in R^{\oplus}$ with core inverses $a^{\oplus}$ and $b^{\oplus}$, respectively. If $ab = 0$ and $a^*b = 0$, then $a + b$ is core invertible with*

$$
(a+b)^{\oplus} = (1 - b^{\oplus}b)a^{\oplus} + b^{\oplus}.
$$

***Proof*** Suppose $a, b \in R^{\oplus}$. Then by Lemma 5.2.4, we have $a, b \in R^{\#}$ with $a^{\#} = (a^{\oplus})^2 a$ and $b^{\#} = (b^{\oplus})^2 b$, respectively. Since $ab = 0$, it follows that $a + b \in R^{\#}$ by Corollary 4.8.2 and

$$
\begin{aligned}
(a+b)^{\#} &\\
&= (1 - bb^{\#})a^{\#} + b^{\#}(1 - aa^{\#}) \\
&= (1 - b^{\#}b)a^{\#} + b^{\#}(1 - a^{\#}a) \\
&= (1 - (b^{\oplus})^2 b^2)(a^{\oplus})^2 a + (b^{\oplus})^2 b(1 - (a^{\oplus})^2 a^2) \\
&= (1 - b^{\oplus}b)(a^{\oplus})^2 a + (b^{\oplus})^2 b(1 - a^{\oplus}a).
\end{aligned}
$$

Since $ab = 0$ and $a^*b = 0$, we obtain

$$ab^{\circledcirc} = ab(b^{\circledcirc})^2 = 0,$$

$$b^{\circledcirc}a = b^{\circledcirc}bb^{\circledcirc}a = b^{\circledcirc}(bb^{\circledcirc})^*a = b^{\circledcirc}(b^{\circledcirc})^*b^*a = b^{\circledcirc}(b^{\circledcirc})^*(a^*b)^* = 0,$$

$$a^{\circledcirc}b = a^{\circledcirc}aa^{\circledcirc}b = a^{\circledcirc}(aa^{\circledcirc})^*b = a^{\circledcirc}(a^{\circledcirc})^*a^*b = 0.$$

Let $x = (1 - b^{\circledcirc}b)a^{\circledcirc} + b^{\circledcirc}$. Since

$$(a + b)x$$

$$= (a + b)[(1 - b^{\circledcirc}b)a^{\circledcirc} + b^{\circledcirc}]$$

$$= a(1 - b^{\circledcirc}b)a^{\circledcirc} + b(1 - b^{\circledcirc}b)a^{\circledcirc} + ab^{\circledcirc} + bb^{\circledcirc}$$

$$= a(1 - b^{\circledcirc}b)a^{\circledcirc} + bb^{\circledcirc}$$

$$= aa^{\circledcirc} + bb^{\circledcirc} \text{ is Hermitian, and}$$

$$(a + b)x(a + b)$$

$$= (aa^{\circledcirc} + bb^{\circledcirc})(a + b) = aa^{\circledcirc}a + aa^{\circledcirc}b + bb^{\circledcirc}a + bb^{\circledcirc}b$$

$$= aa^{\circledcirc}a + bb^{\circledcirc}b = a + b,$$

we see that $x$ is a $\{1, 3\}$-inverse of $a + b$. Hence by Theorem 5.2.5, we have

$$(a + b)^{\circledcirc}$$

$$= (a + b)^{\#}(a + b)(a + b)^{(1,3)}$$

$$= [(1 - b^{\circledcirc}b)(a^{\circledcirc})^2a + (b^{\circledcirc})^2b(1 - a^{\circledcirc}a)](a + b)[(1 - b^{\circledcirc}b)a^{\circledcirc} + b^{\circledcirc}]$$

$$= [(1 - b^{\circledcirc}b)(a^{\circledcirc})^2a^2 + (1 - b^{\circledcirc}b)(a^{\circledcirc})^2ab + (b^{\circledcirc})^2b(1 - a^{\circledcirc}a)a$$

$$\quad + (b^{\circledcirc})^2b(1 - a^{\circledcirc}a)b][(1 - b^{\circledcirc}b)a^{\circledcirc} + b^{\circledcirc}]$$

$$= [(1 - b^{\circledcirc}b)a^{\circledcirc}a + b^{\circledcirc}b][(1 - b^{\circledcirc}b)a^{\circledcirc} + b^{\circledcirc}]$$

$$= (1 - b^{\circledcirc}b)a^{\circledcirc}a(1 - b^{\circledcirc}b)a^{\circledcirc} + b^{\circledcirc}b(1 - b^{\circledcirc}b)a^{\circledcirc} + (1 - b^{\circledcirc}b)a^{\circledcirc}ab^{\circledcirc} + b^{\circledcirc}bb^{\circledcirc}$$

$$= (1 - b^{\circledcirc}b)a^{\circledcirc} + b^{\circledcirc}.$$

$\square$

**Example 5.4.2** The condition $ab = 0$ in Theorem 5.4.1 cannot be dropped. Indeed, let $R = F^{2\times2}$, where $F$ is a field, and let $*$ be the transpose map. Take $a = \begin{bmatrix} 1 & 0 \\ -1 & 0 \end{bmatrix}$ and $b = \begin{bmatrix} -1 & 0 \\ -1 & 0 \end{bmatrix}$. Then it is easy to see that $a, b \in R^{\circledcirc}$ and $a^*b = 0$. But, $a + b = \begin{bmatrix} 0 & 0 \\ -2 & 0 \end{bmatrix}$ is not group invertible and therefore not core invertible by Theorem 5.2.5.

For Moore-Penrose inverses, it is direct to check that if $a, b \in R^{\dagger}$ and $a^*b = ab^* = 0$ then $a + b \in R^{\dagger}$ with $(a + b)^{\dagger} = a^{\dagger} + b^{\dagger}$. For core inverses, it is in general not the case.

**Remark 5.4.3** Let $R = \mathbb{Z}_4^{2 \times 2}$ and $*$ be the transpose map. Take $a = \begin{bmatrix} -1 & 1 \\ 0 & 0 \end{bmatrix}$ and $b = \begin{bmatrix} 0 & 0 \\ 1 & 1 \end{bmatrix} \in R$. Then we have $a^*b = 0$ and $ab^* = 0$, and it is easy to see that $a, b \in R^{\circledcirc}$. But $a + b \notin R(a + b)^*(a + b)$, which implies that $a + b \notin R^{\{1,3\}}$ and thus $a + b \notin R^{\circledcirc}$. This also shows that the condition $ab = 0$ in Theorem 5.4.1 cannot be dropped.

**Remark 5.4.4** Let $R = F^{2 \times 2}$ and $*$ be the transpose map, where $F$ is a field. Take $a = \begin{bmatrix} 1 & 0 \\ 0 & 0 \end{bmatrix}$ and $b = \begin{bmatrix} 0 & 0 \\ -1 & 0 \end{bmatrix}$. Then $a, b \in R^{\dagger}$ and $a^*b = ab = 0$, yet $a + b \notin R(a + b)^*(a + b)$, that is $a + b \notin R^{\{1,3\}}$, thus $a + b \notin R^{\circledcirc}$.

**Corollary 5.4.5** *Let $a, b \in R^{\circledcirc}$ with core inverses $a^{\circledcirc}$ and $b^{\circledcirc}$, respectively. If $ab = 0 = ba$ and $a^*b = 0$, then $a + b$ is core invertible with*

$$(a + b)^{\circledcirc} = a^{\circledcirc} + b^{\circledcirc}.$$

Similarly, we have the following results for dual core inverses.

**Theorem 5.4.6** *Let $a, b \in R_{\circledcirc}$ with dual core inverses $a_{\circledcirc}$ and $b_{\circledcirc}$, respectively. If $ab = 0$ and $ab^* = 0$, then $a + b$ is dual core invertible with*

$$(a + b)_{\circledcirc} = a_{\circledcirc} + b_{\circledcirc}(1 - aa_{\circledcirc}).$$

**Corollary 5.4.7** *Let $a, b \in R_{\circledcirc}$ with dual core inverses $a_{\circledcirc}$ and $b_{\circledcirc}$, respectively. If $ab = 0 = ba$ and $ab^* = 0$, then $a + b$ is dual core invertible with*

$$(a + b)_{\circledcirc} = a_{\circledcirc} + b_{\circledcirc}.$$

**Theorem 5.4.8 ([134])** *Let $R$ be a Dedekind-finite ring and $2 \in R^{-1}$. If $a, b \in R^{\circledcirc}$ satisfy $a^2 a^{\circledcirc} b^{\circledcirc} b = baa^{\circledcirc}$ and $ab^{\circledcirc} b = aa^{\circledcirc} b$, then $a + b \in R^{\circledcirc}$ and*

$$(a + b)^{\circledcirc} = a^{\circledcirc} + b^{\circledcirc} - \frac{1}{2}a^{\circledcirc}bb^{\circledcirc} - \frac{1}{2}b^{\circledcirc}ba^{\circledcirc} - \frac{1}{2}a^{\circledcirc}ab^{\circledcirc}.$$

***Proof*** Since $a^2 a^{\circledcirc} b^{\circledcirc} b = baa^{\circledcirc}$, we have

$$a^{\circledcirc}b^{\circledcirc}ba^{\circledcirc}baa^{\circledcirc} = a^{\circledcirc}b^{\circledcirc}ba^{\circledcirc}a^2 a^{\circledcirc}b^{\circledcirc}b = a^{\circledcirc}b^{\circledcirc}baa^{\circledcirc}b^{\circledcirc}b = a^{\circledcirc}b^{\circledcirc}baa^{\circledcirc}$$

$$= (a^{\circledcirc})^2 a^2 a^{\circledcirc}b^{\circledcirc}baa^{\circledcirc} = (a^{\circledcirc})^2 baa^{\circledcirc} = (a^{\circledcirc})^2 a^2 a^{\circledcirc}b^{\circledcirc}b = a^{\circledcirc}b^{\circledcirc}b,$$

$$a^{\circledcirc}ba^{\circledcirc}baa^{\circledcirc} = a^{\circledcirc}ba^{\circledcirc}a^2 a^{\circledcirc}b^{\circledcirc}b = a^{\circledcirc}baa^{\circledcirc}b^{\circledcirc}b = a^{\circledcirc}baa^{\circledcirc},$$

$$a^{\circledcirc}baa^{\circledcirc}baa^{\circledcirc} = a^{\circledcirc}a^2 a^{\circledcirc}b^{\circledcirc}bbaa^{\circledcirc} = aa^{\circledcirc}baa^{\circledcirc} = aa^{\circledcirc}a^2 a^{\circledcirc}b^{\circledcirc}b$$

$$= a^2 a^{\circledcirc}b^{\circledcirc}b = baa^{\circledcirc}.$$

Since $baa^{\circledcirc}(a^{\circledcirc}b^{\circledcirc}b)baa^{\circledcirc} = ba^{\circledcirc}baa^{\circledcirc} = ba^{\circledcirc}a^2a^{\circledcirc}b^{\circledcirc}b = baa^{\circledcirc}b^{\circledcirc}b = baa^{\circledcirc}$,
$a^{\circledcirc}b^{\circledcirc}b$ is a $\{1\}$-inverse of $baa^{\circledcirc}$. Then we have

$$(1 + a^{\circledcirc}b^{\circledcirc}b - a^{\circledcirc}baa^{\circledcirc})(1 + baa^{\circledcirc} - a^{\circledcirc}baa^{\circledcirc})$$

$$= 1 + baa^{\circledcirc} - a^{\circledcirc}baa^{\circledcirc} + a^{\circledcirc}b^{\circledcirc}b + a^{\circledcirc}baa^{\circledcirc} - a^{\circledcirc}b^{\circledcirc}ba^{\circledcirc}baa^{\circledcirc}$$

$$- a^{\circledcirc}baa^{\circledcirc} - a^{\circledcirc}baa^{\circledcirc}baa^{\circledcirc} + a^{\circledcirc}ba^{\circledcirc}baa^{\circledcirc}$$

$$= 1 + baa^{\circledcirc} + a^{\circledcirc}b^{\circledcirc}b - a^{\circledcirc}b^{\circledcirc}ba^{\circledcirc}baa^{\circledcirc} - a^{\circledcirc}baa^{\circledcirc} - a^{\circledcirc}baa^{\circledcirc}baa^{\circledcirc}$$

$$+ a^{\circledcirc}ba^{\circledcirc}baa^{\circledcirc} = 1 + baa^{\circledcirc} + a^{\circledcirc}b^{\circledcirc}b - a^{\circledcirc}b^{\circledcirc}b - a^{\circledcirc}baa^{\circledcirc}$$

$$- baa^{\circledcirc} + a^{\circledcirc}baa^{\circledcirc} = 1.$$

Since $R$ is Dedekind-finite, $(1 + baa^{\circledcirc} - a^{\circledcirc}baa^{\circledcirc})(1 + a^{\circledcirc}b^{\circledcirc}b - a^{\circledcirc}baa^{\circledcirc}) = 1$.
Hence $1 + baa^{\circledcirc} - a^{\circledcirc}baa^{\circledcirc} \in R^{-1}$. According to Theorem 3.2.7, we know that
$baa^{\circledcirc}$ is group invertible. By Proposition 3.2.9 and $b^{\circledcirc}bbaa^{\circledcirc} = a^2a^{\circledcirc}b^{\circledcirc}b$, we obtain
$b^{\circledcirc}b(baa^{\circledcirc})^{\#} = a^{\circledcirc}b^{\circledcirc}b$. Then we have

$$b^{\circledcirc}b(baa^{\circledcirc})^{\#} = b^{\circledcirc}bbaa^{\circledcirc}(baa^{\circledcirc})^{\#}(baa^{\circledcirc})^{\#} = baa^{\circledcirc}((baa^{\circledcirc})^{\#})^2 = (baa^{\circledcirc})^{\#}.$$

So $(baa^{\circledcirc})^{\#} = a^{\circledcirc}b^{\circledcirc}b$. Furthermore, $b^{\circledcirc}ba^{\circledcirc}b^{\circledcirc}b = a^{\circledcirc}b^{\circledcirc}b$. Thus $b^{\circledcirc}ba^{\circledcirc}bb^{\circledcirc} = a^{\circledcirc}bb^{\circledcirc}$. It is easy to obtain that $baa^{\circledcirc}a^{\circledcirc}bb^{\circledcirc} = ba^{\circledcirc}bb^{\circledcirc}$ and $baa^{\circledcirc}a^{\circledcirc}bb^{\circledcirc} = a^2a^{\circledcirc}(b^{\circledcirc}ba^{\circledcirc}bb^{\circledcirc}) = aa^{\circledcirc}bb^{\circledcirc}$. That is, $ba^{\circledcirc}bb^{\circledcirc} = aa^{\circledcirc}bb^{\circledcirc}$. Since

$$(a^{\circledcirc}bb^{\circledcirc})^*(baa^{\circledcirc})^*baa^{\circledcirc} = (baa^{\circledcirc}a^{\circledcirc}bb^{\circledcirc})^*baa^{\circledcirc}$$

$$= (aa^{\circledcirc}bb^{\circledcirc})^*baa^{\circledcirc} = bb^{\circledcirc}aa^{\circledcirc}baa^{\circledcirc} = bb^{\circledcirc}baa^{\circledcirc} = baa^{\circledcirc},$$

$a^{\circledcirc}bb^{\circledcirc}$ is a $\{1, 3\}$-inverse of $baa^{\circledcirc}$ by Lemma 2.2.3. So $baa^{\circledcirc}$ is core invertible and

$$(baa^{\circledcirc})^{\circledcirc} = (baa^{\circledcirc})^{\#}(baa^{\circledcirc})(baa^{\circledcirc})^{(1,3)} = a^{\circledcirc}b^{\circledcirc}bbaa^{\circledcirc}a^{\circledcirc}bb^{\circledcirc} = a^{\circledcirc}baa^{\circledcirc}a^{\circledcirc}bb^{\circledcirc}$$

$$= a^{\circledcirc}a^2a^{\circledcirc}b^{\circledcirc}ba^{\circledcirc}bb^{\circledcirc} = aa^{\circledcirc}(b^{\circledcirc}ba^{\circledcirc}bb^{\circledcirc}) = aa^{\circledcirc}a^{\circledcirc}bb^{\circledcirc} = a^{\circledcirc}bb^{\circledcirc}.$$

We know that $baa^{\circledcirc}a^{\circledcirc}bb^{\circledcirc} = aa^{\circledcirc}bb^{\circledcirc}$ is Hermitian, i.e.,

$$aa^{\circledcirc}bb^{\circledcirc} = bb^{\circledcirc}aa^{\circledcirc}. \tag{5.4.1}$$

Since $b$ is core invertible, $a^2a^{\circledcirc}b^{\circledcirc}b = baa^{\circledcirc}$, $aa^{\circledcirc}a^2a^{\circledcirc}b^{\circledcirc}b = baa^{\circledcirc}$ and $baa^{\circledcirc}$ is group invertible,

$$(a^2a^{\circledcirc}b^{\circledcirc}b)^{\#} = b^{\#}aa^{\circledcirc} = b^{\#}bb^{\#}aa^{\circledcirc} = (b^{\circledcirc})^2baa^{\circledcirc}.$$

Since $a^2a^{⊕}b^{⊕}b(a^2a^{⊕}b^{⊕}b)^{(1,3)} = baa^{⊕}(baa^{⊕})^{(1,3)} = aa^{⊕}bb^{⊕}$,

$$(a^2a^{⊕}b^{⊕}b)^{⊕} = (a^2a^{⊕}b^{⊕}b)^{\#}(a^2a^{⊕}b^{⊕}b)(a^2a^{⊕}b^{⊕}b)^{(1,3)} = (b^{⊕})^2baa^{⊕}aa^{⊕}bb^{⊕}$$
$$= (b^{⊕})^2baa^{⊕}bb^{⊕} = (b^{⊕})^2bbb^{⊕}aa^{⊕} = b^{⊕}aa^{⊕}.$$

Thus

$$a^{⊕}bb^{⊕} = b^{⊕}aa^{⊕}. \qquad (5.4.2)$$

By the condition $ab^{⊕}b = aa^{⊕}b$, we have

$$ab^{⊕} = ab^{⊕}bb^{⊕} = aa^{⊕}bb^{⊕}, \qquad (5.4.3)$$
$$ab^{⊕}ba^{⊕} = aa^{⊕}ba^{⊕} = aa^{⊕}baa^{⊕}a^{⊕} = baa^{⊕}a^{⊕} = ba^{⊕}, \qquad (5.4.4)$$
$$ba^{⊕}ab^{⊕} = ba^{⊕}aa^{⊕}bb^{⊕} = ba^{⊕}bb^{⊕} = bb^{⊕}aa^{⊕} = aa^{⊕}bb^{⊕} = ab^{⊕}, \quad (5.4.5)$$
$$b^{⊕}ba = b^{⊕}ba^{⊕}a^2 = b^{⊕}ab^{⊕}ba^{⊕}a^2 = b^{⊕}ab^{⊕}ba = b^{⊕}aa^{⊕}ba \qquad (5.4.6)$$
$$= a^{⊕}bb^{⊕}ba = a^{⊕}ba. \qquad (5.4.7)$$

By the above equalities, we have the following equalities:

$$b^{⊕}ba^{⊕}ab = b^{⊕}ba^{⊕}ab^{⊕}b^2 = b^{⊕}bb^{⊕}aa^{⊕}b^2 = b^{⊕}aa^{⊕}b^2 = a^{⊕}b^2, \quad (5.4.8)$$
$$b^{⊕}ba^{⊕}ba = b^{⊕}ba^{⊕}bb^{⊕}ba = b^{⊕}aa^{⊕}ba = a^{⊕}ba, \qquad (5.4.9)$$
$$b^{⊕}ba^{⊕}b^2 = b^{⊕}ba^{⊕}bb^{⊕}b^2 = b^{⊕}aa^{⊕}b^2 = a^{⊕}b^2, \qquad (5.4.10)$$
$$a^{⊕}ab = a^{⊕}ab^{⊕}bb = a^{⊕}aa^{⊕}bb^{⊕}b^2 = a^{⊕}b^2, \qquad (5.4.11)$$
$$a^{⊕}ab^{⊕}a^2 = a^{⊕}bb^{⊕}a^2 = b^{⊕}a^2, \ a^{⊕}ab^{⊕}b^2 = a^{⊕}b^2, \qquad (5.4.12)$$
$$a^{⊕}ab^{⊕}ab = a^{⊕}bb^{⊕}ab = b^{⊕}aa^{⊕}ab = b^{⊕}ab, \qquad (5.4.13)$$
$$a^{⊕}ab^{⊕}ba = a^{⊕}aa^{⊕}bb^{⊕}ba = a^{⊕}ba = b^{⊕}ba. \qquad (5.4.14)$$

Let $x = a^{⊕} + b^{⊕} - \frac{1}{2}a^{⊕}bb^{⊕} - \frac{1}{2}b^{⊕}ba^{⊕} - \frac{1}{2}a^{⊕}ab^{⊕}$. Since $ab^{⊕}b = aa^{⊕}b$, by equalities (5.4.2)–(5.4.5), we have

$$(a+b)x = (a+b)\left(a^{⊕} + b^{⊕} - \frac{1}{2}a^{⊕}bb^{⊕} - \frac{1}{2}b^{⊕}ba^{⊕} - \frac{1}{2}a^{⊕}ab^{⊕}\right)$$

$$= aa^{⊕} + ab^{⊕} - \frac{1}{2}aa^{⊕}bb^{⊕} - \frac{1}{2}ab^{⊕}ba^{⊕} - \frac{1}{2}aa^{⊕}ab^{⊕}$$

$$+ ba^{⊕} + bb^{⊕} - \frac{1}{2}ba^{⊕}bb^{⊕} - \frac{1}{2}ba^{⊕} - \frac{1}{2}ba^{⊕}ab^{⊕}$$

$$= aa^{⊕} + \frac{1}{2}ab^{⊕} + \frac{1}{2}ba^{⊕} - \frac{1}{2}aa^{⊕}bb^{⊕} - \frac{1}{2}ab^{⊕}ba^{⊕} - \frac{1}{2}ba^{⊕}bb^{⊕} - \frac{1}{2}ba^{⊕}ab^{⊕} + bb^{⊕}$$

$$= aa^\oplus + bb^\oplus + \frac{1}{2}aa^\oplus bb^\oplus + \frac{1}{2}ba^\oplus - \frac{1}{2}aa^\oplus bb^\oplus - \frac{1}{2}ba^\oplus - \frac{1}{2}bb^\oplus aa^\oplus - \frac{1}{2}bb^\oplus aa^\oplus$$

$$= aa^\oplus + bb^\oplus - bb^\oplus aa^\oplus = aa^\oplus + bb^\oplus - aa^\oplus bb^\oplus.$$

Thus $(a + b)x$ is Hermitian. According to equalities (5.4.1) and (5.4.2), we have

$$(a + b)x^2 = (aa^\oplus + bb^\oplus - aa^\oplus bb^\oplus)(a^\oplus + b^\oplus - \frac{1}{2}a^\oplus bb^\oplus - \frac{1}{2}b^\oplus ba^\oplus - \frac{1}{2}a^\oplus ab^\oplus)$$

$$= a^\oplus + aa^\oplus b^\oplus - \frac{1}{2}a^\oplus bb^\oplus - \frac{1}{2}aa^\oplus b^\oplus ba^\oplus - \frac{1}{2}a^\oplus ab^\oplus$$

$$+ bb^\oplus a^\oplus + b^\oplus - \frac{1}{2}bb^\oplus(a^\oplus bb^\oplus)$$

$$- \frac{1}{2}b^\oplus ba^\oplus - \frac{1}{2}bb^\oplus a^\oplus ab^\oplus - aa^\oplus bb^\oplus a^\oplus - aa^\oplus b^\oplus + \frac{1}{2}aa^\oplus bb^\oplus a^\oplus bb^\oplus$$

$$+ \frac{1}{2}aa^\oplus b^\oplus ba^\oplus + \frac{1}{2}aa^\oplus bb^\oplus a^\oplus ab^\oplus$$

$$= a^\oplus - \frac{1}{2}a^\oplus bb^\oplus - \frac{1}{2}aa^\oplus b^\oplus ba^\oplus - \frac{1}{2}a^\oplus ab^\oplus + bb^\oplus a^\oplus + b^\oplus - \frac{1}{2}bb^\oplus(b^\oplus aa^\oplus)$$

$$- \frac{1}{2}b^\oplus ba^\oplus - \frac{1}{2}bb^\oplus a^\oplus ab^\oplus - (bb^\oplus aa^\oplus)a^\oplus + \frac{1}{2}(bb^\oplus aa^\oplus)a^\oplus bb^\oplus$$

$$+ \frac{1}{2}aa^\oplus b^\oplus ba^\oplus + \frac{1}{2}(bb^\oplus aa^\oplus)a^\oplus ab^\oplus$$

$$= a^\oplus + b^\oplus - \frac{1}{2}a^\oplus bb^\oplus - \frac{1}{2}aa^\oplus b^\oplus ba^\oplus - \frac{1}{2}a^\oplus ab^\oplus + bb^\oplus a^\oplus - \frac{1}{2}b^\oplus aa^\oplus - \frac{1}{2}b^\oplus ba^\oplus$$

$$- \frac{1}{2}bb^\oplus a^\oplus ab^\oplus - bb^\oplus a^\oplus + \frac{1}{2}bb^\oplus a^\oplus bb^\oplus + \frac{1}{2}aa^\oplus b^\oplus ba^\oplus + \frac{1}{2}bb^\oplus a^\oplus ab^\oplus$$

$$= a^\oplus + b^\oplus - \frac{1}{2}a^\oplus bb^\oplus - \frac{1}{2}a^\oplus ab^\oplus - \frac{1}{2}b^\oplus aa^\oplus - \frac{1}{2}b^\oplus ba^\oplus + \frac{1}{2}b^\oplus aa^\oplus$$

$$= a^\oplus + b^\oplus - \frac{1}{2}a^\oplus bb^\oplus - \frac{1}{2}a^\oplus ab^\oplus - \frac{1}{2}b^\oplus ba^\oplus = x.$$

By equalities (5.4.2) and (5.4.6)–(5.4.14), we have

$$x(a + b)(a + b) = (a^\oplus + b^\oplus - \frac{1}{2}a^\oplus bb^\oplus - \frac{1}{2}b^\oplus ba^\oplus - \frac{1}{2}a^\oplus ab^\oplus)$$

$$(a^2 + ab + ba + b^2)$$

$$= a + a^\oplus ab + a^\oplus ba + a^\oplus b^2 + b^\oplus a^2 + b^\oplus ab + b^\oplus ba + b - \frac{1}{2}a^\oplus bb^\oplus a^2$$

$$- \frac{1}{2}a^\oplus bb^\oplus ab - \frac{1}{2}a^\oplus bb^\oplus ba - \frac{1}{2}a^\oplus bb^\oplus b^2$$

$$- \frac{1}{2}b^{\circledast}ba^{\circledast}a^2 - \frac{1}{2}b^{\circledast}ba^{\circledast}ab - \frac{1}{2}b^{\circledast}ba^{\circledast}ba$$

$$- \frac{1}{2}b^{\circledast}ba^{\circledast}b^2 - \frac{1}{2}a^{\circledast}ab^{\circledast}a^2 - \frac{1}{2}a^{\circledast}ab^{\circledast}ab - \frac{1}{2}a^{\circledast}ab^{\circledast}ba - \frac{1}{2}a^{\circledast}ab^{\circledast}b^2$$

$$= a + b + a^{\circledast}ab + a^{\circledast}ba + a^{\circledast}b^2 + b^{\circledast}a^2 + b^{\circledast}ab + b^{\circledast}ba - \frac{1}{2}b^{\circledast}a^2 - \frac{1}{2}b^{\circledast}ab$$

$$- \frac{1}{2}a^{\circledast}ba - \frac{1}{2}a^{\circledast}b^2 - \frac{1}{2}b^{\circledast}ba - \frac{1}{2}b^{\circledast}ba^{\circledast}ab - \frac{1}{2}b^{\circledast}ba^{\circledast}ba$$

$$- \frac{1}{2}b^{\circledast}ba^{\circledast}b^2 - \frac{1}{2}a^{\circledast}ab^{\circledast}a^2$$

$$- \frac{1}{2}a^{\circledast}ab^{\circledast}ab - \frac{1}{2}a^{\circledast}ab^{\circledast}ba - \frac{1}{2}a^{\circledast}ab^{\circledast}b^2$$

$$= a + b + a^{\circledast}b^2 + a^{\circledast}ba + a^{\circledast}b^2 + b^{\circledast}a^2 + b^{\circledast}ab + a^{\circledast}ba$$

$$- \frac{1}{2}b^{\circledast}a^2 - \frac{1}{2}b^{\circledast}ab - \frac{1}{2}a^{\circledast}ba$$

$$- \frac{1}{2}a^{\circledast}b^2 - \frac{1}{2}b^{\circledast}ba - \frac{1}{2}a^{\circledast}b^2 - \frac{1}{2}a^{\circledast}ba - \frac{1}{2}a^{\circledast}b^2 - \frac{1}{2}b^{\circledast}a^2$$

$$- \frac{1}{2}b^{\circledast}ab - \frac{1}{2}a^{\circledast}ba - \frac{1}{2}a^{\circledast}b^2$$

$$= a + b.$$

Hence, $(a + b)^{\circledast} = a^{\circledast} + b^{\circledast} - \frac{1}{2}a^{\circledast}bb^{\circledast} - \frac{1}{2}b^{\circledast}ba^{\circledast} - \frac{1}{2}a^{\circledast}ab^{\circledast}$.                    □

In general, the core inverse of $a - b$ does not exist under the conditions of Theorem 5.4.8. Next, the sufficient condition, which guarantee that $a - b$ has core inverse, is established.

**Theorem 5.4.9 ([134])** *Let $R$ be a Dedekind-finite ring. If $a$, $b \in R^{\circledast}$ satisfy $a^2a^{\circledast}b^{\circledast}b = baa^{\circledast}$ and $ab^{\circledast}b = aa^{\circledast}b = bb^{\circledast}a$, then $a - b \in R^{\circledast}$ and*

$$(a - b)^{\circledast} = (a - b)(a^{\circledast} - b^{\circledast})^2.$$

*Proof* Since $a^2a^{\circledast}b^{\circledast}b = baa^{\circledast}$ and $ab^{\circledast}b = aa^{\circledast}b$, according to the proof of Theorem 5.4.8, we have the following equalities:

$$a^{\circledast}bb^{\circledast} = b^{\circledast}aa^{\circledast}, \ ab^{\circledast} = aa^{\circledast}bb^{\circledast} = bb^{\circledast}aa^{\circledast}, \ a(b^{\circledast})^2 = aa^{\circledast}b^{\circledast}, \quad (5.4.15)$$

$$ba^{\circledast} = ab^{\circledast}ba^{\circledast} = bb^{\circledast}aa^{\circledast} = ba^{\circledast}ab^{\circledast}, \ b(a^{\circledast})^2 = bb^{\circledast}a^{\circledast}, \quad (5.4.16)$$

$$a^{\circledast}b^2 = b^{\circledast}ba^{\circledast}ab = a^{\circledast}ab, \ b^{\circledast}a^2 = a^{\circledast}ab^{\circledast}a^2, \ b^{\circledast}ba^{\circledast}ba = a^{\circledast}ba, \quad (5.4.17)$$

$$a^{\circledast}ab^{\circledast}ab = b^{\circledast}ab, \ a^{\circledast}ab^{\circledast}ba = b^{\circledast}ba, \quad (5.4.18)$$

$$bb^{\circledast}a^{\circledast}ab = a^{\circledast}b^2, \ baa^{\circledast}b^{\circledast} = a^2a^{\circledast}b^{\circledast}. \quad (5.4.19)$$

By the above equalities and $ab^{⊕}b = aa^{⊕}b = bb^{⊕}a$, we have

$$bab^{⊕}a^{⊕} = b^2b^{⊕}a^{⊕} = bbb^{⊕}a(a^{⊕})^2 = baa^{⊕}b(a^{⊕})^2 = a^2b^{⊕}a^{⊕}, \quad (5.4.20)$$

$$a^2b^{⊕}a^{⊕} = abb^{⊕}aa^{⊕}a^{⊕} = abb^{⊕}a^{⊕}, \quad (5.4.21)$$

$$aba^{⊕}b^{⊕} = aaa^{⊕}bb^{⊕}b^{⊕} = a^2a^{⊕}b^{⊕}, \quad (5.4.22)$$

$$b^2a^{⊕}b^{⊕} = baa^{⊕}bb^{⊕}b^{⊕} = baa^{⊕}b^{⊕} = a^2a^{⊕}b^{⊕}, \quad (5.4.23)$$

$$ab(a^{⊕})^2 = abb^{⊕}aa^{⊕}a^{⊕} = aaa^{⊕}bb^{⊕}a^{⊕} = a^2b^{⊕}a^{⊕}. \quad (5.4.24)$$

Let $y = (a - b)(a^{⊕} - b^{⊕})^2$. By equalities (5.4.15), (5.4.16) and (5.4.20)–(5.4.24), then

$$(a - b)y = (a - b)^2(a^{⊕} - b^{⊕})^2$$

$$= (a^2 + b^2 - ab - ba)((a^{⊕})^2 + (b^{⊕})^2 - a^{⊕}b^{⊕} - b^{⊕}a^{⊕})$$

$$= aa^{⊕} + a^2(b^{⊕})^2 - a^2a^{⊕}b^{⊕} - a^2b^{⊕}a^{⊕} + b^2(a^{⊕})^2 + bb^{⊕} - b^2a^{⊕}b^{⊕} - b^2b^{⊕}a^{⊕}$$

$$- ab(a^{⊕})^2 - ab^{⊕} + aba^{⊕}b^{⊕} + abb^{⊕}a^{⊕} - ba^{⊕} - ba(b^{⊕})^2 + baa^{⊕}b^{⊕} + bab^{⊕}a^{⊕}$$

$$= aa^{⊕} + a^2a^{⊕}b^{⊕} - a^2a^{⊕}b^{⊕} - a^2b^{⊕}a^{⊕} + b^2b^{⊕}a^{⊕} + bb^{⊕} - b^2a^{⊕}b^{⊕}$$

$$- b^2b^{⊕}a^{⊕} - a^2b^{⊕}a^{⊕} - ab^{⊕} + a^2a^{⊕}b^{⊕} + abb^{⊕}a^{⊕} - ba^{⊕} - a^2a^{⊕}b^{⊕}$$

$$+ a^2a^{⊕}b^{⊕} + a^2b^{⊕}a^{⊕}$$

$$= aa^{⊕} + bb^{⊕} - aa^{⊕}bb^{⊕} - bb^{⊕}aa^{⊕}.$$

So $(a - b)y$ is Hermitian. Since $aa^{⊕}b = bb^{⊕}a$,

$$(a - b)y^2 = (aa^{⊕} + bb^{⊕} - aa^{⊕}bb^{⊕} - bb^{⊕}aa^{⊕})(a - b)(a^{⊕} - b^{⊕})^2$$

$$= (a + bb^{⊕}a - aa^{⊕}bb^{⊕}a - bb^{⊕}a - aa^{⊕}b - b + aa^{⊕}b + bb^{⊕}aa^{⊕}b)(a^{⊕} - b^{⊕})^2$$

$$= (a - b - bb^{⊕}a + aa^{⊕}b)(a^{⊕} - b^{⊕})^2 = (a - b)(a^{⊕} - b^{⊕})^2 = y.$$

By equalities $aa^{⊕}b = bb^{⊕}a$ and (5.4.15)–(5.4.18), we obtain

$$y(a - b)^2 = (a - b)(a^{⊕} - b^{⊕})^2(a - b)^2$$

$$= (a^{⊕} + a(b^{⊕})^2 - aa^{⊕}b^{⊕} - ab^{⊕}a^{⊕} - b(a^{⊕})^2 - b^{⊕} + ba^{⊕}b^{⊕} + bb^{⊕}a^{⊕})(a - b)^2$$

$$= (a^{⊕} - b^{⊕} - ab^{⊕}a^{⊕} + ba^{⊕}b^{⊕})(a^2 + b^2 - ab - ba)$$

$$= (a^{⊕} - b^{⊕} - bb^{⊕}a^{⊕} + aa^{⊕}b^{⊕})(a^2 + b^2 - ab - ba)$$

$$= a + a^{⊕}b^2 - a^{⊕}ab - a^{⊕}ba - b^{⊕}a^2 - b + b^{⊕}ab + b^{⊕}ba - bb^{⊕}a - bb^{⊕}a^{⊕}b^2$$

$$+ bb^{⊕}a^{⊕}ab + bb^{⊕}a^{⊕}ba + aa^{⊕}b^{⊕}a^2 + aa^{⊕}b - aa^{⊕}b^{⊕}ab - aa^{⊕}b^{⊕}ba$$

$$= a + a^{\oplus}b^2 - a^{\oplus}ab - a^{\oplus}ba - b^{\oplus}a^2 - b + b^{\oplus}ab + b^{\oplus}ba - bb^{\oplus}a - a^{\oplus}b^2 + a^{\oplus}ab$$

$$+ a^{\oplus}ba + b^{\oplus}a^2 + aa^{\oplus}b - b^{\oplus}ab - b^{\oplus}ba = a - b.$$

Therefore, it is shown that $(a - b)^{\oplus} = (a - b)(a^{\oplus} - b^{\oplus})^2$.          □

Finally, sufficient conditions, which ensure that the core inverses of $a + b$ and $a - b$ exist, are presented.

**Theorem 5.4.10 ([134])** *Let* $a$, $b \in R^{\oplus}$ *and* $2 \in R^{-1}$. *If* $ba^{\oplus}a = a$, *then* $a + b \in R^{\oplus}$ *and*

$$(a + b)^{\oplus} = -\frac{1}{2}a^{\oplus} + b^{\oplus} + \frac{1}{2}a^{\oplus}bb^{\oplus} - \frac{1}{2}a^{\oplus}ab^{\oplus}.$$

**Proof** Since $ba^{\oplus}a = a$,

$$ab^{\oplus} = ba^{\oplus}ab^{\oplus}, \ ba^{\oplus} = ba^{\oplus}aa^{\oplus} = aa^{\oplus}. \tag{5.4.25}$$

From

$$aa^{\oplus} = ba^{\oplus}aa^{\oplus} = bb^{\oplus}ba^{\oplus}aa^{\oplus} = bb^{\oplus}aa^{\oplus},$$

we obtain $aa^{\oplus} = (aa^{\oplus})^* = (bb^{\oplus}aa^{\oplus})^* = aa^{\oplus}bb^{\oplus}$. That is

$$aa^{\oplus} = aa^{\oplus}bb^{\oplus} = bb^{\oplus}aa^{\oplus}. \tag{5.4.26}$$

Thus we have the following equalities:

$$bb^{\oplus}a^{\oplus} = bb^{\oplus}a(a^{\oplus})^2 = bb^{\oplus}ba^{\oplus}a(a^{\oplus})^2 = ba^{\oplus}a(a^{\oplus})^2 = a^{\oplus}, \tag{5.4.27}$$

$$a^{\oplus}bb^{\oplus}a^2 = a^{\oplus}bb^{\oplus}ba^{\oplus}aa = a^{\oplus}ba^{\oplus}aa = a, \tag{5.4.28}$$

$$a^{\oplus}bb^{\oplus}ab = a^{\oplus}bb^{\oplus}ba^{\oplus}ab = a^{\oplus}ba^{\oplus}ab = a^{\oplus}ab, \tag{5.4.29}$$

$$a^{\oplus}ab^{\oplus}ba = a^{\oplus}ab^{\oplus}bba^{\oplus}a = a^{\oplus}a^2 = a, \tag{5.4.30}$$

$$b^{\oplus}ba = b^{\oplus}bba^{\oplus}a = ba^{\oplus}a = a, \ b^{\oplus}a^2 = b^{\oplus}ba^{\oplus}a^2 = a, \tag{5.4.31}$$

$$b^{\oplus}ab = b^{\oplus}ba^{\oplus}ab = b^{\oplus}ba(a^{\oplus})^2ab = a^{\oplus}ab, \tag{5.4.32}$$

$$a^{\oplus}ab^{\oplus}ab = a^{\oplus}aa^{\oplus}ab = a^{\oplus}ab. \tag{5.4.33}$$

Let $x = -\frac{1}{2}a^{\oplus} + b^{\oplus} + \frac{1}{2}a^{\oplus}bb^{\oplus} - \frac{1}{2}a^{\oplus}ab^{\oplus}$. By equalities (5.4.25) and (5.4.26), we have

$$(a+b)x = (a+b)(-\frac{1}{2}a^{\oplus} + b^{\oplus} + \frac{1}{2}a^{\oplus}bb^{\oplus} - \frac{1}{2}a^{\oplus}ab^{\oplus})$$

$$= -\frac{1}{2}aa^{\oplus} + ab^{\oplus} + \frac{1}{2}aa^{\oplus}bb^{\oplus} - \frac{1}{2}ab^{\oplus} - \frac{1}{2}ba^{\oplus} + bb^{\oplus}$$

$$+ \frac{1}{2}ba^{\oplus}bb^{\oplus} - \frac{1}{2}ba^{\oplus}ab^{\oplus}$$

$$= -\frac{1}{2}aa^{\oplus} + \frac{1}{2}ab^{\oplus} + \frac{1}{2}aa^{\oplus} - \frac{1}{2}aa^{\oplus} + bb^{\oplus} + \frac{1}{2}aa^{\oplus}bb^{\oplus} - \frac{1}{2}ab^{\oplus}$$

$$= bb^{\oplus}.$$

Therefore, $(a+b)x$ is Hermitian. By equality (5.4.27), we obtain

$$(a+b)x^2 = bb^{\oplus}(-\frac{1}{2}a^{\oplus} + b^{\oplus} + \frac{1}{2}a^{\oplus}bb^{\oplus} - \frac{1}{2}a^{\oplus}ab^{\oplus})$$

$$= -\frac{1}{2}bb^{\oplus}a^{\oplus} + b^{\oplus} + \frac{1}{2}bb^{\oplus}a^{\oplus}bb^{\oplus} - \frac{1}{2}bb^{\oplus}a^{\oplus}ab^{\oplus}$$

$$= -\frac{1}{2}a^{\oplus} + b^{\oplus} + \frac{1}{2}a^{\oplus}bb^{\oplus} - \frac{1}{2}a^{\oplus}ab^{\oplus} = x.$$

By equalities (5.4.28)–(5.4.33), we have

$$x(a+b)^2 = (-\frac{1}{2}a^{\oplus} + b^{\oplus} + \frac{1}{2}a^{\oplus}bb^{\oplus} - \frac{1}{2}a^{\oplus}ab^{\oplus})(a+b)^2$$

$$= (-\frac{1}{2}a^{\oplus} + b^{\oplus} + \frac{1}{2}a^{\oplus}bb^{\oplus} - \frac{1}{2}a^{\oplus}ab^{\oplus})(a^2 + b^2 + ab + ba)$$

$$= -\frac{1}{2}a - \frac{1}{2}a^{\oplus}b^2 - \frac{1}{2}a^{\oplus}ab - \frac{1}{2}a^{\oplus}ba + b^{\oplus}a^2 + b + b^{\oplus}ab$$

$$+ b^{\oplus}ba + \frac{1}{2}a^{\oplus}bb^{\oplus}a^2 + \frac{1}{2}a^{\oplus}b^2$$

$$+ \frac{1}{2}a^{\oplus}bb^{\oplus}ab + \frac{1}{2}a^{\oplus}bb^{\oplus}ba - \frac{1}{2}a^{\oplus}ab^{\oplus}a^2$$

$$- \frac{1}{2}a^{\oplus}ab^{\oplus}b^2$$

$$- \frac{1}{2}a^{\oplus}ab^{\oplus}ab - \frac{1}{2}a^{\oplus}ab^{\oplus}ba$$

$$= -\frac{1}{2}a - \frac{1}{2}a^{\oplus}b^2 - \frac{1}{2}a^{\oplus}ab - \frac{1}{2}a^{\oplus}ba$$

$$+ a + b + a^{\circledast}ab + a + \frac{1}{2}a + \frac{1}{2}a^{\circledast}b^2 + \frac{1}{2}a^{\circledast}ab$$

$$+ \frac{1}{2}a^{\circledast}ba - \frac{1}{2}a - \frac{1}{2}a^{\circledast}ab - \frac{1}{2}a^{\circledast}ab - \frac{1}{2}a$$

$$= a + b.$$

So, we have that $(a + b)^{\circledast} = -\frac{1}{2}a^{\circledast} + b^{\circledast} + \frac{1}{2}a^{\circledast}bb^{\circledast} - \frac{1}{2}a^{\circledast}ab^{\circledast}$. □

According to the Theorem 5.4.10, sufficient condition, which guarantee the core inverse of $a - b$ exists, is given.

**Theorem 5.4.11 ([134])** *Let $a, b \in R^{\circledast}$. If $ba^{\circledast}a = a = aa^{\circledast}b$, then $a - b \in R^{\circledast}$ and*

$$(a - b)^{\circledast} = aa^{\circledast}b^{\circledast} - b^{\circledast}.$$

***Proof*** Since $a = aa^{\circledast}b$, $ab^{\circledast} = aa^{\circledast}bb^{\circledast}$. According to the proof of Theorem 5.4.10, we have the following equalities:

$$aa^{\circledast} = aa^{\circledast}bb^{\circledast}, \; b^{\circledast}ba = a = aa^{\circledast}b^{\circledast}ba = b^{\circledast}a^2,$$

$$b^{\circledast}ab = a^{\circledast}ab = aa^{\circledast}b^{\circledast}ab,$$

$$baa^{\circledast}b^{\circledast} = ba^{\circledast}a^2a^{\circledast}b^{\circledast} = a^2a^{\circledast}b^{\circledast}.$$

Let $x = aa^{\circledast}b^{\circledast} - b^{\circledast}$. Then

$$(a - b)x = (a - b)(aa^{\circledast}b^{\circledast} - b^{\circledast})$$

$$= a^2a^{\circledast}b^{\circledast} - ab^{\circledast} - baa^{\circledast}b^{\circledast} + bb^{\circledast}$$

$$= -aa^{\circledast}bb^{\circledast} + bb^{\circledast}$$

$$= -aa^{\circledast} + bb^{\circledast}.$$

Thus, $(a - b)x$ is Hermitian. Since

$$(a - b)x^2 = (-aa^{\circledast} + bb^{\circledast})(aa^{\circledast}b^{\circledast} - b^{\circledast})$$

$$= -aa^{\circledast}b^{\circledast} + aa^{\circledast}b^{\circledast} + bb^{\circledast}aa^{\circledast}b^{\circledast} - b^{\circledast}$$

$$= aa^{\circledast}bb^{\circledast}b^{\circledast} - b^{\circledast}$$

$$= aa^{\circledast}b^{\circledast} - b^{\circledast} = x$$

and

$$x(a-b)^2 = (aa^{\circledast}b^{\circledast} - b^{\circledast})(a^2 + b^2 - ab - ba)$$

$$= aa^{\circledast}b^{\circledast}a^2 + aa^{\circledast}b - aa^{\circledast}b^{\circledast}ab - aa^{\circledast}b^{\circledast}ba - b^{\circledast}a^2 - b + b^{\circledast}ab + b^{\circledast}ba$$

$$= a + aa^{\circledast}b - a^{\circledast}ab - a - a - b + a^{\circledast}ab + a$$

$$= a - b,$$

then, $(a-b)^{\circledast} = aa^{\circledast}b^{\circledast} - b^{\circledast}$.                                            ⊔

## 5.5   The Core Inverse of a Product $paq$

In this section, the properties of core inverse and dual core inverse of a product are investigated and the formulae are obtained.

In view of Corollary 3.2.8, if $a \in R$ is regular, then $a \in R^{\#}$ if and only if $s = a + 1 - aa^-$ is a unit, independent of the choice of $a^-$, or, equivalently, $t = a + 1 - a^- a$ is a unit, in which case

$$a^{\#} = s^{-2}a = at^{-2}. \tag{5.5.1}$$

Using this remark and Theorem 5.2.5, we have the following result.

**Proposition 5.5.1** *Let $a \in R$. Then:*

(1) $a \in R^{\circledast}$ *if and only if $a\{1,3\} \neq \emptyset$ and $u = a + 1 - aa^{(1,3)}$ is a unit. In this case,*
$a^{\circledast} = u^{-1}aa^{(1,3)}$.

(2) $a \in R_{\circledast}$ *if and only if $a\{1,4\} \neq \emptyset$ and $s = a + 1 - a^{(1,4)}a$ is a unit. In this case,*
$a_{\circledast} = a^{(1,4)}as^{-1}$.

**Proof** (1) From the above remark, we know that $a \in R^{\#}$ if and only if $u = a + 1 - aa^{(1,3)}$ is a unit. By Theorem 5.2.5, $a \in R^{\circledast}$ if and only if $a\{1,3\} \neq \emptyset$ and $u = a + 1 - aa^{(1,3)}$ is a unit.

As $u = a + 1 - aa^{(1,3)}$ is a unit, we have $ua = a^2$, and then $a = u^{-1}a^2$. Thus, by Theorem 5.2.5 and Eq. (5.5.1), we have

$$a^{\circledast} = a^{\#}aa^{(1,3)} = (u^{-2}a)aa^{(1,3)} = u^{-1}(u^{-1}a^2)a^{(1,3)} = u^{-1}aa^{(1,3)}.$$

(2) is similar to (1).                                                            □

Note that $v = a^* + 1 - aa^{(1,3)}$ is a unit if and only if $u = v^* = a + 1 - aa^{(1,3)}$ is a unit.

Now, we investigate the existence of the core inverse of $paq$ with $p, a, q \in R$, and we first let $a\{1,4\} \neq \emptyset$.

**Theorem 5.5.2 ([67])** *Let $a, p, q \in R$ be such that $a\{1, 4\} \neq \emptyset$. Suppose that there exist $p', q' \in R$ such that $p'pa = a = aqq'$. Then the following are equivalent:*

(1) $(paq)^{\circledcirc}$ *exists.*
(2) $u = (pa)^*pa + 1 - a^{(1,4)}a$ *and* $v = paq + 1 - pau^{-1}(pa)^*$ *are units.*

*In this case,*

$$(paq)^{\circledcirc} = v^{-1}pau^{-1}(pa)^*.$$

**Proof** By Proposition 5.5.1, $paq \in R^{\circledcirc}$ if and only if $(paq)\{1, 3\} \neq \emptyset$ and $v = paq + 1 - paq(paq)^{(1,3)}$ is a unit, and in this case $(paq)^{\circledcirc} = v^{-1}paq(paq)^{(1,3)}$. Since $a\{1, 4\} \neq \emptyset$ and $p'pa = a = aqq'$, by Theorem 2.3.8, it follows that $(paq)\{1, 3\} \neq \emptyset$ if and only if $u = (pa)^*pa + 1 - a^{(1,4)}a$ is a unit, and in this case Eq. (2.3.1) holds. Therefore, using Proposition 5.5.1 again, we have $(paq)^{\circledcirc}$ exists if and only if $u$ and $v$ are units. By substituting (2.3.1) into $v$, we obtain

$$v = paq + 1 - paq(q'a^{(1,4)}au^{-1}(pa)^*) = paq + 1 - pau^{-1}(pa)^*.$$

Moreover,

$$(paq)^{\circledcirc} = v^{-1}paq(paq)^{(1,3)} = v^{-1}paq(q'a^{(1,4)}au^{-1}(pa)^*) = v^{-1}pau^{-1}(pa)^*.$$

$\square$

Now, if $a\{1, 3\} \neq \emptyset$, we consider the existence of the core inverse of $paq$.

**Theorem 5.5.3 ([67])** *Let $a, p, q \in R$ be such that $a\{1, 3\} \neq \emptyset$. Suppose that there exist $p', q' \in R$ such that $p'pa = a = aqq'$. Then the following are equivalent:*

(1) $(paq)^{\circledcirc}$ *exists.*
(2) $u = aqpaa^{(1,3)} + 1 - aa^{(1,3)}$ *is a unit and* $(paq)\{1, 3\} \neq \emptyset$.
(3) $v = a^{(1,3)}aqpa + 1 - a^{(1,3)}a$ *is a unit and* $(paq)\{1, 3\} \neq \emptyset$.

*In this case,*

$$(paq)^{\circledcirc} = pu^{-2}aqpaq(paq)^{(1,3)} = pav^{-2}qpaq(paq)^{(1,3)}.$$

**Proof** In view of Corollary 3.3.3, $(paq)^{\#}$ exists if and only if $u = aqpaa^{(1,3)} + 1 - aa^{(1,3)}$ is a unit if and only if $v = a^{(1,3)}aqpa + 1 - a^{(1,3)}a$ is a unit when $p'pa = a = aqq'$, and in this case $(paq)^{\#} = pu^{-2}aq = pav^{-2}q$. Using Theorem 5.2.5, we have

$$(paq)^{\circledcirc} = (paq)^{\#}paq(paq)^{(1,3)} = (pu^{-2}aq)paq(paq)^{(1,3)}$$

$$= (pav^{-2}q)paq(paq)^{(1,3)}.$$

$\square$

If $p, q$ are invertible, we know that $(paq)^{\#}$ exists if and only if $u = aqp + 1 - aa^{(1,3)}$ is invertible if and only if $v = qpa + 1 - a^{(1,3)}a$ is invertible, and in this case $(paq)^{\#} = pu^{-2}aq = pav^{-2}q$. Then using Theorem 5.2.5, we have the following corollary, which will be used.

**Corollary 5.5.4** *Let* $a, p, q \in R$ *be such that* $a\{1, 3\} \neq \emptyset$ *and* $p, q$ *be invertible. Then* $(paq)^{\oplus}$ *exists if and only if* $u = aqp + 1 - aa^{(1,3)}$ *is a unit and* $(paq)\{1, 3\} \neq \emptyset$, *or, equivalently,* $v = qpa + 1 - a^{(1,3)}a$ *is a unit and* $(paq)\{1, 3\} \neq \emptyset$. *In this case,*

$$(paq)^{\oplus} = pu^{-2}aqpaq(paq)^{(1,3)} = pav^{-2}qpaq(paq)^{(1,3)}.$$

For the dual core inverse, we have the following analogous results.

**Theorem 5.5.5 ([67])** *Let* $a, p, q \in R$ *be such that* $a\{1, 3\} \neq \emptyset$. *Suppose that there exist* $p', q' \in R$ *such that* $p'pa = a = aqq'$. *Then the following are equivalent:*

(1) $(paq)_{\oplus}$ *exists.*
(2) $s = aq(aq)^* + 1 - aa^{(1,3)}$ *and* $t = paq + 1 - (aq)^*s^{-1}aq$ *are units.*

*In this case,*

$$(paq)_{\oplus} = t^{-1}(aq)^*s^{-1}aq.$$

**Theorem 5.5.6 ([67])** *Let* $a, p, q \in R$ *be such that* $a\{1, 4\} \neq \emptyset$. *Suppose that there exist* $p', q' \in R$ *such that* $p'pa = a = aqq'$. *Then the following are equivalent:*

(1) $(paq)_{\oplus}$ *exists.*
(2) $s = aqpaa^{(1,4)} + 1 - aa^{(1,4)}$ *is a unit and* $(paq)\{1, 4\} \neq \emptyset$.
(3) $t = a^{(1,4)}aqpa + 1 - a^{(1,4)}a$ *is a unit and* $(paq)\{1, 4\} \neq \emptyset$.

*In this case,*

$$(paq)_{\oplus} = (paq)^{(1,4)}paqps^{-2}aq = (paq)^{(1,4)}paqpat^{-2}q.$$

**Corollary 5.5.7** *Let* $a, p, q \in R$ *be such that* $a\{1, 4\} \neq \emptyset$ *and* $p, q$ *be invertible. Then* $(paq)_{\oplus}$ *exists if and only if* $s = aqp + 1 - aa^{(1,4)}$ *is a unit and* $(paq)\{1, 4\} \neq \emptyset$, *or, equivalently,* $t = qpa + 1 - a^{(1,4)}a$ *is a unit and* $(paq)\{1, 4\} \neq \emptyset$. *In this case,*

$$(paq)_{\oplus} = (paq)^{(1,4)}paqps^{-2}aq = (paq)^{(1,4)}paqpat^{-2}q.$$

It is well known that $a \in R$ is Moore-Penrose invertible if and only if it is both $\{1, 3\}$ and $\{1, 4\}$-invertible. Based on previous Theorems 5.5.2 and 5.5.5, we derive a characterization of the existence of the (dual) core inverse of $paq$ in the case that $a^{\dagger}$ exists.

**Theorem 5.5.8 ([67])** *Let* $a, p, q \in R$ *be such that* $a \in R^{\dagger}$. *Suppose that there exist* $p', q' \in R$ *such that* $p'pa = a = aqq'$. *Then the following are equivalent:*

(1)  $(paq)^{\circledR}$ and $(paq)_{\circledR}$ exist.
(2)  $u = (pa)^* pa + 1 - a^\dagger a$, $s = aq(aq)^* + 1 - aa^\dagger$ and $w = paq + 1 - paq(aq)^* s^{-1} au^{-1}(pa)^*$ are invertible.
(3)  $u = (pa)^* pa + 1 - a^\dagger a$, $s = aq(aq)^* + 1 - aa^\dagger$ and $w' = paq + 1 - (aq)^* s^{-1} au^{-1}(pa)^* paq$ are invertible.
(4)  $(paq)^{\#}$ and $(paq)^\dagger$ exist.

*In this case,*

$$(paq)^{\circledR} = w^{-2} paqpau^{-1}(pa)^* = paqw'^{-2} pau^{-1}(pa)^*,$$

$$(paq)_{\circledR} = (aq)^* s^{-1} aqw^{-2} paq = (aq)^* s^{-1} aqpaqw'^{-2}.$$

Next, we use Theorem 5.2.5 to consider the existence of the core inverse of $paq$ when $a$ is regular.

**Theorem 5.5.9 ([67])**  *Suppose $a \in R$ is regular, $aa^- a = a$ and $p, q \in R$,*

$$u = aq(paq)^* paa^- + 1 - aa^-, \quad v = a^- aq(paq)^* pa + 1 - a^- a,$$

$$s = aqpaa^- + 1 - aa^-, \quad t = a^- aqpa + 1 - a^- a.$$

*Then the following are equivalent:*

(1)  $Ru = R$ and $s$ is a unit.
(2)  $Raq(paq)^* pa = Ra = Raqpa$, $aqpaR = aR$.
(3)  $Rv = R$ and $t$ is a unit.

*If in addition $aR = aqR$, then (1)–(3) are also equivalent to*

(4)  $(paq)^{\#}$ exists and $Raq(paq)^* paq = Rpaq$, $Ra = Rpa$.

*In this case, $(paq)^{\circledR}$ exists and there exists $x \in R$ such that*

$$(paq)^{\circledR} = ps^{-1} at^{-1} qpaqq^* a^* x = ps^{-2} aqpaqq^* a^* x = pat^{-2} qpaqq^* a^* x.$$

**Proof**  From Theorem 3.3.1, $sR = R$ if and only if $aqpaR = aR$ if and only if $tR = R$. By symmetry it follows that $Rs = R$ is equivalent to $Raqpa = Ra$, or equivalently, $Rt = R$. So $s$ is a unit if and only if $t$ is a unit if and only if $Ra = Raqpa$, $aqpaR = aR$. Similarly, we know that $Ru = R$ if and only if $Raq(paq)^* pa = Ra$ if and only if $Rv = R$. Therefore, the statements (1)–(3) are equivalent.

Now suppose $aR = aqR$. By Corollary 3.3.3, $s$ is a unit if and only if "$(paq)^{\#}$ exists and $Ra = Rpa$, $aR = aqR$". Notice that $Ru = R$ is also equivalent to "$Raq(paq)^* paq = Rpaq$ and $Ra = Rpa$", and in this case, $paq$ has a $\{1, 3\}$-inverse $q^* a^* x$ for some $x \in R$. Hence, (1)–(3) are also equivalent to (4). Using

Theorem 5.2.5, $(paq)^{\circledR}$ exists and

$$(paq)^{\circledR} = (paq)^{\#} paq(paq)^{(1,3)}$$

$$= ps^{-1}at^{-1}qpaqq^*a^*x = ps^{-2}aqpaqq^*a^*x$$

$$= pat^{-2}qpaqq^*a^*x,$$

for some $x \in R$.                                                                                               □

We note that the existence of $(paq)\{1,3\}$ cannot imply $Raq(paq)^*paq = Rpaq$ in general. For example, suppose that $R = M_2(\mathbb{C})$ and the involution is a transpose. Let $P = Q = I$, $A = \left[\begin{smallmatrix} 1 & i \\ 0 & 0 \end{smallmatrix}\right]$. Then $A^* = \left[\begin{smallmatrix} 1 & 0 \\ i & 0 \end{smallmatrix}\right]$, $RA^*A = RA$ since $rank(A^*A) = rank(A) = 1$. Consequently, $A^{(1,3)}$ exists. However, $RAA^*A = 0 \neq RA$ since $AA^* = 0$.

By symmetry, we have the following result for the dual core inverse.

**Theorem 5.5.10 ([67])** *Let* $p, a, q \in R$, *where* $a$ *is regular. Suppose*

$$u = aq(paq)^*paa^- + 1 - aa^-, \quad v = a^-aq(paq)^*pa + 1 - a^-a,$$

$$s = aqpaa^- + 1 - aa^-, \quad t = a^-aqpa + 1 - a^-a.$$

*Then the following statements are equivalent:*

(1) $uR = R$ *and* $s$ *is a unit.*
(2) $aq(paq)^*paR = aR = aqpaR$, $Ra = Raqpa$.
(3) $vR = R$ *and* $t$ *is a unit.*

*If in addition* $Ra = Rpa$, *then (1)–(3) are also equivalent to*

(4) $(paq)^{\#}$ *exists and* $paq(paq)^*paR = paqR$, $aR = aqR$.

*In this case,* $(paq)_{\circledR}$ *exists and there exists* $x \in R$ *such that*

$$(paq)_{\circledR} = xa^*p^*paqps^{-1}at^{-1}q = xa^*p^*paqps^{-2}aq = xa^*p^*paqpat^{-2}q.$$

Now, we consider the core inverse and dual core inverse of the product of two elements.

**Theorem 5.5.11 ([67])** *Let* $a, p \in R$ *be such that* $a^{(1,3)}$ *exists and* $Rpa = Ra$. *Let* $e = aa^{(1,3)}$ *and* $p(1-e) = 1 - e$. *Suppose* $u = p^*pe + 1 - e$, $s = ape + 1 - e$, *then the following are equivalent:*

(1) $(pa)^{\circledR}$ *exists.*
(2) $Ru = R$ *and* $s$ *is a unit.*
(3) $uR = R$ *and* $t = ap + 1 - e$ *is a unit.*

*In this case,* $(pa)^{\circledR} = ps^{-2}apeu^{-1}p^* = p(1 - (ap-1)t^{-1}e)^{-1}apeu^{-1}p^*$.

**Proof** By Jacobson's lemma, $s = ape + 1 - e = 1 + (ap - 1)e$ is a unit if and only if $t = 1 + e(ap - 1) = ap + 1 - e$ is a unit. In this case $s^{-1} = 1 - (ap - 1)t^{-1}e$. If $Rpa = Ra$, then $(pa)^{\#}$ exists if and only if $s$ is a unit. In this case, $(pa)^{\#} = ps^{-2}a$. Also, $pa$ has a $\{1, 3\}$-inverse if and only if $Ru = R$ if and only if $uR = R$. In this case, $(pa)^{(1,3)} = a^{(1,3)}u^{-1}p^{*}$. Hence, using Theorem 5.2.5, the equivalence of (1)–(3) are satisfied. Moreover,

$$(pa)^{\circledR} = (pa)^{\#}pa(pa)^{(1,3)} = ps^{-2}apaa^{(1,3)}u^{-1}p^{*} = ps^{-2}apeu^{-1}p^{*}$$

$$= p(1 - (ap - 1)t^{-1}e)^{-1}apeu^{-1}p^{*}.$$

$\square$

Similarly, we get the following result for the dual core inverse.

**Theorem 5.5.12 ([67])** Let $a, q \in R$ be such that $a^{(1,4)}$ exists and $aR = aqR$. Let $f = a^{(1,4)}a$ and $(1 - f)q = 1 - f$. Suppose $v = fqq^{*} + 1 - f$, $s' = fqa + 1 - f$, then the following are equivalent:

(1) $(aq)_{\circledR}$ exists.
(2) $Rv = R$ and $s'$ is a unit.
(3) $vR = R$ and $t' = qa + 1 - f$ is a unit.

In this case, $(aq)_{\circledR} = q^{*}v^{-1}fqas'^{-2}q = q^{*}v^{-1}fqa(1 - ft'^{-1}(qa - 1))^{-1}q$.

## 5.6   Core Inverses of Companion Matrices

In this section, we will investigate the core inverse of a companion matrix. Recall that the lower companion matrix $L \in R^{n \times n}$ associated with the monic polynomial $p(\lambda) = p_0 + p_1\lambda + \cdots + p_{n-1}\lambda^{n-1} + \lambda^n$ is denoted by

$$L = \begin{bmatrix} 0 & & & & -p_0 \\ 1 & 0 & & & -p_1 \\ & 1 & 0 & & -p_2 \\ & & \ddots & \ddots & \vdots \\ & & & 1 & 0 & -p_{n-2} \\ & & & & 1 & -p_{n-1} \end{bmatrix}_{n \times n}.$$

**Theorem 5.6.1 ([82])** *Let $L \in R^{n \times n}$ be the lower companion matrix of $p(\lambda)$. Then $L$ is core invertible if and only if there exist solutions $x$ and $y$ to*

$$(1)\ (p_0 x)^* = p_0 x,$$

$$(2)\ y p_0 = 0 = p_0 y,$$

$$(3)\ (x,\ y) \begin{pmatrix} p_0 \\ p_1 \end{pmatrix} = -1, \tag{5.6.1}$$

$$(4)\ (1 + p_0 x)(p_1 x,\ 1 + p_1 y) = (0, 0).$$

*In either case,*

$$L^{\oplus} = \left[ \begin{array}{cc|ccc} x_1 & y_1 & 0 & \cdots & 0 \\ \vdots & \vdots & & I_{n-2} & \\ x_{n-1} & y_{n-1} & & & \\ \hline x & y & 0 & \cdots & 0 \end{array} \right],$$

*where $x_i = p_i x$ and $y_i = \delta_{i1} + p_i y$, $i = 1, 2, \ldots, n - 1$.*

**Proof** Suppose that $L^{\oplus}$ exists and has columns $(\alpha_1, \cdots, \alpha_n)$, then by the routine calculation, we can obtain

$$L^{\oplus} L = (\alpha_2, \alpha_3, \cdots, \alpha_n, L^{\oplus} L e_n).$$

If we set $p = -(p_0, p_1, \cdots, p_{n-1})^{\mathrm{T}}$, then $L = (e_2, \cdots, e_n, p)$ and

$$L^{\oplus} L^2 = (L^{\oplus} L) L = (\alpha_3, \cdots, \alpha_n, L^{\oplus} L e_n, L^{\oplus} L p),$$

thus from $L^{\oplus} L^2 = L$ we have

$$\alpha_3 = e_2, \quad \ldots, \quad \alpha_n = e_{n-1}$$

and

$$L^{\oplus} L e_n = e_n, \quad L^{\oplus} L p = p. \tag{5.6.2}$$

Therefore,

$$L^{\oplus} = (\alpha_1, \alpha_2, e_2, \cdots, e_{n-1}). \tag{5.6.3}$$

Writing $\alpha_1 = (x_1, \cdots, x_n)^T$ and $\alpha_2 = (y_1, \cdots, y_n)^T$, then

$$LL^{\circledR} = \begin{bmatrix} 0 & & & & -p_0 \\ 1 & 0 & & & -p_1 \\ & 1 & 0 & & -p_2 \\ & & \ddots & \ddots & \vdots \\ & & & 1 & 0 & -p_{n-2} \\ & & & & 1 & -p_{n-1} \end{bmatrix} \left[\begin{array}{cc|ccc} x_1 & y_1 & 0 & \cdots & 0 \\ \vdots & \vdots & & I_{n-2} & \\ x_{n-1} & y_{n-1} & & & \\ \hline x_n & y_n & 0 & \cdots & 0 \end{array}\right]$$

$$= \left[\begin{array}{cc|ccc} -p_0 x_n & -p_0 y_n & 0 & \cdots & 0 \\ x_1 - p_1 x_n & y_1 - p_1 y_n & 0 & \cdots & 0 \\ \hline \vdots & \vdots & & I_{n-2} & \\ x_{n-1} - p_{n-1}x_n & y_{n-1} - p_{n-1}y_n & & & \end{array}\right].$$

From $(LL^{\circledR})^* = LL^{\circledR}$ we obtain

$$(p_0 x_n)^* = p_0 x_n, \quad (y_1 - p_1 y_n)^* = y_1 - p_1 y_n, \tag{5.6.4}$$

$$x_1 - p_1 x_n = (-p_0 y_n)^*, \tag{5.6.5}$$

and

$$x_i - p_i x_n = 0, \quad y_i - p_i y_n = 0, \quad i = 2, 3, \ldots, n - 1. \tag{5.6.6}$$

Similarly, $LL^{\circledR}L^{\circledR} = L^{\circledR}$ follows that

$$\left[\begin{array}{ccc|ccc} -p_0 x_n x_1 - p_0 y_n x_2 & -p_0 x_n y_1 - p_0 y_n y_2 & -p_0 y_n & 0 & \cdots & 0 \\ (x_1 - p_1 x_n)x_1 + (y_1 - p_1 y_n)x_2 & (x_1 - p_1 x_n)y_1 + (y_1 - p_1 y_n)y_2 & y_1 - p_1 y_n & 0 & \cdots & 0 \\ \hline x_3 & y_3 & 0 & & & \\ \vdots & \vdots & \vdots & & I_{n-3} & \\ x_{n-1} & y_{n-1} & 0 & & & \\ \hline x_n & y_n & 0 & 0 & \cdots & 0 \end{array}\right]$$

$$= \left[\begin{array}{cc|ccc} x_1 & y_1 & 0 & \cdots & 0 \\ \vdots & \vdots & & I_{n-2} & \\ x_{n-1} & y_{n-1} & & & \\ \hline x_n & y_n & 0 & \cdots & 0 \end{array}\right],$$

which shows that

$$p_0 y_n = 0, \quad y_1 - p_1 y_n = 1, \tag{5.6.7}$$

thus

$$x_1 = -p_0 x_n x_1 - p_0 y_n x_2 = -p_0 x_n x_1,$$
$$y_1 = -p_0 x_n y_1 - p_0 y_n y_2 = -p_0 x_n y_1.$$
(5.6.8)

Moreover, from (5.6.5) and (5.6.7) we know that

$$r_1 - p_1 x_n = (-p_0 y_n)^* = 0,$$
(5.6.9)

and (5.6.8) yields respectively by using (5.6.9) and (5.6.7)

$$(1 + p_0 x_n) p_1 x_n = 0,$$
$$(1 + p_0 x_n)(1 + p_1 y_n) = 0.$$
(5.6.10)

Therefore, (5.6.3) with identities (5.6.6), (5.6.7) and (5.6.9) together yield that

$$L^{\oplus} = \begin{bmatrix} x_1 & y_1 & 0 & \cdots & 0 \\ \vdots & \vdots & & I_{n-2} & \\ x_{n-1} & y_{n-1} & & & \\ x & y & 0 & \cdots & 0 \end{bmatrix},$$

where $x_i = p_i x$ and $y_i = \delta_{i1} + p_i y$, $i = 1, 2, \ldots, n-1$. Using this form of $L^{\oplus}$ again in (5.6.2) we arrive at the following conditions:

$$\begin{bmatrix} 0 \\ 0 \\ \vdots \\ 0 \\ 1 \end{bmatrix} = e_n = L^{\oplus} L e_n = \begin{bmatrix} -(x_1 p_0 + y_1 p_1) \\ -(x_2 p_0 + y_2 p_1 + p_2) \\ \vdots \\ -(x_{n-1} p_0 + y_{n-1} p_1 + p_{n-1}) \\ -(x_n p_0 + y_n p_1) \end{bmatrix}$$

and

$$\begin{bmatrix} -p_0 \\ -p_1 \\ \vdots \\ -p_{n-2} \\ -p_{n-1} \end{bmatrix} = p = L^{\oplus} L p = \begin{bmatrix} -y_1 p_0 \\ -y_2 p_0 - p_1 \\ \vdots \\ -y_{n-1} p_0 - p_{n-2} \\ -y_n p_0 - p_{n-1} \end{bmatrix},$$

which respectively follows that

$$x_n p_0 + y_n p_1 = -1 , \quad y_n p_0 = 0.$$
(5.6.11)

As a consequence, Eqs. (5.6.4), (5.6.7), (5.6.10), and (5.6.11) together show that $x_n$ and $y_n$ satisfy the conditions (1)–(4) of (5.6.1).

Conversely, suppose that $x_n$ and $y_n$ satisfy the conditions (1)–(4) of (5.6.1) and suppose that $x_i = p_i x$ and $y_i = \delta_{i1} + p_i y$, $i = 1, 2, \ldots, n - 1$. Let

$$
X = \left[
\begin{array}{cc|ccc}
x_1 & y_1 & 0 & \cdots & 0 \\
\vdots & \vdots & & I_{n-2} & \\
x_{n-1} & y_{n-1} & & & \\
\hline
x & y & 0 & \cdots & 0
\end{array}
\right],
$$

then all that remains is to verify that $X$ is the core inverse of $L$, that is to say, $(LX)^* = LX$, $LX^2 = X$ and $XL^2 = L$. This verification is lengthy but straightforward and will be omitted.                                      □

**Remark 5.6.2**  In particular, there are three special cases of Theorem 5.6.1.

(1) If $p_0(\neq 0)$ is not a divisor of zero, then $y = 0$ and $p_0$ is invertible with $p_0^{-1} = -x$ follows from the system (5.6.1). Therefore, one can easily obtain that $L^\oplus$ exists if and only if $p_0$ is invertible, and hence $L$ is invertible. In this case,

$$
L^{-1} = L^\oplus = \left[
\begin{array}{c|ccc}
-p_1 p_0^{-1} & & & \\
-p_2 p_0^{-1} & & & \\
\vdots & & I_{n-1} & \\
-p_{n-1} p_0^{-1} & & & \\
\hline
-p_0^{-1} & 0 & \cdots & 0
\end{array}
\right].
$$

(2) If $p_0 = 0$, then $x = 0$ and $p_1$ is invertible with $p_1^{-1} = -y$ follows from the system (5.6.1). Therefore, $L^\oplus$ exists if and only if $p_1$ is invertible. In this case,

$$
L^\oplus = \left[
\begin{array}{cc|ccc}
0 & 0 & 0 & \cdots & 0 \\
0 & -p_2 p_1^{-1} & & & \\
\vdots & \vdots & & I_{n-2} & \\
0 & -p_{n-1} p_1^{-1} & & & \\
\hline
0 & -p_1^{-1} & 0 & \cdots & 0
\end{array}
\right].
$$

(3) If $p_1 = 0$, then $y = 0$ and $p_0$ is invertible with $p_0^{-1} = -x$ follows from the system (5.6.1). Therefore, $L^\oplus$ exists if and only if $p_0$ is invertible. In this case,

$$L^\oplus = \begin{bmatrix} -p_1 p_0^{-1} & \\ -p_2 p_0^{-1} & \\ \vdots & I_{n-1} \\ -p_{n-1} p_0^{-1} & \\ \hline -p_0^{-1} & 0 \cdots 0 \end{bmatrix}.$$

Let $a = a_0$, $\beta = [a_1, a_2, \cdots, a_{n-1}]^T$. Then

**Theorem 5.6.3 ([46])**   *If $L = \begin{bmatrix} 0 & a \\ I_{n-1} & \beta \end{bmatrix}$ with $\beta = [a_1, a_2, \cdots, a_{n-1}]^T$ is a companion matrix over R. Then the following conditions are equivalent:*

(1)   $L^\oplus$ *exists.*
(2)   $a^{(1,3)}$ *exists and $h = a - (1 - aa^{(1,3)})a_1$ is invertible.*
(3)   $a^{(1,3)}$ *exists and $k = a - a_1(1 - a^{(1,3)}a)$ is invertible.*

*In this case,*

$$L^\oplus = \begin{bmatrix} \underline{x} aa^{(1,3)} & \underline{y} & 0 & \cdots & 0 \\ & & I_{n-2} & \\ \underline{x} aa^{(1,3)} & \underline{y} & 0 & \cdots & 0 \end{bmatrix},$$

*where $\underline{x} = [x_1, \cdots, x_{n-1}]^T$, $\underline{y} = [y_1, \cdots, y_{n-1}]^T$ and*

$$y_i = \delta_{i1} + a_i y, \quad x_i = y_{i+1} + a_i x, \quad i = 1, \cdots, n-1.$$

$$\begin{cases} x = -\alpha_1 a_2 \alpha_1 + \gamma_1 \beta_1, \ y = \alpha_1, \quad n > 2 \\ x = \alpha_1^2 + \gamma_1 \beta_1, \ y = [\alpha_1 \ \gamma_1] \begin{bmatrix} \alpha_1 & \gamma_1 \\ \beta_1 & \delta_1 \end{bmatrix} \begin{bmatrix} a_1 \\ a \end{bmatrix}, \quad n = 2 \end{cases}$$

$\alpha_1 = -h^{-1}[1 - aa^{(1,3)}]$, $\gamma_1 = h^{-1}aa^{(1,3)}$, $\beta_1 = 1 + (a + a_1)h^{-1}[1 - aa^{(1,3)}]$ and $\delta_1 = 1 - (a + a_1)h^{-1}aa^{(1,3)}$.

**Proof**   Consider the factorization of $L$:

$$L = \begin{bmatrix} 0 & a \\ I_{n-1} & \beta \end{bmatrix} = \begin{bmatrix} 0 & 1 \\ I_{n-1} & 0 \end{bmatrix} \begin{bmatrix} I_{n-1} & 0 \\ 0 & a \end{bmatrix} \begin{bmatrix} I_{n-1} & \beta \\ 0 & 1 \end{bmatrix} := PAQ.$$

Notice that $P$ is a unitary matrix and $Q$ is an invertible matrix. Then $L^{(1,3)} = (PAQ)^{(1,3)}$ exists if and only if $(AQ)^{(1,3)}$ exists if and only if $A^{(1,3)}$ exists if and only if $a^{(1,3)}$ exists. Indeed, if $(AQ)^{(1,3)}$ exists, then $Q(AQ)^{(1,3)}$ is a $\{1, 3\}$-inverse of $A$; conversely, if $A^{(1,3)}$ exists, then $Q^{-1}A^{(1,3)}$ is a $\{1, 3\}$-inverse of $AQ$. In this case,

$$\begin{bmatrix} -\beta a^{(1,3)} & I_{n-1} \\ a^{(1,3)} & 0 \end{bmatrix} \in L\{1,3\} \text{ and } LL^{(1,3)} = \begin{bmatrix} aa^{(1,3)} & 0 \\ 0 & I_{n-1} \end{bmatrix}.$$

From Theorem 3.8.1, together with, $L^{\oplus}$ exists if and only if both $L^{\#}$ and $L^{(1,3)}$ exist, the equivalences of (1)-(3) follow. Moreover,

$$L^{\oplus} = L^{\#}LL^{(1,3)} = \begin{bmatrix} \underline{x} & \underline{y} & 0 & \cdots & 0 \\ & & I_{n-2} & \\ x & y & 0 & \cdots & 0 \end{bmatrix} \begin{bmatrix} aa^{(1,3)} & 0 \\ 0 & I_{n-1} \end{bmatrix} = \begin{bmatrix} \underline{xaa}^{(1,3)} & \underline{y} & 0 & \cdots & 0 \\ & & I_{n-2} & \\ xaa^{(1,3)} & y & 0 & \cdots & 0 \end{bmatrix}.$$

$\square$

## 5.7 The Core Inverse of a Sum of Morphisms

Let $\mathscr{C}$ be an additive category with an involution $*$. Suppose that both $\varphi : X \to Y$ and $\eta : X \to Y$ are morphisms of $\mathscr{C}$. We use the following notations:

$$\alpha = (1_X + \varphi^{\tau}\eta)^{-1},$$
$$\beta = (1_Y + \eta\varphi^{\tau})^{-1},$$
$$\varepsilon = (1_Y - \varphi\varphi^{\tau})\eta\alpha(1_X - \varphi^{\tau}\varphi),$$
$$\gamma = \alpha(1_X - \varphi^{\tau}\varphi)\eta\varphi^{\tau}\beta,$$
$$\delta = \alpha\varphi^{\tau}\eta(1_Y - \varphi\varphi^{\tau})\beta,$$
$$\lambda = \alpha(1_X - \varphi^{\tau}\varphi)\eta^*(\varphi^{\tau})^*\alpha^*,$$
$$\mu = \beta^*(\varphi^{\tau})^*\eta^*(1_Y - \varphi\varphi^{\tau})\beta,$$

where $\tau \in \{\#, (1,2,3), (1,2,4), \dagger\}$.

The $\{1,2,4\}$-inverse, $\{1,2,3\}$-inverse, Moore-Penrose inverse and group inverse of a sum of morphisms in an additive category are considered in previous chapters. Now we consider corresponding versions for core inverses and dual core inverses.

**Theorem 5.7.1 ([83])** *Let $\mathscr{C}$ be an additive category with an involution $*$. Suppose that $\varphi : X \to X$ is a morphism of $\mathscr{C}$ with core inverse $\varphi^{\oplus}$ and $\eta : X \to X$ is a* *morphism of $\mathscr{C}$ such that $1_X + \varphi^\circledast\eta$ is invertible. Let*

$$\alpha = (1_X + \varphi^\circledast\eta)^{-1},$$

$$\beta = (1_X + \eta\varphi^\circledast)^{-1},$$

$$\varepsilon = (1_X - \varphi\varphi^\circledast)\eta\alpha(1_X - \varphi^\circledast\varphi),$$

$$\gamma = \alpha(1_X - \varphi^\circledast\varphi)\beta^{-1}\varphi\varphi^\circledast\beta,$$

$$\sigma = \alpha\varphi^\circledast\varphi\alpha^{-1}(1_X - \varphi\varphi^\circledast)\beta,$$

$$\delta = \beta^*(\varphi^\circledast)^*\eta^*(1_X - \varphi\varphi^\circledast)\beta.$$

*Then the following conditions are equivalent:*

(1) $f = \varphi + \eta - \varepsilon$ *has a core inverse.*
(2) $1_X - \gamma$, $1_X - \sigma$ *and* $1_X - \delta$ *are invertible.*
(3) $1_X - \gamma$ *is left invertible, both* $1_X - \sigma$ *and* $1_X - \delta$ *are right invertible.*

*In this case,*

$$f^\circledast = (1_X - \gamma)^{-1}\alpha\varphi^\circledast(1_X - \delta)^{-1},$$

$$(1_X - \gamma)^{-1} = 1_X - \varphi\varphi^\circledast + f^\circledast f\varphi\varphi^\circledast,$$

$$(1_X - \sigma)^{-1} = 1_X - \varphi\varphi^\circledast + \varphi\varphi^\circledast f^\circledast f,$$

$$(1_X - \delta)^{-1} = 1_X - \varphi\varphi^\circledast + \varphi\varphi^\circledast ff^\circledast.$$

**Proof** By Lemma 2.8.1, $(1_X + \varphi^\circledast\eta)^{-1}\varphi^\circledast \in f\{1, 2\}$.
Let $f_0 = (1_X + \varphi^\circledast\eta)^{-1}\varphi^\circledast = \alpha\varphi^\circledast = \varphi^\circledast\beta$, then

$$\varphi^\circledast f = \varphi^\circledast(\varphi + \eta - \varepsilon) = \varphi^\circledast\varphi + \varphi^\circledast\eta = \varphi^\circledast\varphi(1_X + \varphi^\circledast\eta) = \varphi^\circledast\varphi\alpha^{-1},$$

and $f\varphi^\circledast = \beta^{-1}\varphi\varphi^\circledast$. So $f_0f = \alpha\varphi^\circledast f = \alpha\varphi^\circledast\varphi\alpha^{-1}$, and

$$1_X - f_0f = 1_X - \alpha\varphi^\circledast\varphi\alpha^{-1} = \alpha(1_X - \varphi^\circledast\varphi)\alpha^{-1}$$
$$= \alpha(1_X - \varphi^\circledast\varphi)(1_X + \varphi^\circledast\eta) = \alpha(1_X - \varphi^\circledast\varphi).$$

Similarly, we have $ff_0 = \beta^{-1}\varphi\varphi^\circledast\beta$ and $1_X - ff_0 = (1_X - \varphi\varphi^\circledast)\beta$.
Further,

$$(1_X - f_0f)ff_0 = \alpha(1_X - \varphi^\circledast\varphi)\beta^{-1}\varphi\varphi^\circledast\beta = \gamma,$$

$$f_0f(1_X - ff_0) = \alpha\varphi^\circledast\varphi\alpha^{-1}(1_X - \varphi\varphi^\circledast)\beta = \sigma,$$

$$(ff_0)^*(1_X - ff_0) = (\beta^{-1}\varphi\varphi^\circledR\beta)^*(1_X - \varphi\varphi^\circledR)\beta$$

$$= \beta^*\varphi\varphi^\circledR(\beta^{-1})^*(1_X - \varphi\varphi^\circledR)\beta$$

$$= \beta^*[(1_X - \varphi\varphi^\circledR)\beta^{-1}\varphi\varphi^\circledR]^*\beta$$

$$= \beta^*[(1_X - \varphi\varphi^\circledR)(1_X + \eta\varphi^\circledR)\varphi\varphi^\circledR]^*\beta$$

$$= \beta^*[(1_X - \varphi\varphi^\circledR)(\varphi\varphi^\circledR + \eta\varphi^\circledR)]^*\beta$$

$$= \beta^*[(1_X - \varphi\varphi^\circledR)\eta\varphi^\circledR]^*\beta$$

$$= \delta.$$

Therefore, we obtain

$$f\gamma = 0, \quad \sigma f = 0, \quad \delta f = 0,$$

moreover,

$$f_0 f^2 = f - f + f_0 f^2 = f - (1_X - f_0 f)f$$
$$= [1_X - (1_X - f_0 f)ff_0]f = (1_X - \gamma)f,$$

$$f^2 f_0 = f - f + f^2 f_0 = f - f(1_X - ff_0)$$
$$= f[1_X - f_0 f(1_X - ff_0)] = f(1_X - \sigma),$$

$$f^* ff_0 = f^* - f^* + f^* ff_0 = f^* - f^*(1_X - ff_0)$$
$$= f^*[1_X - (ff_0)^*(1_X - ff_0)] = f^*(1_X - \delta).$$

Now we are ready to show the equivalence of three conditions.

(1) $\Rightarrow$ (2). The first step is to show that $1_X - \varphi\varphi^\circledR + f^\circledR f\varphi\varphi^\circledR$ is the inverse of $1_X - \gamma$.

Note that

$$(1_X - \gamma)f^\circledR f = (1_X - \gamma)ff^\circledR f^\circledR f = f_0 f^2 f^\circledR f^\circledR f = f_0 f$$

$$= \alpha\varphi^\circledR\varphi\alpha^{-1} = \alpha\varphi^\circledR\varphi(1_X + \varphi^\circledR\eta) = \alpha(\varphi^\circledR\varphi + \psi^\circledR\eta)$$

$$= \alpha(1_X + \varphi^\circledR\eta + \varphi^\circledR\varphi - 1_X) = \alpha(\alpha^{-1} + \varphi^\circledR\varphi - 1_X)$$

$$= 1_X + \alpha(\varphi^\circledR\varphi - 1_X).$$

Post-multiplication $\varphi\varphi^\circledR$ on the equality above yields

$$(1_X - \gamma)f^\circledR f\varphi\varphi^\circledR = \varphi\varphi^\circledR.$$

As

$$\gamma(1_X - \varphi\varphi^\circledR) = \alpha(1_X - \varphi^\circledR\varphi)\beta^{-1}\varphi\varphi^\circledR\beta(1_X - \varphi\varphi^\circledR)$$
$$= \alpha(1_X - \varphi^\circledR\varphi)\beta^{-1}\varphi\alpha\varphi^\circledR(1_X - \varphi\varphi^\circledR)$$
$$= 0,$$

we obtain

$$(1_X - \gamma)(1_X - \varphi\varphi^\circledR + f^\circledR f\varphi\varphi^\circledR)$$
$$= 1_X - \varphi\varphi^\circledR - \gamma(1_X - \varphi\varphi^\circledR) + (1_X - \gamma)f^\circledR f\varphi\varphi^\circledR$$
$$= 1_X - \varphi\varphi^\circledR - 0 + \varphi\varphi^\circledR$$
$$= 1_X.$$

So $1_X - \varphi\varphi^\circledR + f^\circledR f\varphi\varphi^\circledR$ is the right inverse of $1_X - \gamma$. Next, we prove that $1_X - \varphi\varphi^\circledR + f^\circledR f\varphi\varphi^\circledR$ is also the left inverse of $1_X - \gamma$.

Note that

$$\gamma = \alpha(1_X - \varphi^\circledR\varphi)\beta^{-1}\varphi\varphi^\circledR\beta$$
$$= \alpha(1_X - \varphi^\circledR\varphi)(1_X + \eta\varphi^\circledR)\varphi\varphi^\circledR\beta$$
$$= \alpha(1_X - \varphi^\circledR\varphi)(\varphi\varphi^\circledR + \eta\varphi^\circledR)\beta$$
$$= \alpha(1_X - \varphi^\circledR\varphi)\eta\varphi^\circledR\beta,$$

$1_X - \varphi^\circledR\eta\alpha = \alpha$ and $1_X - \eta\varphi^\circledR\beta = \beta$, thus we have

$$\varphi\varphi^\circledR\gamma = \varphi\varphi^\circledR\alpha(1_X - \varphi^\circledR\varphi)\eta\varphi^\circledR\beta$$
$$= \varphi\varphi^\circledR(1_X - \varphi^\circledR\eta\alpha)(1_X - \varphi^\circledR\varphi)\eta\varphi^\circledR\beta$$
$$= \varphi\varphi^\circledR(1_X - \varphi^\circledR\varphi)\eta\varphi^\circledR\beta - \varphi^\circledR\eta\alpha(1_X - \varphi^\circledR\varphi)\eta\varphi^\circledR\beta$$
$$= \varphi\varphi^\circledR(1_X - \varphi^\circledR\varphi)\eta\varphi^\circledR\beta + (\alpha - 1_X)(1_X - \varphi^\circledR\varphi)\eta\varphi^\circledR\beta$$
$$= \varphi\varphi^\circledR(1_X - \varphi^\circledR\varphi)\eta\varphi^\circledR\beta + \alpha(1_X - \varphi^\circledR\varphi)\eta\varphi^\circledR\beta - (1_X - \varphi^\circledR\varphi)\eta\varphi^\circledR\beta$$
$$= \alpha(1_X - \varphi^\circledR\varphi)\eta\varphi^\circledR\beta - (1_X - \varphi\varphi^\circledR)(1_X - \varphi^\circledR\varphi)\eta\varphi^\circledR\beta$$
$$= \gamma - (1_X - \varphi\varphi^\circledR)\eta\varphi^\circledR\beta$$
$$= \gamma - (1_X - \varphi\varphi^\circledR)(1_X - \beta).$$

So $(1_X - \varphi\varphi^\circledR)\gamma = (1_X - \varphi\varphi^\circledR)(1_X - \beta)$, which implies

$$(1_X - \varphi\varphi^\circledR)(1_X - \gamma) = (1_X - \varphi\varphi^\circledR)\beta.$$

Furthermore,

$$
\begin{aligned}
f^{\oplus} f \varphi \varphi^{\oplus} \gamma &= f^{\oplus} f [\gamma - (1_X - \varphi \varphi^{\oplus})(1_X - \beta)] \\
&= f^{\oplus} f \gamma - f^{\oplus} f (1_X - \varphi \varphi^{\oplus})(1_X - \beta), \\
&= -f^{\oplus} f (1_X - \varphi \varphi^{\oplus})(1_X - \beta).
\end{aligned}
$$

In addition,

$$
\begin{aligned}
&1_X - \varphi \varphi^{\oplus} + f^{\oplus} f \varphi \varphi^{\oplus} \\
&= 1_X + \eta \varphi^{\oplus} - f \varphi^{\oplus} + f^{\oplus} f \varphi \varphi^{\oplus} \\
&= 1_X + \eta \varphi^{\oplus} - f^{\oplus} f (f \varphi^{\oplus} - \varphi \varphi^{\oplus}) \\
&= 1_X + \eta \varphi^{\oplus} - f^{\oplus} f \eta \varphi^{\oplus} \\
&= f^{\oplus} f + (1_X - f^{\oplus} f)(1_X + \eta \varphi^{\oplus}) \\
&= f^{\oplus} f + (1_X - f^{\oplus} f) \beta^{-1}.
\end{aligned}
$$

Therefore,

$$
\begin{aligned}
&(1_X - \varphi \varphi^{\oplus} + f^{\oplus} f \varphi \varphi^{\oplus})(1_X - \gamma) \\
&= (1_X - \varphi \varphi^{\oplus})(1_X - \gamma) + f^{\oplus} f \varphi \varphi^{\oplus} - f^{\oplus} f \varphi \varphi^{\oplus} \gamma \\
&= (1_X - \varphi \varphi^{\oplus}) \beta + f^{\oplus} f \varphi \varphi^{\oplus} + f^{\oplus} f (1_X - \varphi \varphi^{\oplus})(1_X - \beta) \\
&= (1_X - \varphi \varphi^{\oplus}) \beta + f^{\oplus} f \varphi \varphi^{\oplus} + f^{\oplus} f (1_X - \beta) - f^{\oplus} f \varphi \varphi^{\oplus} + f^{\oplus} f \varphi \varphi^{\oplus} \beta \\
&= f^{\oplus} f (1_X - \beta) + (1_X - \varphi \varphi^{\oplus} + f^{\oplus} f \varphi \varphi^{\oplus}) \beta \\
&= f^{\oplus} f (1_X - \beta) + [f^{\oplus} f + (1_X - f^{\oplus} f) \beta^{-1}] \beta \\
&= f^{\oplus} f - f^{\oplus} f \beta + f^{\oplus} f \beta + 1_X - f^{\oplus} f \\
&= 1_X.
\end{aligned}
$$

Hence $1_X - \gamma$ is invertible with inverse $(1_X - \gamma)^{-1} = 1_X - \varphi \varphi^{\oplus} + f^{\oplus} f \varphi \varphi^{\oplus}$.
    The second step is to prove that $1_X - \varphi \varphi^{\oplus} + \varphi \varphi^{\oplus} f^{\oplus} f$ is the inverse of $1_X - \sigma$. On the one hand,

$$
\begin{aligned}
f^{\oplus} f (1_X - \sigma) &= f^{\oplus} f^2 f_0 = f f_0 = \beta^{-1} \varphi \varphi^{\oplus} \beta = (1_X + \eta \varphi^{\oplus}) \varphi \varphi^{\oplus} \beta \\
&= (\varphi \varphi^{\oplus} + \eta \varphi^{\oplus}) \beta = (1_X + \eta \varphi^{\oplus} + \varphi \varphi^{\oplus} - 1_X) \beta \\
&= 1_X + (\varphi \varphi^{\oplus} - 1_X) \beta.
\end{aligned}
$$

Pre-multiplication $\varphi \varphi^{\oplus}$ on the equality above yields

$$
\varphi \varphi^{\oplus} f^{\oplus} f (1_X - \sigma) = \varphi \varphi^{\oplus}.
$$

And because

$$(1_X - \varphi\varphi^{\circledR})\sigma = (1_X - \varphi\varphi^{\circledR})\alpha\varphi^{\circledR}\varphi\alpha^{-1}(1_X - \varphi\varphi^{\circledR})\beta$$
$$= (1_X - \varphi\varphi^{\circledR})\varphi^{\circledR}\beta\varphi\alpha^{-1}(1_X - \varphi\varphi^{\circledR})\beta$$
$$= 0,$$

we have

$$(1_X - \varphi\varphi^{\circledR} + \varphi\varphi^{\circledR}f^{\circledR}f)(1_X - \sigma)$$
$$= (1_X - \varphi\varphi^{\circledR})(1_X - \sigma) + \varphi\varphi^{\circledR}f^{\circledR}f(1_X - \sigma)$$
$$= 1_X - \varphi\varphi^{\circledR} + \varphi\varphi^{\circledR}$$
$$= 1_X.$$

On the other hand, as $1_X - \beta\eta\varphi^{\circledR} = \beta$, we obtain

$$\sigma\varphi\varphi^{\circledR} = \alpha\varphi^{\circledR}\varphi\alpha^{-1}(1_X - \varphi\varphi^{\circledR})\beta\varphi\varphi^{\circledR}$$
$$= \alpha\varphi^{\circledR}\varphi\alpha^{-1}(1_X - \varphi\varphi^{\circledR})(1_X - \beta\eta\varphi^{\circledR})\varphi\varphi^{\circledR}$$
$$= \alpha\varphi^{\circledR}\varphi\alpha^{-1}(1_X - \varphi\varphi^{\circledR})\varphi\varphi^{\circledR} - \alpha\varphi^{\circledR}\varphi\alpha^{-1}(1_X - \varphi\varphi^{\circledR})\beta\eta\varphi^{\circledR}\varphi\varphi^{\circledR}$$
$$= -\alpha\varphi^{\circledR}\varphi\alpha^{-1}(1_X - \varphi\varphi^{\circledR})\beta\eta\varphi^{\circledR}$$
$$= -\alpha\varphi^{\circledR}\varphi\alpha^{-1}(1_X - \varphi\varphi^{\circledR})(1_X - \beta)$$
$$= \sigma - \alpha\varphi^{\circledR}\varphi\alpha^{-1}(1_X - \varphi\varphi^{\circledR}).$$

Thus we have

$$\sigma(1_X - \varphi\varphi^{\circledR}) = \alpha\varphi^{\circledR}\varphi\alpha^{-1}(1_X - \varphi\varphi^{\circledR}),$$

and

$$\sigma\varphi\varphi^{\circledR}f^{\circledR}f = [\sigma - \alpha\varphi^{\circledR}\varphi\alpha^{-1}(1_X - \varphi\varphi^{\circledR})]f^{\circledR}f$$
$$= \sigma f^{\circledR}f - \alpha\varphi^{\circledR}\varphi\alpha^{-1}(1_X - \varphi\varphi^{\circledR})f^{\circledR}f$$
$$= \sigma ff^{\circledR}f^{\circledR}f - \alpha\varphi^{\circledR}\varphi\alpha^{-1}(1_X - \varphi\varphi^{\circledR})f^{\circledR}f$$
$$= -\alpha\varphi^{\circledR}\varphi\alpha^{-1}(1_X - \varphi\varphi^{\circledR})f^{\circledR}f.$$

In addition,

$$\varphi\varphi^{\oplus} + \alpha\varphi^{\oplus}\varphi\alpha^{-1}(1_X - \varphi\varphi^{\oplus})$$

$$= \varphi\varphi^{\oplus} + \alpha\varphi^{\oplus}\varphi(1_X + \varphi^{\oplus}\eta)(1_X - \varphi\varphi^{\oplus})$$

$$= \alpha[\alpha^{-1}\varphi\varphi^{\oplus} + \varphi^{\oplus}\varphi(1_X + \varphi^{\oplus}\eta)(1_X - \varphi\varphi^{\oplus})]$$

$$= \alpha[(1_X + \varphi^{\oplus}\eta)\varphi\varphi^{\oplus} + (\varphi^{\oplus}\varphi + \varphi^{\oplus}\eta)(1_X - \varphi\varphi^{\oplus})]$$

$$= \alpha[(1_X + \varphi^{\oplus}\eta)\varphi\varphi^{\oplus} + \varphi^{\oplus}\varphi + \varphi^{\oplus}\eta - (\varphi^{\oplus}\varphi + \varphi^{\oplus}\eta)\varphi\varphi^{\oplus}]$$

$$= \alpha[\varphi^{\oplus}\varphi + \varphi^{\oplus}\eta + (1_X + \varphi^{\oplus}\eta - \varphi^{\oplus}\varphi - \varphi^{\oplus}\eta)\varphi\varphi^{\oplus}]$$

$$= \alpha(\varphi^{\oplus}\varphi + \varphi^{\oplus}\eta)$$

$$= \alpha\varphi^{\oplus}f.$$

Therefore

$$(1_X - \sigma)(1_X - \varphi\varphi^{\oplus} + \varphi\varphi^{\oplus}f^{\oplus}f)$$

$$= (1_X - \sigma)(1_X - \varphi\varphi^{\oplus}) + (1_X - \sigma)\varphi\varphi^{\oplus}f^{\oplus}f$$

$$= 1_X - \varphi\varphi^{\oplus} - \sigma(1_X - \varphi\varphi^{\oplus}) + \varphi\varphi^{\oplus}f^{\oplus}f - \sigma\varphi\varphi^{\oplus}f^{\oplus}f$$

$$= 1_X - \varphi\varphi^{\oplus} - \alpha\varphi^{\oplus}\varphi\alpha^{-1}(1_X - \varphi\varphi^{\oplus}) + \varphi\varphi^{\oplus}f^{\oplus}f + \alpha\varphi^{\oplus}\varphi\alpha^{-1}(1_X - \varphi\varphi^{\oplus})f^{\oplus}f$$

$$= 1_X - \varphi\varphi^{\oplus} + \varphi\varphi^{\oplus}f^{\oplus}f - \alpha\varphi^{\oplus}\varphi\alpha^{-1}(1_X - \varphi\varphi^{\oplus})(1_X - f^{\oplus}f)$$

$$= 1_X - \varphi\varphi^{\oplus}(1_X - f^{\oplus}f) - \alpha\varphi^{\oplus}\varphi\alpha^{-1}(1_X - \varphi\varphi^{\oplus})(1_X - f^{\oplus}f)$$

$$= 1_X - [\varphi\varphi^{\oplus} + \alpha\varphi^{\oplus}\varphi\alpha^{-1}(1_X - \varphi\varphi^{\oplus})](1_X - f^{\oplus}f)$$

$$= 1_X - \alpha\varphi^{\oplus}f(1_X - f^{\oplus}f)$$

$$= 1_X.$$

Thus $1_X - \sigma$ is invertible with inverse $(1_X - \sigma)^{-1} = 1_X - \varphi\varphi^{\oplus} + \varphi\varphi^{\oplus}f^{\oplus}f$.

In the end, $1_X - \delta$ is invertible with inverse $(1_X - \delta)^{-1} = 1_X - \varphi\varphi^{\oplus} + \varphi\varphi^{\oplus}ff^{\oplus}$ can be deduced immediately from Proposition 2.8.5 and the fact that the core inverse is a $\{1, 2, 3\}$-inverse.

(2) $\Rightarrow$ (3). Obviously.

(3) $\Rightarrow$ (1). Assume that $\omega$ is the left inverse of $1_X - \gamma$. $\nu$ and $\lambda$ are the right inverses of $1_X - \delta$ and $1_X - \sigma$, respectively. Then we have

$$f = \omega(1_X - \gamma)f = \omega f_0 f^2, \tag{5.7.1}$$

$$f = f(1_X - \sigma)\lambda = f^2 f_0 \lambda, \tag{5.7.2}$$

and

$$f = (f^*)^* = (f^*(1_X - \delta)\nu)^* = (f^* f f_0 \nu)^* = \nu^* f_0^* f^* f. \qquad (5.7.3)$$

By Theorem 3.2.2, equalities (5.7.1) and (5.7.2) tell us that $f$ is group invertible with group inverse $f^{\#} = \omega f_0 f f_0 \lambda = \omega f_0 \lambda$. And by Lemma 2.2.3, equality (5.7.3) shows that $f$ is $\{1, 3\}$-invertible with $f_0 \nu \in f\{1, 3\}$. Hence, $f$ is core invertible by Theorem 5.2.5. Moreover,

$$f^{\circledR} = f^{\#} f f^{(1,3)} = \omega f_0 \lambda f f_0 \nu.$$

From (1) $\Rightarrow$ (2), we know that $1_X - \gamma$, $1_X - \sigma$ and $1_X - \delta$ are invertible with inverse $\omega = (1_X - \gamma)^{-1}$, $\lambda = (1_X - \sigma)^{-1}$ and $\nu = (1_X - \delta)^{-1}$, respectively. As $\sigma f = 0$, we have $(1_X - \sigma) f = f$, that is to say $f = (1_X - \sigma)^{-1} f = \lambda f$. Therefore,

$$f^{\circledR} = \omega f_0 \lambda f f_0 \nu = \omega f_0 f f_0 \nu = \omega f_0 \nu = (1_X - \gamma)^{-1} \alpha \varphi^{\circledR} (1_X - \delta)^{-1}.$$

□

**Remark 5.7.2 ([83])** If the conditions as in Theorem 5.7.1 hold, and if $1_X - \delta$ is invertible, then $f f_0$ is core invertible with $(f f_0)^{\circledR} = f f_0 (1_X - \delta)^{-1}$.

**Proof** Since $1_X - f f_0 = (1_X - \varphi \varphi^{\circledR})\beta$,

$$\begin{aligned}
1_X - \delta - f f_0 &= (1_X - f f_0) - \delta \\
&= (1_X - \varphi \varphi^{\circledR})\beta - \beta^*(\varphi^{\circledR})^* \eta^*(1_X - \varphi \varphi^{\circledR})\beta \\
&= (1_X - \eta \varphi^{\circledR} \beta)^*(1_X - \varphi \varphi^{\circledR})\beta \\
&= \beta^*(1_X - \varphi \varphi^{\circledR})\beta.
\end{aligned}$$

Set $q = \beta^*(1_X - \varphi \varphi^{\circledR})\beta = 1_X - \delta - f f_0$, then $q = q^*$, $q f f_0 = 0$ and $q + f f_0 = 1_X - \delta$ is invertible. Thus $f f_0$ is core invertible and $(f f_0)^{\circledR} = (1_X - \delta)^{-1} f f_0 (1_X - \delta)^{-1}$ by Lemma 5.3.2. Since $\delta f = 0$, $(1_X - \delta) f = f$, which implies $f = (1_X - \delta)^{-1} f$. Hence

$$(f f_0)^{\circledR} = (1_X - \delta)^{-1} f f_0 (1_X - \delta)^{-1} = f f_0 (1_X - \delta)^{-1}.$$

□

There is a result for the dual core inverse, which corresponds to Theorem 5.7.1, as follows.

**Theorem 5.7.3 ([83])** *Let $\mathscr{C}$ be an additive category with an involution $*$. Suppose that $\varphi : X \to X$ is a morphism of $\mathscr{C}$ with dual core inverse $\varphi_{\circledR} : X \to X$ and*

$\eta : X \to X$ is a morphism of $\mathscr{C}$ such that $1_X + \varphi_{\circledast}\eta$ is invertible. Let

$$\alpha = (1_X + \varphi_{\circledast}\eta)^{-1},$$

$$\beta = (1_X + \eta\varphi_{\circledast})^{-1},$$

$$\varepsilon = (1_X - \varphi\varphi_{\circledast})\eta\alpha(1_X - \varphi_{\circledast}\varphi),$$

$$\rho = \alpha\varphi_{\circledast}\varphi\alpha^{-1}(1_X - \varphi\varphi_{\circledast})\beta,$$

$$\zeta = \alpha(1_X - \varphi_{\circledast}\varphi)\beta^{-1}\varphi\varphi_{\circledast}\beta,$$

$$\xi = \alpha(1_X - \varphi_{\circledast}\varphi)\eta^*(\varphi_{\circledast})^*\alpha^*.$$

*Then the following conditions are equivalent:*

(1)   $f = \varphi + \eta - \varepsilon$ *has a dual core inverse.*
(2)   $1_X - \rho$, $1_X - \zeta$ *and* $1_X - \xi$ *are invertible.*
(3)   $1_X - \rho$ *is right invertible, both* $1_X - \zeta$ *and* $1_X - \xi$ *are left invertible.*

*In this case,*

$$f_{\circledast} = (1_X - \xi)^{-1}\alpha\varphi_{\circledast}(1_X - \rho)^{-1},$$

$$(1_X - \rho)^{-1} = 1_X - \varphi_{\circledast}\varphi + \varphi_{\circledast}\varphi f f_{\circledast},$$

$$(1_X - \zeta)^{-1} = 1_X - \varphi_{\circledast}\varphi + f f_{\circledast}\varphi_{\circledast}\varphi,$$

$$(1_X - \xi)^{-1} = 1_X - \varphi_{\circledast}\varphi + f_{\circledast}f\varphi_{\circledast}\varphi.$$

Let $J(R)$ be Jacobson radical of $R$. As a matter of convenience, we use the following notation:

$$\varepsilon_\tau = (1 - aa^\tau)j(1 + a^\tau j)^{-1}(1 - a^\tau a),$$

where $\tau \in \{(1), (1, 2, 3), (1, 2, 4), \dagger, \#\}$ and $j \in J(R)$.
From the previous chapters, we know that:

(I)   If $a \in R^{\{1\}}$, then $a + j \in R^{\{1\}}$ if and only if $\varepsilon_{(1)} = 0$.
(II)   If $a \in R^\dagger$, then $a + j \in R^\dagger$ if and only if $\varepsilon_\dagger = 0$.
(III)   If $a \in R^\#$, then $a + j \in R^\#$ if and only if $\varepsilon_\# = 0$.
(IV)   If $a \in R^{\{1,2,3\}}$, then $a + j \in R^{\{1,2,3\}}$ if and only if $\varepsilon_{(1,2,3)} = 0$.
(V)   If $a \in R^{\{1,2,4\}}$, then $a + j \in R^{\{1,2,4\}}$ if and only if $\varepsilon_{(1,2,4)} = 0$.

Moreover, the expressions of $(a + j)^{(1,2)}$, $(a + j)^\dagger$, $(a + j)^\#$, $(a + j)^{(1,2,3)}$, $(a + j)^{(1,2,4)}$ are presented, respectively.

Next, we will show that the core invertible element has a similar result as the above.

**Theorem 5.7.4 ([83])** *Let $R$ be a unital $*$-ring and $J(R)$ its Jacobson radical. If $a \in R^{\circledcirc}$ with core inverse $a^{\circledcirc}$ and $j \in J(R)$, then*

$$a + j \in R^{\circledcirc} \ \text{if and only if} \ \varepsilon = (1 - aa^{\circledcirc})j(1 + a^{\circledcirc}j)^{-1}(1 - a^{\circledcirc}a) = 0.$$

*In this case,*

$$(a + j)^{\circledcirc} = (1 - \gamma)^{-1}(1 + a^{\circledcirc}j)^{-1}a^{\circledcirc}(1 - \delta)^{-1},$$

*where*

$$\gamma = (1 + a^{\circledcirc}j)^{-1}(1 - a^{\circledcirc}a)(1 + ja^{\circledcirc})aa^{\circledcirc}(1 + ja^{\circledcirc})^{-1},$$

$$\delta = (1 + (a^{\circledcirc})^* j^*)^{-1}(a^{\circledcirc})^* j^*(1 - aa^{\circledcirc})(1 + ja^{\circledcirc})^{-1}.$$

***Proof*** Remark first that, if $j \in J(R)$, then $1 + a^{\circledcirc}j \in R^{-1}$ and $j^* \in J(R)$.

Set $\phi = (a + j)^{\circledcirc}$, then $\phi \in \varepsilon\{1\}$ by Lemma 2.8.1. This shows that the element $\varepsilon \in J(R)$ is von Neumann regular, so it must be zero, that is to say

$$\varepsilon = (1 - aa^{\circledcirc})j(1 + a^{\circledcirc}j)^{-1}(1 - a^{\circledcirc}a) = 0.$$

On the contrary, suppose that $\varepsilon = (1 - aa^{\circledcirc})j(1 + a^{\circledcirc}j)^{-1}(1 - a^{\circledcirc}a) = 0$. Then it is easy to see that $(1 + a^{\circledcirc}j)^{-1}a^{\circledcirc} \in (a + j)\{1, 2\}$ by Lemma 2.8.1 and the fact that the core inverse is a $\{1, 2\}$-inverse. Thus, we have

$$(a + j)(1 + a^{\circledcirc}j)^{-1}a^{\circledcirc}(a + j) = (a + j),$$

which implies

$$[1 - (a + j)(1 + a^{\circledcirc}j)^{-1}a^{\circledcirc}](a + j) = 0, \tag{5.7.4}$$

where

$$1 - (a + j)(1 + a^{\circledcirc}j)^{-1}a^{\circledcirc}$$

$$= 1 - (a + j)a^{\circledcirc}(1 + ja^{\circledcirc})^{-1}$$

$$= (1 + ja^{\circledcirc})(1 + ja^{\circledcirc})^{-1} - (a + j)a^{\circledcirc}(1 + ja^{\circledcirc})^{-1}$$

$$= [(1 + ja^{\circledcirc}) - (a + j)a^{\circledcirc}](1 + ja^{\circledcirc})^{-1}$$

$$= (1 - aa^{\circledcirc})(1 + ja^{\circledcirc})^{-1}.$$

Hence the equality (5.7.4) can be written as

$$(1 - aa^{\circledcirc})(1 + ja^{\circledcirc})^{-1}(a + j) = 0. \tag{5.7.5}$$

Since $a \in R^{\circledcirc}$, set $p = 1 - aa^{\circledcirc}$. Then $p$ is an Hermitian element such that $pa = 0$, $a + p \in R^{-1}$ by the proof of Lemma 5.3.2. Let

$$q = [(1 + ja^{\circledcirc})^{-1}]^* p(1 + ja^{\circledcirc})^{-1} = [(1 + ja^{\circledcirc})^{-1}]^*(1 - aa^{\circledcirc})(1 + ja^{\circledcirc})^{-1}.$$

Then $q = q^*$ and

$$q(a + j) = [(1 + ja^{\circledcirc})^{-1}]^*(1 - aa^{\circledcirc})(1 + ja^{\circledcirc})^{-1}(a + j) \overset{(5.7.5)}{=} 0.$$

Moreover, we have

$$a + q = a + [(1 + ja^{\circledcirc})^{-1}]^* p(1 + ja^{\circledcirc})^{-1}$$
$$= [(1 + ja^{\circledcirc})^{-1}]^*[(1 + ja^{\circledcirc})^* a(1 + ja^{\circledcirc}) + p](1 + ja^{\circledcirc})^{-1},$$

where

$$(1 + ja^{\circledcirc})^* a(1 + ja^{\circledcirc}) + p$$
$$= a + aja^{\circledcirc} + (a^{\circledcirc})^* j^* a + (a^{\circledcirc})^* j^* aja^{\circledcirc} + p$$
$$= (a + p) + aja^{\circledcirc} + (a^{\circledcirc})^* j^* a + (a^{\circledcirc})^* j^* aja^{\circledcirc}$$
$$= (a + p)[1 + (a + p)^{-1}(aja^{\circledcirc} + (a^{\circledcirc})^* j^* a + (a^{\circledcirc})^* j^* aja^{\circledcirc})]$$

is invertible which follows from the property of Jacobson radical and the fact that $a + p \in R^{-1}$. Thus $a + q \in R^{-1}$. Therefore,

$$a + j + q = (a + q)[1 + (a + q)^{-1} j] \in R^{-1}.$$

In conclusion, $q$ is an Hermitian element such that $q(a + j) = 0$, $a + j + q \in R^{-1}$. Applying Lemma 5.3.2, we can get that $a + j \in R^{\circledcirc}$. Therefore, $1_X - \gamma$, $1_X - \sigma$ and $1_X - \delta$ are invertible by Theorem 5.7.1, where

$$\gamma = (1 + a^{\circledcirc} j)^{-1}(1 - a^{\circledcirc} a)(1 + ja^{\circledcirc})aa^{\circledcirc}(1 + ja^{\circledcirc})^{-1},$$
$$\sigma = (1 + a^{\circledcirc} j)^{-1} a^{\circledcirc} a(1 + a^{\circledcirc} j)(1 - aa^{\circledcirc})(1 + ja^{\circledcirc})^{-1},$$
$$\delta = (1 + (a^{\circledcirc})^* j^*)^{-1}(a^{\circledcirc})^* j^*(1 - aa^{\circledcirc})(1 + ja^{\circledcirc})^{-1}.$$

Hence we obtain

$$(a + j)^{\circledcirc} = (1 - \gamma)^{-1}(1 + a^{\circledcirc} j)^{-1} a^{\circledcirc}(1 - \delta)^{-1}.$$

$\square$

Similar to Theorem 5.7.4, we have

**Theorem 5.7.5 ([83])** *Let $R$ be a unital $*$-ring and $J(R)$ its Jacobson radical. If $a \in R_{\oplus}$ with dual core inverse $a_{\oplus}$ and $j \in J(R)$, then*

$$a + j \in R_{\oplus} \text{ if and only if } \varepsilon = (1 - aa_{\oplus})j(1 + a_{\oplus}j)^{-1}(1 - a_{\oplus}a) = 0.$$

*In this case,*

$$(a + j)_{\oplus} = (1 - \xi)^{-1}(1 + a_{\oplus}j)^{-1}a_{\oplus}(1 - \rho)^{-1},$$

*where*

$$\rho = (1 + a_{\oplus}j)^{-1}a_{\oplus}a(1 + a_{\oplus}j)(1 - aa_{\oplus})(1 + ja_{\oplus})^{-1},$$

$$\xi = (1 + a_{\oplus}j)^{-1}(1 - a_{\oplus}a)j^*(a_{\oplus})^*(1 + j^*a_{\oplus}^*)^{-1}.$$

# Chapter 6
# Pseudo Core Inverses

For any square complex matrix $A$ with index 1, we know that the core inverse of $A$ exists and it is equal to $A^{\#} P_{A^{\#}}$. For a square complex matrix $A$ of an arbitrary index, by noting that the Drazin inverse $A^D$ always exists, it is natural to consider the generalized inverse $A^D P_{A^D}$ so as to generalize the core inverse. To this end, K. Manjunatha Prasad and K. S. Mohana [103] proposed in 2014 the notion of core-EP inverses. Let $A$ have index $k$. Then the core-EP inverse of $A$ was defined by $A^k[(A^k)^* A^{k+1}]^- (A^k)^*$ and exactly equals $A^D P_{A^D}$. The term "core-EP inverse" is used because such a generalized inverse is an EP matrix (and hence a core matrix). In [44], by making use of three algebraic equations, Yuefeng Gao and Jianlong Chen introduced the notion of pseudo core inverses in an arbitrary $*$-ring which generalizes core-EP inverses of complex matrices.

## 6.1 Core-EP Inverses of Complex Matrices

In this section, the core-EP inverse of a complex matrix is discussed.

**Definition 6.1.1 ([103])** Let $A \in \mathbb{C}^{n \times n}$ have index $k$. Then $X \in \mathbb{C}^{n \times n}$ is called a core-EP inverse of $A$ and denoted by $A^{\oplus}$ if

$$XAX = X \quad \text{and} \quad \mathcal{R}(X) = \mathcal{R}(X^*) = \mathcal{R}(A^k).$$

**Theorem 6.1.2 ([103])** *Let $A \in \mathbb{C}^{n \times n}$. Then the core-EP inverse of $A$ is unique. If $\mathrm{ind}(A) = 1$, the core-EP inverse of $A$ reduces to the core inverse of $A$.*

**Proof** Suppose that $A$ is with index $k$. Let $X = A^D A^k (A^k)^{\dagger}$. Since $X$ satisfies

$$XAX = X, \mathcal{R}(X) = \mathcal{R}(X^*) = \mathcal{R}(A^k),$$

it follows that $X$ is a core-EP inverse of $A$. For the uniqueness of the core-EP inverse, we can see that if $Y$ is another core-EP inverse of $A$ then

$$\mathcal{R}(X) = \mathcal{R}(X^*) = \mathcal{R}(Y) = \mathcal{R}(Y^*),$$

which implies that $X = YT$ and $Y = SX$ for some $S, T \in \mathbb{C}^{n \times n}$. Thus

$$X = YT = (YAY)T = YAX = (SX)AX = S(XAX) = SX = Y,$$

as needed.                                                                                                       □

Next, several methods are provided to compute the core-EP inverse of $A$. Let $A \in \mathbb{C}^{n \times n}$ and $\mathrm{rank}(A) = r$. Consider the Hartwig-Spidelböck decomposition

$$A = U \begin{bmatrix} \Sigma K & \Sigma L \\ 0 & 0 \end{bmatrix} U^* \tag{6.1.1}$$

for $A$, where $U \in \mathbb{C}^{n \times n}$ is unitary, $\Sigma = \mathrm{diag}(\sigma_1 I_{r_1}, \sigma_2 I_{r_2}, \cdots, \sigma_t I_{r_t})$ is a diagonal matrix, the diagonal entries $\sigma_i$ being singular values of $A$, $\sigma_1 > \sigma_2 > \cdots > \sigma_t > 0$, $r_1 + r_2 + \cdots + r_t = r$ and $K \in \mathbb{C}^{r \times r}$, $L \in \mathbb{C}^{r \times (n-r)}$ satisfying $KK^* + LL^* = I_r$.

**Theorem 6.1.3 ([44])**  *Let $A \in \mathbb{C}^{n \times n}$ be of the form (6.1.1). Then*

$$A^{\oplus} = U \begin{bmatrix} (\Sigma K)^{\oplus} & 0 \\ 0 & 0 \end{bmatrix} U^*.$$

***Proof***  Suppose that $A$ is of index $m$ and $A^{\oplus} = X$. Then

$$XA^{m+1} = A^m, \quad AX^2 = X, \quad (AX)^* = AX.$$

Since $XA^{m+1} = A^m$, there exists $X = \begin{bmatrix} X_1 & X_2 \\ X_3 & X_4 \end{bmatrix} \in \mathbb{C}^{n \times n}$ such that

$$\begin{bmatrix} X_1 & X_2 \\ X_3 & X_4 \end{bmatrix} \begin{bmatrix} (\Sigma K)^{m+1} & (\Sigma K)^m \Sigma L \\ 0 & 0 \end{bmatrix} = \begin{bmatrix} (\Sigma K)^m & (\Sigma K)^{m-1} \Sigma L \\ 0 & 0 \end{bmatrix}.$$

So

$$\begin{cases} (\Sigma K)^m & = X_1 (\Sigma K)^{m+1}, \\ (\Sigma K)^{m-1} \Sigma L & = X_1 (\Sigma K)^m \Sigma L. \end{cases}$$

Hence $(\Sigma K)^{m-1}\Sigma(K, L) = X_1(\Sigma K)^m \Sigma(K, L)$. Multiplying $\begin{pmatrix} K^* \\ L^* \end{pmatrix}$ on the right side of this equality, we get

$$(\Sigma K)^{m-1}\Sigma \begin{pmatrix} K & L \end{pmatrix} \begin{pmatrix} K^* \\ L^* \end{pmatrix} = X_1(\Sigma K)^m \Sigma \begin{pmatrix} K & L \end{pmatrix} \begin{pmatrix} K^* \\ L^* \end{pmatrix}.$$

Since $KK^* + LL^* = I$, we have $(\Sigma K)^{m-1}\Sigma = X_1(\Sigma K)^m \Sigma$. Hence $(\Sigma K)^{m-1} = X_1(\Sigma K)^m$. Thus $\text{rank}((\Sigma K)^{m-1}) = \text{rank}((\Sigma K)^m)$ and $ind(\Sigma K) \leq m - 1$. We know that

$$(\Sigma K)^{\textcircled{\dagger}} = (\Sigma K)^D (\Sigma K)^{m-1}((\Sigma K)^{m-1})^\dagger. \tag{6.1.2}$$

A direct calculation shows

$$A^D = U \begin{bmatrix} (\Sigma K)^D & ((\Sigma K)^D)^2 \Sigma L \\ 0 & 0 \end{bmatrix} U^*.$$

Since

$$A^m = (U \begin{bmatrix} \Sigma K & \Sigma L \\ 0 & 0 \end{bmatrix} U^*)^m = U \begin{pmatrix} \Sigma K & \Sigma L \\ 0 & 0 \end{pmatrix}^m U^*$$

$$= U \begin{bmatrix} (\Sigma K)^m & (\Sigma K)^{m-1}\Sigma L \\ 0 & 0 \end{bmatrix} U^*,$$

we have $(A^m)^\dagger = U \begin{bmatrix} (\Sigma K)^m & (\Sigma K)^{m-1}\Sigma L \\ 0 & 0 \end{bmatrix}^\dagger U^*$. So

$$A^m(A^m)^\dagger = U \begin{bmatrix} (\Sigma K)^m & (\Sigma K)^{m-1}\Sigma L \\ 0 & 0 \end{bmatrix} \begin{bmatrix} (\Sigma K)^m & (\Sigma K)^{m-1}\Sigma L \\ 0 & 0 \end{bmatrix}^\dagger U^*$$

$$= U \begin{bmatrix} B & C \\ 0 & 0 \end{bmatrix} \begin{bmatrix} B & C^\dagger \\ 0 & 0 \end{bmatrix} U^*,$$

where $B = (\Sigma K)^m$, $C = (\Sigma K)^{m-1}\Sigma L$. Since $\begin{bmatrix} 0 & 0 \\ B & C \end{bmatrix}^\dagger = \begin{bmatrix} 0 & B^*L^\dagger \\ 0 & C^*L^\dagger \end{bmatrix}$, where $L = BB^* + CC^*$, it follows that

$$\begin{bmatrix} B & C \\ 0 & 0 \end{bmatrix}^\dagger = \begin{bmatrix} 0 & 1 \\ 1 & 0 \end{bmatrix}\begin{bmatrix} 0 & 0 \\ B & C \end{bmatrix}^\dagger = \begin{bmatrix} 0 & 0 \\ B & C \end{bmatrix}^\dagger \begin{bmatrix} 0 & 1 \\ 1 & 0 \end{bmatrix} = \begin{bmatrix} B^*L^\dagger & 0 \\ C^*L^\dagger & 0 \end{bmatrix},$$

$$\begin{bmatrix} B & C \\ 0 & 0 \end{bmatrix}\begin{bmatrix} B & C \\ 0 & 0 \end{bmatrix}^\dagger = \begin{bmatrix} B & C \\ 0 & 0 \end{bmatrix}\begin{bmatrix} B^*L^\dagger & 0 \\ C^*L^\dagger & 0 \end{bmatrix} = \begin{bmatrix} BB^*L^\dagger + CC^*L^\dagger & 0 \\ 0 & 0 \end{bmatrix} = \begin{bmatrix} LL^\dagger & 0 \\ 0 & 0 \end{bmatrix}.$$

So $A^m(A^m)^\dagger = U\begin{bmatrix} LL^\dagger & 0 \\ 0 & 0 \end{bmatrix}U^*$, where

$$L = BB^* + CC^*$$
$$= (\Sigma K)^m((\Sigma K)^m)^* + (\Sigma K)^{m-1}\Sigma L((\Sigma K)^{m-1}\Sigma L)^*$$
$$= (\Sigma K)^{m-1}[\Sigma K(\Sigma K)^* + \Sigma L(\Sigma L)^*]((\Sigma K)^{m-1})^*$$
$$= (\Sigma K)^{m-1}\Sigma\Sigma^*((\Sigma K)^{m-1})^*$$
$$= (\Sigma K)^{m-1}\Sigma((\Sigma K)^{m-1}\Sigma)^*$$
$$= TT^*, \text{ where} T = (\Sigma K)^{m-1}\Sigma.$$

So $LL^\dagger = TT^*(TT^*)^\dagger = TT^\dagger = (\Sigma K)^{m-1}\Sigma((\Sigma K)^{m-1}\Sigma)^\dagger = T_1\Sigma(T_1\Sigma)^\dagger$, where $T_1 = (\Sigma K)^{m-1}$. Also, we know that

$$(T_1\Sigma)^\dagger = (T_1\Sigma)^*[(T_1\Sigma(T_1\Sigma)^* + I - T_1T_1^\dagger]^{-1}.$$

Thus $T_1\Sigma(T_1\Sigma)^\dagger = T_1\Sigma(T_1\Sigma)^*[(T_1\Sigma(T_1\Sigma)^*+I-T_1T_1^\dagger]^{-1} = T_1T_1^\dagger[T_1\Sigma(T_1\Sigma)^*+ I - T_1T_1^\dagger][(T_1\Sigma(T_1\Sigma)^* + I - T_1T_1^\dagger]^{-1} = T_1T_1^\dagger$.

So $A^m(A^m)^\dagger = U\begin{bmatrix} T_1T_1^\dagger & 0 \\ 0 & 0 \end{bmatrix}U^*$, where $T_1 = (\Sigma K)^{m-1}$.

Hence

$$A^\oplus = A^D A^m(A^m)^\dagger = U\begin{bmatrix} (\Sigma K)^D & ((\Sigma K)^D)^2\Sigma L \\ 0 & 0 \end{bmatrix}\begin{bmatrix} T_1T_1^\dagger & 0 \\ 0 & 0 \end{bmatrix}U^*$$

$$= U\begin{bmatrix} (\Sigma K)^D T_1T_1^\dagger & 0 \\ 0 & 0 \end{bmatrix}U^* = U\begin{bmatrix} (\Sigma K)^D(\Sigma K)^{m-1}((\Sigma K)^{m-1})^\dagger & 0 \\ 0 & 0 \end{bmatrix}U^*$$

$$= U\begin{bmatrix} (\Sigma K)^\oplus & 0 \\ 0 & 0 \end{bmatrix}U^*.$$

$\square$

In view of Theorem 1.1.3, for any matrix $A \in \mathbb{C}^{n \times n}$, there exists unitary matrix $U$ such that

$$A = U \begin{bmatrix} D & L \\ 0 & N \end{bmatrix} U^*, \tag{6.1.3}$$

where $D$ is invertible and $N$ is nilpotent. The next result gives a representation of the core-EP inverse by making use of this decomposition (6.1.3).

**Theorem 6.1.4** *Let $A$ be written as in (6.1.3). Then*

$$A^{\oplus} = U \begin{bmatrix} D^{-1} & 0 \\ 0 & 0 \end{bmatrix} U^*.$$

**Proof** Let $A$ have index $k$. Then $A^k = U \begin{bmatrix} D^k & \hat{L} \\ 0 & 0 \end{bmatrix} U^*$, from which it can be seen that $A^k(A^k)^{\dagger} = U \begin{bmatrix} I & 0 \\ 0 & 0 \end{bmatrix} U^*$. Since $A^D = U \begin{bmatrix} D^{-1} & \star \\ 0 & 0 \end{bmatrix} U^*$, it follows from Theorem 6.1.2 that $A^{\oplus} = A^D A^k (A^k)^{\dagger} = U \begin{bmatrix} D^{-1} & 0 \\ 0 & 0 \end{bmatrix} U^*$. $\square$

Let $A \in \mathbb{C}^{n \times n}$ be written in the Jordan normal form

$$A = P^{-1} \begin{bmatrix} D & 0 \\ 0 & N \end{bmatrix} P, \tag{6.1.4}$$

where $P \in \mathbb{C}^{n \times n}$ and $D \in \mathbb{C}^{t \times t}$ are nonsingular, and $N \in \mathbb{C}^{(n-t) \times (n-t)}$ is nilpotent. Suppose that $P = \begin{bmatrix} P_1 \\ P_2 \end{bmatrix}$ and $P^{-1} = [Q_1, Q_2]$.

**Theorem 6.1.5 ([44])** *Let $A \in \mathbb{C}^{n \times n}$ be of the form (6.1.4). Then*

$$A^{\oplus} = Q_1 D^{-1} (Q_1^* Q_1)^{-1} Q_1^*.$$

**Proof** From $P = \begin{bmatrix} P_1 \\ P_2 \end{bmatrix}$, $P^{-1} = (Q_1, Q_2)$ and $A = P^{-1} \begin{bmatrix} D & O \\ O & N \end{bmatrix} P$, we have

$$A^D = P^{-1} \begin{bmatrix} D^{-1} & O \\ O & O \end{bmatrix} P = Q_1 D^{-1} P_1 \text{ and } P_1 Q_1 = I.$$

Let the nilpotent index of $N$ be $m$. Then $N^m = 0$. So

$$A^m = (Q_1, Q_2) \begin{bmatrix} D^m & O \\ O & O \end{bmatrix} \begin{bmatrix} P_1 \\ P_2 \end{bmatrix} = Q_1 D^m P_1.$$

Set $B = Q_1(D^{-1})^m(Q_1^*Q_1)^{-1}Q_1^*$. Then

$$A^m B = Q_1 D^m P_1 Q_1(D^{-1})^m(Q_1^*Q_1)^{-1}Q_1^* = Q_1(Q_1^*Q_1)^{-1}Q_1^*.$$

Thus $(A^m B)^* = A^m B$ and

$$A^m B A^m = Q_1(Q_1^*Q_1)^{-1}Q_1^*Q_1 D^m P_1 = Q_1 D^m P_1 = A^m,$$

i.e., $B$ is a $\{1, 3\}$-inverse of $A^m$. So

$$A^{\oplus} = A^D A^m (A^m)^{(1,3)} = Q_1 D^{-1} P_1 Q_1(Q_1^*Q_1)^{-1}Q_1^*$$
$$= Q_1 D^{-1}(Q_1^*Q_1)^{-1}Q_1^*.$$

$\square$

Let $A \in \mathbb{C}^{n \times n}$ and $\mathrm{ind}(A) = k$. If $A = B_1 G_1$ is a full-rank decomposition and

$$G_i B_i = B_{i+1}G_{i+1}, \quad i = 1, 2, \ldots, k - 1. \tag{6.1.5}$$

are also full-rank decomposition, then $G_k B_k$ is nonsingular or zero by Theorem 4.1.4, and $A^{\dagger} = G_1^*(G_1 G_1^*)^{-1}(B_1^* B_1)^{-1}B_1^*$ by Theorem 2.1.2. Moreover, $A^k = B_1 B_2 \cdots B_k G_k \cdots G_2 G_1$.

**Theorem 6.1.6** *Let $A \in \mathbb{C}^{n \times n}$, $\mathrm{ind}(A) = k$ and the full-rank decomposition of $A$ be as in (6.1.5). Then*

$$A^{\oplus} = \begin{cases} B(G_k B_k)^{-1}(B^* B)^{-1}B^*, & G_k B_k \text{ is nonsingular,} \\ 0, & G_k B_k = 0, \end{cases}$$

*where $B = B_1 B_2 \cdots B_k$ and $G = G_k \cdots G_2 G_1$.*

**Proof** Suppose that $A \in \mathbb{C}^{n \times n}$, $\mathrm{ind}(A) = k$ and the full-rank decomposition of $A$ be as in (6.1.5). Set $B = B_1 B_2 \cdots B_k$ and $G = G_k \cdots G_2 G_1$. Then

$$A^{\oplus} = A^D A^k (A^k)^{\dagger} = B(G_k B_k)^{-k-1}GBGG^*(GG^*)^{-1}(B^* B)^{-1}B^*$$
$$= B(G_k B_k)^{-k-1}GB(B^* B)^{-1}B^*$$
$$= B(G_k B_k)^{-k-1}(G_k B_k)^k(B^* B)^{-1}B^*$$
$$= B(G_k B_k)^{-1}(B^* B)^{-1}B^*.$$

If $G_k B_k = 0$, then $BG = 0$. So, $A^{\oplus} = 0$.

$\square$

## 6.2   Pseudo Core Inverses of Elements in Rings with Involution

In this section we introduce the notion of pseudo core inverses which serves as a common generalization for core inverses of ring elements and core-EP inverses for complex matrices.

**Definition 6.2.1 ([44])**   An element $a \in R$ is said to be pseudo core invertible if there exists $x \in R$ such that

$$(I)\ xa^{m+1} = a^m \text{ for some nonnegative integer } m,$$

$$(II)\ ax^2 = x,\ (III)\ (ax)^* = ax,$$

in which case such an $x$ is called a pseudo core inverse of $a$ and denoted by $a^{\circledD}$. Dually, $a$ is said to be dual pseudo core invertible if there is $y \in R$ such that

$$(I')\ a^{m+1}y = a^m \text{ for some nonnegative integer } m,$$

$$(II')\ x^2a = x,\ (III')\ (xa)^* = xa,$$

in which case such an $y$ is called a dual pseudo core inverse of $a$ and denoted by $a_{\circledD}$.

Since only multiplication is required in Definition 6.2.1, the above two definitions hold, without modification, in a $*$-semigroup $S$.

In the following, the sets of all pseudo core invertible elements and dual pseudo core invertible elements in ring $R$ are denoted by $R^{\circledD}$ and $R_{\circledD}$, respectively.

**Lemma 6.2.2 ([44])**   *For $a \in R$, if there exist $x \in R$ and $k \in \mathbb{N}$ such that $xa^{k+1} = a^k$ and $ax^2 = x$, then*

(1) $ax = a^m x^m$ *for arbitrary positive integer $m$.*
(2) $xax = x$.
(3) $a^m x^m a^m = a^m$ *for any $m \geq k$.*
(4) $a \in R^D$, $a^D = x^{k+1}a^k$ *with* $\mathrm{ind}(a) \leq k$, *and* $xa^{\mathrm{ind}(a)+1} = a^{\mathrm{ind}(a)}$.

*Proof*

(1) Since $ax^2 = x$, we have $ax = a(ax^2) = a^2x^2 = a^2(ax^2)x = a^3x^3 = \cdots = a^m x^m$ for arbitrary positive integer $m$.
(2) As $ax = a^{k+1}x^{k+1} = a^k x^k$ follows from (1), we have

$$xax = xa^{k+1}x^{k+1} = a^k x^{k+1} = (ax)x = ax^2 = x.$$

(3) Let $m \geq k$. Then

$$a^m = a^k a^{m-k} = x a^{k+1} a^{m-k} = x a^{m+1} = a x^2 a^{m+1}$$
$$= (a^m x^m) x a^{m+1} = a^m (x^{m+1} a^{m+1})$$

Since $x^{m+1} a^{m+1} = x^m a^m$, $a^m = a^m x^m a^m$.

(4) Let $b = x^{k+1} a^k$. Then we get

$$a^k b a = a^k (x^{k+1} a^k) a = a^k (x^{k+1} a^{k+1}) = a^k x^k a^k = a^k;$$
$$bab = (x^{k+1} a^k) a (x^{k+1} a^k) = x^{k+1} a^{k+1} x^{k+1} a^k$$
$$= x^k a^k x^{k+1} a^k = x^k (a x^2) a^k = x^{k+1} a^k = b;$$

$ab = a x^{k+1} a^k = x^k a^k = x^{k+1} a^{k+1} = (x^{k+1} a^k) a = ba$. Hence

$$a^D = b = x^{k+1} a^k,$$

with $\mathrm{ind}(a) \leq k$. Moreover, multiplying $x a^{k+1} = a^k$ by $(a^D)^{k-\mathrm{ind}(a)}$ from the right side gives $x a^{\mathrm{ind}(a)+1} = a^{\mathrm{ind}(a)}$.

□

From the above result it can be seen that if $a$ is (dual) pseudo core invertible then it is Drazin invertible and the smallest nonnegative integer $m$ satisfying Eqs. $(I)$–$(III)$ (resp., $(I')$–$(III')$) is exactly the Drazin index of $a$. In what follows, we shall indicate the fact that $a$ is pseudo core invertible with Drazin index $m$ by saying $a$ is pseudo core invertible with index $m$.

First and most fundamentally, we have the following theorem.

**Theorem 6.2.3 ([44])** *Let $a \in R$. Then $a$ has at most one pseudo core inverse in $R$.*

***Proof*** Suppose that $a$ is pseudo core invertible with index $k$ and let $x$ and $y$ be two pseudo core inverses of $a$. Then by Lemma 6.2.2, we have

$$x^{k+1} a^k = x^{m+1} a^m = a^D = y^{n+1} a^n = y^{k+1} a^k,$$

and

$$a^k x^k a^k = a^k, \ (a^k x^k)^* = (ax)^* = ax = a^k x^k$$

which yield that $x^k$ is a $\{1, 3\}$-inverse of $a^k$. Likewise, $y^k$ is one of the $\{1, 3\}$-inverse of $a^k$. So $a^k x^k = a^k (a^k)^{(1,3)} = a^k y^k$. Then,

$$x = x(ax) = x(a^k x^k) = x^2 a^{k+1} x^k = \cdots = x^{k+1} a^{2k} x^k = (x^{k+1} a^k)(a^k x^k)$$
$$= a^D(a^k (a^k)^{(1,3)}) = (y^{k+1} a^k)(a^k y^k) = y^{k+1} a^{2k} y^k = y.$$

Thus $a$ has at most one pseudo core inverse. □

The following result gives an equivalent characterization for the existence of the pseudo core inverse as well as the expression in terms of Drazin inverse and $\{1, 3\}$-inverse.

**Theorem 6.2.4 ([44])** *The following statements are equivalent for any $a \in R$:*

(1) $a \in R^{\circledD}$.
(2) $a \in R^D$ and $a^m \in R^{\{1,3\}}$ for any integer $m \geq \text{ind}(a)$.
(3) $a \in R^D$ and $a^m \in R^{\{1,3\}}$ for some integer $m \geq \text{ind}(a)$.

*In this case, $a^{\circledD} = a^D a^m (a^m)^{(1,3)}$ for any integer $m \geq \text{ind}(a)$.*

**Proof** (1)$\Rightarrow$(2). It follows by Lemma 6.2.2.
(2)$\Rightarrow$(3). It is clear.
(3)$\Rightarrow$(1). Suppose $a \in R^D$ with $\text{ind}(a) = k$ and suppose $a^m \in R^{\{1,3\}}$ for some $m \geq k$. Setting $x = a^D a^m (a^m)^{(1,3)}$, next we prove $a^{\circledD} = x$. In fact,

$$xa^{k+1} = a^D a^m (a^m)^{(1,3)} a^{k+1} = a^D a^m (a^m)^{(1,3)} a^m a^{k+1} (a^D)^m$$
$$= a^D a^m a^{k+1} (a^D)^m = a^k;$$
$$ax^2 = aa^D a^m (a^m)^{(1,3)} a^D a^m (a^m)^{(1,3)}$$
$$= a^m (a^m)^{(1,3)} a^D a^m (a^m)^{(1,3)} = a^D a^m (a^m)^{(1,3)} = x;$$

$ax = aa^D a^m (a^m)^{(1,3)} = a^m (a^m)^{(1,3)}$, so $(ax)^* = ax$. Hence $a \in R^{\circledD}$. □

For Drazin inverses, we have seen that $a \in R^D$ if and only if $a^k \in R^\#$ for some positive integer $k$; $a \in R^D$ if and only if $a^k \in R^D$ for arbitrary positive integer $k$ if and only if $a^k \in R^D$ for some positive integer $k$; and if $a \in R^D$ then $a^D \in R^D$ with $(a^D)^D = a^2 a^D$.

Now, for pseudo core inverses, we have the following results.

**Theorem 6.2.5 ([44])** *Let $a \in R$. Then $a \in R^{\circledD}$ if and only if $a^m \in R^{\circledast}$ for some positive integer $m$. In this case, $(a^m)^{\circledast} = (a^{\circledD})^m$ and $a^{\circledD} = a^{m-1}(a^m)^{\circledast}$.*

**Proof** Suppose $a^{\text{\textcircled{D}}} = x$ with $\text{ind}(a) = m$. Setting $y = x^m$, by the definition of pseudo core inverse and by Lemma 6.2.2, we can check that

$$y(a^m)^2 = x^m(a^m)^2 = (x^m a^m)a^m = (x^{m+1}a^{m+1})a^m$$

$$= (x^{m+1}a^m)a^{m+1} = a^D a^{m+1} = a^m;$$

$$a^m y^2 = a^m(x^m)^2 = (a^m x^m)x^m = a x x^m = a x^{m+1} = x^m = y;$$

$$(a^m y)^* = (a^m x^m)^* = (ax)^* = ax = a^m x^m = a^m y.$$

Therefore $(a^m)^{\text{\textcircled{+}}} = y = (a^{\text{\textcircled{D}}})^m$.

On the contrary, since $a^m \in R^{\text{\textcircled{+}}}$, by the notion of core inverse, we have

$$(a^m)^{\text{\textcircled{+}}}(a^m)^2 = a^m, \quad a^m((a^m)^{\text{\textcircled{+}}})^2 = (a^m)^{\text{\textcircled{+}}} \text{ and } (a^m(a^m)^{\text{\textcircled{+}}})^* = a^m(a^m)^{\text{\textcircled{+}}}.$$

Let $x = a^{m-1}(a^m)^{\text{\textcircled{+}}}$. Then we notice

$$xa^{m+1} = a^{m-1}(a^m)^{\text{\textcircled{+}}}a^{m+1} = a^{m-1}((a^m)^{\#}a^m)a = a^m;$$

$$ax^2 = a(a^{m-1}(a^m)^{\text{\textcircled{+}}})^2 = a^m(a^m)^{\text{\textcircled{+}}}a^{m-1}(a^m)^{\text{\textcircled{+}}} = a^m(a^m)^{\text{\textcircled{+}}}a^{m-1}a^m((a^m)^{\text{\textcircled{+}}})^2 = a^{m-1}(a^m)^{\text{\textcircled{+}}} = x;$$

$$(ax)^* = (aa^{m-1}(a^m)^{\text{\textcircled{+}}})^* = (a^m(a^m)^{\text{\textcircled{+}}})^* = a^m(a^m)^{\text{\textcircled{+}}} = a(a^{m-1}(a^m)^{\text{\textcircled{+}}}) = ax.$$

Hence $a^{\text{\textcircled{D}}} = x = a^{m-1}(a^m)^{\text{\textcircled{+}}}$.  □

**Theorem 6.2.6 ([44])** *Let $a \in R$ and $k \in \mathbb{N}^+$. Then $a \in R^{\text{\textcircled{D}}}$ if and only if $a^k \in R^{\text{\textcircled{D}}}$. In this case, $(a^k)^{\text{\textcircled{D}}} = (a^{\text{\textcircled{D}}})^k$ and $a^{\text{\textcircled{D}}} = a^{k-1}(a^k)^{\text{\textcircled{D}}}$.*

**Proof** Suppose $a^{\text{\textcircled{D}}} = x$ with $\text{ind}(a) = m$. Then we have

$$xa^{m+1} = a^m, \quad ax^2 = x, \quad (ax)^* = ax.$$

For arbitrary positive integer $k$, let $n$ be the unique integer satisfying $0 \le kn - m < k$. Then

$$(a^k)^n = a^{kn} = a^m a^{kn-m} = xa^{m+1}a^{kn-m} = xa^{kn+1}, \text{ by induction,}$$

$$(a^k)^n = x^k(a^k)^{n+1}; \; a^k(x^k)^2 = (a^k x^k)x^k = (ax)x^k = ax^{k+1} = x^k;$$

$$a^k x^k = ax, \text{ so } (a^k x^k)^* = a^k x^k.$$

Thus $(a^k)^{\text{\textcircled{D}}} = x^k = (a^{\text{\textcircled{D}}})^k$, with $\text{ind}(a^k) \le n$. $\text{ind}(a^k) < n$ clearly forces that $x^k(a^k)^n = (a^k)^{n-1}$, and, since $x^k a^{kn} = (a^k)^{\text{\textcircled{D}}}a^{kn} = (a^k)^D a^{kn} = (a^D)^k a^{kn} = a^D a^{kn-k+1}$, which implies $a^D a^{kn-k+1} = a^{kn-k}$, whence, by $\text{ind}(a) = m$, we should have $kn - k \ge m$, contrary to our definition of $n$. Hence $\text{ind}(a^k) = n$.

Conversely, suppose $(a^k)^{\circledcirc} = y$ with $\text{ind}(a^k) = n$. Then we have

$$y(a^k)^{n+1} = (a^k)^n, \quad a^k y^2 = y, \quad (a^k y)^* = a^k y.$$

Set $x = a^{k-1}y$. In what follows, we prove $a^{\circledcirc} = x$.
    Since $xa^{kn+1} = a^{k-1}ya^{kn+1} = a^{k-1}(a^k y^2)a^{kn+1}$,

$$xa^{kn+1} = a^{k-1}(a^{kn}y^{n+1})a^{kn+1}$$

by induction. So, we get

$$xa^{kn+1} = a^{kn+k-1}(y^{n+1}a^{kn})a = a^{kn+k-1}(a^k)^D a = (a^k)^D a^{kn+k}$$
$$= y^{n+1}(a^k)^{n+1}a^{kn} = y^n a^{kn} a^{kn} = y^n (a^k)^{2n} = a^{kn};$$
$$ax^2 = a(a^{k-1}y)^2 = a^k y a^{k-1} y = a^k y a^{k-1}(a^k y^2) = a^k y a^{k-1}((a^k)^{n+1} y^{n+2})$$
$$= a^k(y(a^k)^{n+1})a^{k-1}y^{n+2} = a^k a^{kn} a^{k-1} y^{n+2} = a^{k-1}a^{kn}y^{n+1} = a^{k-1}y = x;$$
$$ax = aa^{k-1}y = a^k y, \quad \text{so } (ax)^* = ax.$$

From the above, $a^{\circledcirc} = x = a^{k-1}(a^k)^{\circledcirc}$ with $\text{ind}(a) \le kn$. □

**Theorem 6.2.7 ([44])** *Let $a \in R$. If $a \in R^{\circledcirc}$, then $a^{\circledcirc} \in R^{\oplus}$. In fact $a^{\circledcirc}$ is core invertible whenever it exists, and $(a^{\circledcirc})^{\circledcirc} = (a^{\circledcirc})^{\oplus} = a^2 a^{\circledcirc}$.*

**Proof** To prove this, one has merely to verify that if $x$ satisfies (I), (II) and (III) which defines the pseudo core inverse, then $y = a^2 x$ satisfies

$$yx^2 = x, \quad xy^2 = y, \quad (xy)^* = xy.$$

Here we omit the details. □

**Proposition 6.2.8 ([44])** *Let $a \in R^{\circledcirc}$. Then $((a^{\circledcirc})^{\circledcirc})^{\circledcirc} = a^{\circledcirc}$.*

**Proof** Suppose $a \in R^{\circledcirc}$ with $\text{ind}(a) = m$. By Theorem 6.2.7, we have

$$((a^{\circledcirc})^{\circledcirc})^{\circledcirc} = (a^{\circledcirc})^2(a^{\circledcirc})^{\circledcirc} = (a^{\circledcirc})^2 a^2 a^{\circledcirc}.$$

Since $(a^{\circledcirc})^2 a^2 a^{\circledcirc} = (a^{\circledcirc})^2 a^{m+1}(a^{\circledcirc})^m = a^{\circledcirc} a^m (a^{\circledcirc})^m = a^{\circledcirc} a a^{\circledcirc} = a^{\circledcirc}$, we get

$$((a^{\circledcirc})^{\circledcirc})^{\circledcirc} = a^{\circledcirc}.$$

□

Let $a \in R^D$ with $\text{ind}(a) = m$ and let sum $a = c_a + n_a$ be the core nilpotent decomposition of $a$.

**Theorem 6.2.9 ([44])** *An element $a \in R^{\circledD}$ if and only if $a \in R^D$ and $c_a \in R^{\circledast}$. In this case, $a^{\circledD}$ coincides with $c_a^{\circledast}$.*

**Proof** Supposing $a \in R^{\circledD}$ with $\mathrm{ind}(a) = m$, we have $a \in R^D$ by Theorem 6.2.4 and

$$a^{\circledD} c_a^2 = a^{\circledD}(aa^D a)^2 = a^{\circledD} a^3 a^D = a^{\circledD} a^{m+2}(a^D)^m$$

$$= a^{m+1}(a^D)^m = a^2 a^D = aa^D a = c_a,$$

$$c_a(a^{\circledD})^2 = aa^D a(a^{\circledD})^2 = aa^D a^{\circledD} = a^{\circledD},$$

$$c_a a^{\circledD} = aa^D aa^{\circledD} = aa^{\circledD}, \text{ which implies } (c_a a^{\circledD})^* = c_a a^{\circledD}.$$

Thus $c_a^{\circledast} = a^{\circledD}$.

Conversely, suppose $a \in R^D$ with $\mathrm{ind}(a) = m$, and suppose $c_a \in R^{\circledast}$ which gives $c_a \in R^{\{1,3\}}$ and $c_a^m \in R^{\{1,3\}}$. Since $a = c_a + n_a$, $c_a n_a = n_a c_a = 0$, and $n_a^m = 0$, we get $a^m = c_a^m$. So we get $a^m \in R^{\{1,3\}}$, and $a^m(a^m)^{(1,3)} = c_a^m(c_a^m)^{(1,3)} = c_a c_a^{(1,3)}$. From Theorem 6.2.4, it follows that $a \in R^{\circledD}$, $a^{\circledD} = a^D a^m(a^m)^{(1,3)} = c_a^{\#} c_a c_a^{(1,3)} = c_a^{\circledast}$ and $\mathrm{ind}(a) = m$.  $\square$

**Theorem 6.2.10 ([44])** *Let $a, \ x \in R$. Then the following conditions are equivalent:*

(1) $a^{\circledD} = x$.

(2) $xax = x$, and $xR = x^*R = a^m R$ for some positive integer $m$.

(3) $xax = x$, $xR = a^m R$ and $a^m R \subseteq x^*R$ for some positive integer $m$.

(4) $xax = x$, $xR = a^m R$ and $^{\circ}(x^*) \subseteq {}^{\circ}(a^m)$ for some positive integer $m$.

(5) $xax = x$, $^{\circ}x = {}^{\circ}(a^m)$ and $^{\circ}(x^*) \subseteq {}^{\circ}(a^m)$ for some positive integer $m$.

**Proof** (1) $\Rightarrow$ (2). Suppose $a^{\circledD} = x$ with $\mathrm{ind}(a) = m$. Then by the definition of pseudo core inverse, we have $xax = x$.

$$xR \subseteq x^*R \text{ since } x = ax^2 = (ax)x = (ax)^*x = x^*a^*x \in x^*R;$$

$$x^*R \subseteq a^m R \text{ since } x^* = (xax)^* = axx^* = a^m x^m x^* \in a^m R;$$

$$a^m R \subseteq xR \text{ since } a^m = xa^{m+1} \in xR.$$

Thus, $xR = x^*R = a^m R$.

It is easy to check that (2) $\Rightarrow$ (3), (3) $\Rightarrow$ (4) and (4) $\Rightarrow$ (5) hold.

(5) $\Rightarrow$ (1). Note that $xa - 1 \in {}^{\circ}x$ and $^{\circ}x = {}^{\circ}(a^m)$. We have $xa^{m+1} = a^m$. From $x^*a^* - 1 \in {}^{\circ}(x^*)$ and $^{\circ}(x^*) \subseteq {}^{\circ}(a^m)$, it follows that $(x^*a^* - 1)a^m = 0$, i.e., $(ax)^*a^m = a^m$. Post-multiply this equality by $a$, then we get $(ax)^*aa^m = aa^m$, which implies $(ax)^*a - a \in {}^{\circ}(a^m) = {}^{\circ}(x)$. Thus $(ax)^*ax = ax$, and so $(ax)^* = ax$. The equalities $(ax)^*a^m = a^m$, $(ax)^* = ax$ and $^{\circ}(x) = {}^{\circ}(a^m)$ yield that $ax^2 = x$. Hence, we get $a^{\circledD} = x$.  $\square$

If we particularize Theorem 6.2.10 to the ring $\mathbb{C}^{n \times n}$, then (1) $\Leftrightarrow$ (2) indicates that the pseudo core inverse of a complex matrix coincides with its core-EP inverse. In other words, the notion of pseudo core inverse generalizes the notion of core-EP inverse from matrices to an arbitrary $*$-ring, in terms of equations.

**Theorem 6.2.11 ([44])** *Let $a \in R$. Then $a \in R^{\circledD}$ if and only if there exist $u, v \in R$ and positive integers $p, q$ such that*

$$a^p = u(a^*)^{p+1}a^p, \quad a^q = va^{q+1}.$$

***Proof*** Suppose $a^{\circledD} = x$ with $\mathrm{ind}(a) = m$. Then

$$xa^{m+1} = a^m, \quad ax^2 = x, \quad (ax)^* = ax.$$

By Lemma 6.2.2, we have $a^m x^m a^m = a^m$, $(a^m x^m)^* = a^m x^m$, which yield $a^m = (a^m x^m)^* a^m$. Therefore $a^m = (x^m)^* (a^m)^* a^m = (ax^{m+1})^* (a^m)^* a^m = (x^{m+1})^* (a^*)^{m+1} a^m$. Consequently, the necessity holds.

Conversely, suppose that $u, v \in R$ and $p, q$ are positive integers such that $a^p = u(a^*)^{p+1}a^p$, $a^q = va^{q+1}$. Since $a^p = u(a^*)^{p+1}a^p \in R(a^p)^* a^p$, $au^*$ is a $\{1, 3\}$-inverse of $a^p$. So $a^p = a^p au^* a^p = a^{p+1}u^* a^p$, together with $a^q = va^{q+1}$, which implies that $a \in R^D$ with $ind(a) \leq p$. By Theorem 6.2.4, we get $a \in R^{\circledD}$, moreover, $a^{\circledD} = a^D a^p (a^p)^{(1,3)} = a^D a^p au^* = a^p u^*$. □

From the proof of Theorem 6.2.11 and its dual case, we have the following result.

**Theorem 6.2.12 ([44])** *Let $a \in R$. Then the following conditions are equivalent:*

(1) $a \in R^D$ and $a^m \in R^\dagger$ for some positive integer $m \geq \mathrm{ind}(a)$.
(2) $a \in R^{\circledD} \cap R_{\circledD}$.
(3) $a^m \in a^m (a^*)^{m+1} R \cap R(a^*)^{m+1} a^m$ for some positive integer $m$.

Add one more equation $axa = a$ (resp. $a^2 x = a$) to the three equations which exactly define the pseudo core inverse, then we can observe $a \in R^{\circledast}$ with $a^{\circledast} = a^{\circledD}$.

**Proposition 6.2.13 ([44])** *Let $a, x \in R$. Then (1) $\Leftrightarrow$ (2), (3) $\Rightarrow$ (1), where*

(1) $a^{\circledast} = x$;
(2) $xa^{m+1} = a^m$, $ax^2 = x$, $(ax)^* = ax$, $axa = a$ *for some positive integer $m$;*
(3) $xa^{m+1} = a^m$, $ax^2 = x$, $(ax)^* = ax$, $a^2 x = a$ *for some positive integer $m$.*

***Proof*** (1) $\Rightarrow$ (2). It is clear.
(2) $\Rightarrow$ (1). By Theorem 6.2.2, we have

$$x = xax = xa^m x^m = x(xa^{m+1})x^m = x^2 a^{m+1} x^m,$$

then $x = x^{m+1}a^{2m}x^m$ by induction. So $x = (x^{m+1}a^m)a^m x^m = a^D(a^m x^m) = a^D ax$. Thus

$$a = axa = a(a^D ax)a = aa^D(axa) = aa^D a,$$

which implies $a \in R^{\#}$. Hence $xa^{m+1} = a^m$ becomes $xa^2 = a$. From $a^2x^2a = a = xa^2$, we get $a^{\#} = x^2a$. Therefore $a \in R^{\circledcirc}$ with $a^{\circledcirc} = a^{\#}aa^{(1,3)} = (x^2a)ax = xax = x$.

(3) $\Rightarrow$ (1). Since $a^2x = a$, $a = a^{m+1}x^m$ by induction. So we have $axa = a(xa^{m+1})x^m = a^{m+1}x^m = a$. Therefore (2) holds, then (1) holds.                          □

**Remark 6.2.14 ([44])** In Proposition 6.2.13, (1) may not imply (3). For example: take $R = \mathbb{C}^{2 \times 2}$ with transpose as involution and let $a = \begin{bmatrix} 1 & i \\ 0 & 0 \end{bmatrix} \in R$. By a simple calculation, $a^{\circledcirc} = a^{\#}aa^{(1,3)} = \begin{bmatrix} 1 & 0 \\ 0 & 0 \end{bmatrix}$, but $a^2a^{\circledcirc} = \begin{bmatrix} 1 & 0 \\ 0 & 0 \end{bmatrix} \neq a$.

Next, new characterizations for pseudo core inverses are given.

**Lemma 6.2.15** Let $a \in R$ be regular and $b \in R$. If $aR = bR$, then $b$ is regular.

**Lemma 6.2.16 ([108])** Let $a, b \in R$. Then

(1) If $aR \subseteq bR$, then $^{\circ}b \subseteq {}^{\circ}a$.
(2) If $Ra \subseteq Rb$, then $b^{\circ} \subseteq a^{\circ}$.

We present some equivalent conditions of the pseudo core inverse in the following proposition.

**Proposition 6.2.17 ([131])** Let $a, x \in R$ and $k \in \mathbb{N}^{+}$. If $a^k R = a^{k+1} R$, then the following conditions are equivalent:

(1) $a \in R^{\circledcirc}$, $a^{\circledcirc} = x$ and $\mathrm{ind}(a) \leq k$.
(2) $xa^{k+1} = a^k$, $(a^k x^k)^* = a^k x^k$ and $xa^k x^k = x$.
(3) $xa^{k+1} = a^k$, $(ax)^* = ax$ and $xR \subseteq a^k R$.
(4) $a^k \in R^{(1,3)}$, $xa^{k+1} = a^k$ and $x = xa^k \widetilde{a^k}$ for some $\widetilde{a^k} \in a^k\{1,3\}$.

**Proof** (1)$\Rightarrow$(2). Suppose that $a \in R^{\circledcirc}$ with $\mathrm{ind}(a) \leq k$. By Lemma 6.2.2, (1) and (2), we know that $a^k x^k = ax$ and $xax = x$. So the rest proof is obvious.

(2) $\Rightarrow$ (1). Since $a^k R = a^{k+1} R$, there exists $z$ such that $a^{k+1}z = a^k$. Combining the equality $xa^{k+1} = a^k$, we have $a \in R^D$ with $\mathrm{ind}(a) \leq k$ by Theorem 4.2.3. So we have $xa^k = xa^{k+1}a^D = a^k a^D$. Therefore, we have

$$a^k x^k a^k = a^k x^{k-1}xa^k = a^k x^{k-1}a^k a^D = a^k x^{k-2}xa^k a^D = a^k x^{k-2}a^k(a^D)^2$$

$$= \cdots = a^k xa^k(a^D)^{k-1} = a^k a^k(a^D)^k = a^k aa^D = a^k.$$

Since $(a^k x^k)^* = a^k x^k$, $x^k \in a^k\{1,3\}$. By Theorem 6.2.4, we have

$$a^{\circledcirc} = a^D a^k(a^k)^{(1,3)} = xa^k(a^k)^{(1,3)} = xa^k x^k = x.$$

(1) $\Rightarrow$ (3). Suppose that $a^{\circledcirc} = x$ and $\mathrm{ind}(a) \leq k$. Then by Lemma 6.2.2 (1), $x = ax^2 = a^k x^{k+1} \Rightarrow xR \subseteq a^k R$. The rest proof is obvious.

$(3) \Rightarrow (1)$. According to the definition of the pseudo core inverse, it is sufficient to verify that $ax^2 = x$. Since $xa^{k+1} = a^k$ and $a^k R = a^{k+1} R$, by Theorem 4.2.3, we have $a \in R^D$ with $\mathrm{ind}(a) \le k$. So $xR \subseteq a^k R = a^{k+1} a^D R \subseteq a^{k+1} R$. By Lemma 6.2.16, we have $^\circ(a^{k+1}) \subseteq {}^\circ x$. Since $axa^{k+1} = a^{k+1}$,

$$(ax - 1) \in {}^\circ(a^{k+1}) \subseteq {}^\circ x \Rightarrow ax^2 = x.$$

$(1) \Rightarrow (4)$. From Theorem 6.2.4, we know that $x = a^\circledD = a^D a^k (a^k)^{(1,3)}$. So

$$x = (a^D a^k (a^k)^{(1,3)}) a^k (a^k)^{(1,3)} = xa^k (a^k)^{(1,3)} = xa^k \widetilde{a^k},$$

where $\widetilde{a^k} \in a^k\{1, 3\}$.

$(4) \Rightarrow (1)$. By Theorem 4.2.3, we know that $a \in R^D$ with $\mathrm{ind}(a) \le k$. Hence, by Theorem 6.2.4,

$$x = xa^k \widetilde{a^k} = xa^{k+1} a^D \widetilde{a^k} = a^k a^D \widetilde{a^k} = a^D a^k \widetilde{a^k} = a^\circledD.$$

$\square$

In particular, let $k = 1$, we have

**Corollary 6.2.18 ([124, Theorem 3.5, Proposition 3.11])** *Let $a$, $x \in R$. If $aR = a^2 R$, then the following conditions are equivalent:*

(1) $a \in R^\circledast$ *and* $a^\circledast = x$.
(2) $xa^2 = a$, $(ax)^* = ax$ *and* $xax = x$.
(3) $xa^2 = a$, $(ax)^* = ax$ *and* $xR \subseteq aR$.
(4) $a \in R^{(1,3)}$, $xa^2 = a$ *and* $x = xa\tilde{a}$ *for some* $\tilde{a} \in a\{1, 3\}$.

From above proposition, we know that $a^k (a^k)^{(1,3)} = aa^\circledD$ if $a \in R^\circledD$ with $\mathrm{ind}(a) \le k$. We also know that $aa^\circledD$ is a projection. So, we present the equivalent conditions of pseudo core inverse by a unique projection.

**Theorem 6.2.19 ([131])** *Let $a \in R$ and $k \in \mathbb{N}^+$. Then the following conditions are equivalent:*

(1) $a \in R^\circledD$ *with* $\mathrm{ind}(a) \le k$.
(2) *There exists a projection $p \in R$ such that $pR = a^k R = a^{k+1} R$ and $Ra^k \subseteq Ra^{k+1}$.*
(3) $a^{k+1} \in R^{(1)}$ *and there exists a projection $p \in R$ such that $^\circ p = {}^\circ(a^k) = {}^\circ(a^{k+1})$ and $(a^{k+1})^\circ \subseteq (a^k)^\circ$.*

*If the previous conditions are true, then conditions (2) and (3) deal with the same unique projection $p$. Moreover, $a^k (a^{k+1})^{(1)} p$ is invariant under the choice of $(a^{k+1})^{(1)} \in (a^{k+1})\{1\}$ and $a^\circledD = a^k (a^{k+1})^{(1)} p$.*

**Proof** $(1) \Rightarrow (2)$. Assume that $a \in R$ is pseudo core invertible and $\mathrm{ind}(a) \le k$. Set $p = aa^\circledD$. By Lemma 6.2.2 (1), we have $p = aa^\circledD = a^k (a^\circledD)^k = a^{k+1} (a^\circledD)^{k+1}$. So

$pR \subseteq a^k R$ and $pR \subseteq a^{k+1} R$. Since

$$a^{k+1} = aa^{\circledD} a^{k+1} \Rightarrow a^{k+1} R \subseteq pR,$$

$$a^k = a^{\circledD} a^{k+1} = a(a^{\circledD})^2 a^{k+1} = aa^{\circledD} a^k \Rightarrow a^k R \subseteq pR, \ Ra^k \subseteq Ra^{k+1},$$

$pR = a^k R = a^{k+1} R$ and $Ra^k \subseteq Ra^{k+1}$.

(2) $\Rightarrow$ (3). It is clear by Lemmas 6.2.15 and 6.2.16.

(3) $\Rightarrow$ (1). Since $a^{k+1} \in R^{(1)}$ and $^{\circ} p = {}^{\circ}(a^k) = {}^{\circ}(a^{k+1})$,

$$(1 - p) \in {}^{\circ} p = {}^{\circ}(a^k) \Rightarrow pa^k = a^k,$$

$$(1 - a^{k+1}(a^{k+1})^{(1)}) \in {}^{\circ}(a^{k+1}) = {}^{\circ} p \Rightarrow a^{k+1}(a^{k+1})^{(1)} p = p,$$

and

$$(1 - a^{k+1}(a^{k+1})^{(1)}) \in {}^{\circ}(a^{k+1}) = {}^{\circ}(a^k) \Rightarrow a^{k+1}(a^{k+1})^{(1)} a^k = a^k.$$

From $(a^{k+1})^{\circ} \subseteq (a^k)^{\circ}$, we have

$$(1 - (a^{k+1})^{(1)} a^{k+1}) \in (a^{k+1})^{\circ} \subseteq (a^k)^{\circ} \Rightarrow a^k = a^k (a^{k+1})^{(1)} a^{k+1}.$$

Taking $x = a^k (a^{k+1})^{(1)} p$, we have

$$xa^{k+1} = a^k (a^{k+1})^{(1)} pa^{k+1} = a^k (a^{k+1})^{(1)} a^{k+1} = a^k,$$

$$ax^2 = aa^k (a^{k+1})^{(1)} pa^k (a^{k+1})^{(1)} p$$

$$= a^{k+1} (a^{k+1})^{(1)} a^k (a^{k+1})^{(1)} p = a^k (a^{k+1})^{(1)} p = x$$

and

$$ax = aa^k (a^{k+1})^{(1)} p = a^{k+1} (a^{k+1})^{(1)} p = p.$$

Since $p$ is a projection, $ax$ is Hermitian. Thus, $a$ is pseudo core invertible with ind$(a) \leq k$. For arbitrary $\overline{a^{k+1}} \in a^{k+1}\{1\}$, we have

$$a^k \overline{a^{k+1}} p = a^k (a^{k+1})^{(1)} a^{k+1} \overline{a^{k+1}} a^{k+1} (a^{k+1})^{(1)} p$$

$$= a^k (a^{k+1})^{(1)} a^{k+1} (a^{k+1})^{(1)} p = a^k (a^{k+1})^{(1)} p.$$

Finally, it is obvious that $p$ is unique by Lemma 1.3.3.                              $\square$

**Remark 6.2.20 ([131])** If $k = 1$ in Theorem 6.2.19, an analogous result of core inverse are obtained.

In the following theorem, we prove that the pseudo core inverse of an element is EP.

**Theorem 6.2.21 ([131])** *Let $a \in R^{\circledD}$ with $\mathrm{ind}(a) \leq k$. Then $a^{\circledD}$ is EP and $(a^{\circledD})^{\dagger} = (a^{\circledD})^{\#} = a^{k+1}(a^k)^{(1,3)}$.*

**Proof** Suppose that $x = a^{k+1}(a^k)^{(1,3)}$. From Theorem 6.2.4, we know that $a^{\circledD} = a^D a^k (a^k)^{(1,3)}$. Then we have

$$a^{\circledD} x a^{\circledD} = a^{\circledD} a^{k+1} (a^k)^{(1,3)} a^{\circledD} = a^k (a^k)^{(1,3)} a^D a^k (a^k)^{(1,3)} = a^k a^D (a^k)^{(1,3)} = a^{\circledD},$$

$$x a^{\circledD} x = a^{k+1} (a^k)^{(1,3)} a^D a^k (a^k)^{(1,3)} a^{k+1} (a^k)^{(1,3)}$$
$$= a^{k+1} a^D (a^k)^{(1,3)} a^{k+1} (a^k)^{(1,3)} = a^{k+1} (a^k)^{(1,3)} = x$$

and $a^{\circledD} x = a^{\circledD} a^{k+1} (a^k)^{(1,3)} = a^k (a^k)^{(1,3)}$ is Hermitian. Since

$$x a^{\circledD} = a^{k+1} (a^k)^{(1,3)} a^D a^k (a^k)^{(1,3)} = a^{k+1} a^D (a^k)^{(1,3)} = a^k (a^k)^{(1,3)} = a^{\circledD} x,$$

$x a^{\circledD}$ is Hermitian. Therefore, $a^{\circledD} \in R^{\dagger} \cap R^{\#}$ and $(a^{\circledD})^{\dagger} = (a^{\circledD})^{\#} = x$. Hence, $a^{\circledD}$ is EP. $\qquad\square$

**Corollary 6.2.22 ([131])** *Let $a \in R^{\circledD}$ with $\mathrm{ind}(a) \leq k$. Then $a^{\circledD} = (a^{k+1}(a^k)^{(1,3)})^{\dagger}$.*

In the following theorem, we present relations between the pseudo core inverse and an EP element.

**Theorem 6.2.23 ([131])** *Let $a \in R^{\circledD}$ with $\mathrm{ind}(a) = k$. Then the following conditions are equivalent:*

(1) *$a^k$ is EP.*
(2) *There exists a unit $u \in R$ such that $a^{\circledD} = u a^k$.*
(3) *There exists an element $b \in R$ such that $a^{\circledD} = b a^k$.*

**Proof** (1) $\Rightarrow$ (2). Suppose that $a^k$ is EP. We know that $a \in R^{\circledD}$ with $\mathrm{ind}(a) = k$. By Theorem 6.2.4 and Lemma 4.2.5, we have

$$a^{\circledD} = a^D a^k (a^k)^{(1,3)} = a^D a^k (a^k)^{\dagger} = a^D a^k (a^k)^{\#} = a^D a^k (a^D)^k = a^D.$$

Take $u = (a^D)^{k+1} + 1 - a a^D$. By a direct computation, we have

$$u(a^{k+1} + 1 - a a^D) = (a^{k+1} + 1 - a a^D)u = 1.$$

So $u$ is a unit. Thus

$$u a^k = ((a^D)^{k+1} + 1 - a a^D) a^k = (a^D)^{k+1} a^k = a^D = a^{\circledD}.$$

(2) $\Rightarrow$ (3). It is obvious.

(3) $\Rightarrow$ (1). Set $x = a^{\circledD}$ with $\mathrm{ind}(a) = k$. By Theorem 6.2.4, we know that $a \in R^D$ with $\mathrm{ind}(a) = k$ and $a^k \in R^{(1,3)}$. Since $a^{\circledD} = ba^k$ and $aa^{\circledD}$ is Hermitian,

$$\begin{aligned} x^k = (a^{\circledD})^k &= (aa^{\circledD})^*(a^{\circledD})^k \\ &= (a^{\circledD})^*a^*(a^{\circledD})^k = (ba^k)^*a^*(a^{\circledD})^k \\ &= (a^k)^*(ab)^*(a^{\circledD})^k. \end{aligned}$$

So $x^k R \subseteq (a^k)^* R$. By Lemma 6.2.16, we have $^{\circ}((a^k)^*) \subseteq {}^{\circ}(x^k)$. By Lemma 6.2.2 (3), we obtain

$$(1 - (x^ka^k)^*) \in {}^{\circ}((a^k)^*) \subseteq {}^{\circ}(x^k) \Rightarrow x^k = (x^ka^k)^*x^k.$$

Since $x^ka^k = (x^ka^k)^*x^ka^k$, $x^ka^k$ is Hermitian, that is, $a^k \in R^{(1,4)}$. Therefore, $a^k \in R^{\dagger}$. By Lemma 6.2.2 (4), we know that $a^k(a^D)^k = (a^D)^ka^k = aa^D = ax^{k+1}a^k = x^ka^k$ is Hermitian. Since

$$a^k(a^D)^ka^k = a^kaa^D = a^k \text{ and } (a^D)^ka^k(a^D)^k = a^Da(a^D)^k = (a^D)^k,$$

$(a^D)^k = (a^k)^{\dagger}$. Hence, $(a^k)^{\#} = (a^D)^k = (a^k)^{\dagger}$, i.e., $a^k$ is EP. $\qquad\square$

## 6.3   Additive and Multiplicative Properties

In this section, we show that the reverse order law for pseudo core inverse holds under certain conditions and investigate the pseudo core invertibility of the sum of two pseudo core invertible elements.

First, we give a crucial lemma.

**Lemma 6.3.1 ([41])** *Let $a_i, b_i, c_i, y_i \in R$ ($i = 1, 2$), and suppose that*

$$y_ia_iy_i = y_i, \quad y_iR = b_iR, \quad Ry_i = Rc_i.$$

*Then, for arbitrary $d \in R$, $da_1 = a_2d$ and $db_1 = b_2d$, $dc_1 = c_2d$ together imply that $y_2d = dy_1$.*

**Proof** Let the conditions on $a_i, b_i, c_i, y_i$ be satisfied. Then $y_i = b_ir_ic_i$ for some $r_i \in R$, $y_ia_ib_i = b_i$ and $c_ia_iy_i = c_i$. Now if $da_1 = a_2d$ and $db_1 = b_2d$, $dc_1 = c_2d$, then we have

$$y_2d = b_2r_2c_2d = b_2r_2dc_1 = b_2r_2d(c_1a_1y_1) = (b_2r_2dc_1)a_1y_1 = y_2da_1y_1,$$

and dually $dy_1 = y_2ady_1$. Thus, $y_2d = y_2da_1y_1 = y_2a_2dy_1 = dy_1$. $\qquad\square$

Applying the above lemma, we obtain the following result.

**Proposition 6.3.2 ([44])** *Let $a$, $x \in R$ with $ax = xa$, $a^*x = xa^*$. If $a \in R^{\scriptsize{\textcircled{D}}}$, then $a^{\scriptsize{\textcircled{D}}}x = xa^{\scriptsize{\textcircled{D}}}$.*

**Proof** Suppose that $a$ is pseudo core invertible with index $m$. From the condition $ax = xa$, $a^*x = xa^*$, we have $a^m x = xa^m$, $(a^m)^*x = x(a^m)^*$. According to Theorem 6.2.10 and Lemma 6.3.1, we get $a^{\scriptsize{\textcircled{D}}}x = xa^{\scriptsize{\textcircled{D}}}$.       □

Applying Proposition 6.3.2, we obtain the following theorem.

**Theorem 6.3.3 ([44])** *Let $a$, $b \in R^{\scriptsize{\textcircled{D}}}$ with $ab = ba$ and $ab^* = b^*a$. Then $(ab)^{\scriptsize{\textcircled{D}}} = a^{\scriptsize{\textcircled{D}}}b^{\scriptsize{\textcircled{D}}} = b^{\scriptsize{\textcircled{D}}}a^{\scriptsize{\textcircled{D}}}$.*

**Proof** From Proposition 6.3.2, it follows that

$$b^{\scriptsize{\textcircled{D}}}a = ab^{\scriptsize{\textcircled{D}}} \text{ and } a^{\scriptsize{\textcircled{D}}}b = ba^{\scriptsize{\textcircled{D}}}.$$

The condition $b^*a = ab^*$, $a^*b^* = b^*a^*$ ensures that $b^*a^{\scriptsize{\textcircled{D}}} = a^{\scriptsize{\textcircled{D}}}b^*$, which together with $a^{\scriptsize{\textcircled{D}}}b = ba^{\scriptsize{\textcircled{D}}}$, implies that $a^{\scriptsize{\textcircled{D}}}b^{\scriptsize{\textcircled{D}}} = b^{\scriptsize{\textcircled{D}}}a^{\scriptsize{\textcircled{D}}}$.

Let $t = \max\{\text{ind}(a), \text{ind}(b)\}$. Then we have

$$b^{\scriptsize{\textcircled{D}}}a^{\scriptsize{\textcircled{D}}}(ab)^{t+1} = b^{\scriptsize{\textcircled{D}}}a^{\scriptsize{\textcircled{D}}}a^{t+1}b^{t+1} = b^{\scriptsize{\textcircled{D}}}a^t b^{t+1} = a^t b^{\scriptsize{\textcircled{D}}}b^{t+1} = a^t b^t = (ab)^t;$$

$$ab(b^{\scriptsize{\textcircled{D}}}a^{\scriptsize{\textcircled{D}}})^2 = ab(b^{\scriptsize{\textcircled{D}}})^2(a^{\scriptsize{\textcircled{D}}})^2 = ab^{\scriptsize{\textcircled{D}}}(a^{\scriptsize{\textcircled{D}}})^2 = b^{\scriptsize{\textcircled{D}}}a(a^{\scriptsize{\textcircled{D}}})^2 = b^{\scriptsize{\textcircled{D}}}a^{\scriptsize{\textcircled{D}}};$$

$$(abb^{\scriptsize{\textcircled{D}}}a^{\scriptsize{\textcircled{D}}})^* = (aa^{\scriptsize{\textcircled{D}}}bb^{\scriptsize{\textcircled{D}}})^* = (bb^{\scriptsize{\textcircled{D}}})^*(aa^{\scriptsize{\textcircled{D}}})^* = bb^{\scriptsize{\textcircled{D}}}aa^{\scriptsize{\textcircled{D}}} = abb^{\scriptsize{\textcircled{D}}}a^{\scriptsize{\textcircled{D}}}.$$

Thus $(ab)^{\scriptsize{\textcircled{D}}} = b^{\scriptsize{\textcircled{D}}}a^{\scriptsize{\textcircled{D}}}$.       □

Next, we explore the pseudo core invertibility of the sum of two pseudo core invertible elements.

**Theorem 6.3.4** *Let $a$, $b \in R^{\scriptsize{\textcircled{D}}}$ with $ab = ba = 0$, $a^*b = 0$. Then $a + b \in R^{\scriptsize{\textcircled{D}}}$ with $(a + b)^{\scriptsize{\textcircled{D}}} = a^{\scriptsize{\textcircled{D}}} + b^{\scriptsize{\textcircled{D}}}$.*

**Proof** Since $a$, $b \in R^{\scriptsize{\textcircled{D}}}$, by Lemma 6.2.2, they are Drazin invertible, and $(a + b)^D = a^D + b^D$ under the condition $ab = ba = 0$. Again by the hypothesis $ab = ba = 0$, $a^*b = 0$, we find

$$ab^{\scriptsize{\textcircled{D}}} = ab(b^{\scriptsize{\textcircled{D}}})^2 = 0, \ ba^{\scriptsize{\textcircled{D}}} = ba(a^{\scriptsize{\textcircled{D}}})^2 = 0,$$

$$b^{\scriptsize{\textcircled{D}}}a = b^{\scriptsize{\textcircled{D}}}(b^{\scriptsize{\textcircled{D}}})^*b^*a = 0, \ a^{\scriptsize{\textcircled{D}}}b = a^{\scriptsize{\textcircled{D}}}(a^{\scriptsize{\textcircled{D}}})^*a^*b = 0,$$

$$a^{\scriptsize{\textcircled{D}}}b^{\scriptsize{\textcircled{D}}} = a^{\scriptsize{\textcircled{D}}}(a^{\scriptsize{\textcircled{D}}})^*a^*b(b^{\scriptsize{\textcircled{D}}})^2 = 0, \ b^{\scriptsize{\textcircled{D}}}a^{\scriptsize{\textcircled{D}}} = b^{\scriptsize{\textcircled{D}}}(b^{\scriptsize{\textcircled{D}}})^*b^*a(a^{\scriptsize{\textcircled{D}}})^2 = 0.$$

Let $m = \max\{\text{ind}(a), \text{ind}(b)\}$, then $a^m(a^{\scriptsize{\textcircled{D}}})^m a^m = a^m$ and $b^m(b^{\scriptsize{\textcircled{D}}})^m b^m = b^m$, so

$$(a + b)^m((a^{\scriptsize{\textcircled{D}}})^m + (b^{\scriptsize{\textcircled{D}}})^m) = (a^m + b^m)((a^{\scriptsize{\textcircled{D}}})^m + (b^{\scriptsize{\textcircled{D}}})^m)$$

$$= a^m(a^{\scriptsize{\textcircled{D}}})^m + b^m(b^{\scriptsize{\textcircled{D}}})^m = aa^{\scriptsize{\textcircled{D}}} + bb^{\scriptsize{\textcircled{D}}}.$$

Then, $((a+b)^m((a^{\circledD})^m + (b^{\circledD})^m))^* = (a+b)^m((a^{\circledD})^m + (b^{\circledD})^m)$. Further,

$$(a+b)^m((a^{\circledD})^m + (b^{\circledD})^m)(a+b)^m = (a^m(a^{\circledD})^m + b^m(b^{\circledD})^m)(a^m + b^m)$$
$$= a^m(a^{\circledD})^m a^m + b^m(b^{\circledD})^m b^m$$
$$= a^m + b^m = (a+b)^m.$$

Hence $(a^{\circledD})^m + (b^{\circledD})^m$ is a $\{1, 3\}$-inverse of $(a+b)^m$. Therefore, we get

$$(a+b)^{\circledD} = (a+b)^D(a+b)^m((a+b)^m)^{(1,3)}$$
$$= (a^D + b^D)(a^m + b^m)((a^{\circledD})^m + (b^{\circledD})^m)$$
$$= a^D a^m(a^m)^{(1,3)} + b^D b^m(b^m)^{(1,3)}$$
$$= a^{\circledD} + b^{\circledD}.$$

□

**Remark 6.3.5 ([44])** It is noteworthy that condition $ab = 0$, $a^*b = 0$ (without $ba = 0$) is not sufficient to show the pseudo core invertibility of $a+b$ although both $a$ and $b$ are pseudo core invertible.

For example: by setting $R = \mathbb{C}^{2 \times 2}$ with transpose as its involution,

$$a = \begin{bmatrix} i & 0 \\ 0 & 0 \end{bmatrix}, \quad b = \begin{bmatrix} 0 & 0 \\ -1 & 0 \end{bmatrix},$$

we have $ab = a^*b = 0$, but $ba \neq 0$.

Observe that $a^{\#} = -a$ and $aa^{(1,3)} = \begin{bmatrix} 1 & 0 \\ 0 & 0 \end{bmatrix}$, which imply

$$a^{\circledD} = a^{\circledcirc} = a^{\#}aa^{(1,3)} = \begin{bmatrix} -i & 0 \\ 0 & 0 \end{bmatrix}.$$

It is obvious that $b^{\circledD} = 0$. As for $a+b = \begin{bmatrix} i & 0 \\ -1 & 0 \end{bmatrix}$, by calculation, we find that neither $a+b$ nor $(a+b)^2$ has any $\{1, 3\}$-inverse. Since $(a+b)^m = \begin{cases} (-1)^{\frac{m-1}{2}}(a+b) & m \text{ is odd} \\ (-1)^{\frac{m}{2}+1}(a+b)^2 & m \text{ is even} \end{cases}$, we conclude that $(a+b)^m$ has no $\{1, 3\}$-inverses for arbitrary positive integer $m$. Hence $a+b$ is not pseudo core invertible.

## 6.4  Pseudo Core Inverses of Jacobson Pairs

By Lemma 6.2.2, a pseudo core invertible element is always Drazin invertible. An interesting question is under what condition a Drazin invertible element is pseudo core invertible. We give a new equivalent condition here.

**Lemma 6.4.1 ([111])** *Let* $e \in R$ *be an idempotent. Then* $e \in R^{\{1,3\}}$ *if and only if* $1 - e \in R^{\{1,4\}}$. *In this case,* $1 - ee^{(1,3)} \in (1 - e)\{1, 4\}$ *and* $1 - (1 - e)^{(1,4)}(1 - e) \in e\{1, 3\}$.

**Proof** If $e \in R^{\{1,3\}}$, in order to prove that $1 - e \in R^{\{1,4\}}$, we only need to check that $1 - ee^{(1,3)} \in (1 - e)\{1, 4\}$. Indeed, we have

$$(1 - ee^{(1,3)})(1 - e) = 1 - ee^{(1,3)} - e + ee^{(1,3)}e = 1 - ee^{(1,3)}$$

and

$$(1-e)(1-ee^{(1,3)})(1-e) = (1-e)(1-ee^{(1,3)}) = 1-e-ee^{(1,3)}+eee^{(1,3)} = 1-e.$$

The converse statement can be proved similarly by verifying that $1 - (1-e)^{(1,4)}(1 - e) \in e\{1, 3\}$. The details of verification will be omitted.  □

When $a \in R^D$, the idempotent $e = 1 - aa^D$ is the spectral idempotent of $a$, denoted by $a^\pi$.

**Theorem 6.4.2 ([111])** *If* $a \in R^D$, *then the following conditions are equivalent:*

(1) $a \in R^{\circledD}$.
(2) $aa^D \in R^{\{1,3\}}$.
(3) $a^\pi \in R^{\{1,4\}}$.

*In this case,* $aa^{\circledD} \in (aa^D)\{1, 3\}$, $1 - aa^{\circledD} \in a^\pi\{1, 4\}$ *and*

$$a^{\circledD} = a^D(aa^D)^{(1,3)} = a^D(1 - (a^\pi)^{(1,4)}a^\pi),$$

*for any* $(aa^D)^{(1,3)} \in (aa^D)\{1, 3\}$ *and* $(a^\pi)^{(1,4)} \in a^\pi\{1, 4\}$.

**Proof** (1) $\Rightarrow$ (2). Suppose that $\text{ind}(a) = m$. Then we have

$$a^D aa^{\circledD} = a^D a^{m+1}(a^{\circledD})^{m+1} = a^m(a^{\circledD})^{m+1} = a^{\circledD}, \tag{6.4.1}$$

and

$$a^{\circledD} aa^D = a^{\circledD} a^{m+1}(a^D)^{m+1} = a^m(a^D)^{m+1} = a^D. \tag{6.4.2}$$

So $aa^D aa^{\circledD} = aa^{\circledD}$ and $aa^D aa^{\circledD} aa^D = aa^{\circledD} aa^D = aa^D$, which proves that $aa^{\circledD} \in (aa^D)\{1, 3\}$.

$(2) \Rightarrow (1)$. For any $(aa^D)^{(1,3)} \in (aa^D)\{1, 3\}$, let $x = a^D(aa^D)^{(1,3)}$ and suppose that $ind(a) = m$, next we check that $x$ is the pseudo core inverse of $a$. Noting that $ax = aa^D(aa^D)^{(1,3)}$, clearly $ax$ is Hermitian. Then we have

$$ax^2 = aa^D(aa^D)^{(1,3)}a^D(aa^D)^{(1,3)} = aa^D(aa^D)^{(1,3)}a^D aa^D(aa^D)^{(1,3)}$$
$$= a^D(aa^D)^{(1,3)} = x,$$

and

$$xa^{m+1} = a^D(aa^D)^{(1,3)}a^{m+1} = a^D aa^D(aa^D)^{(1,3)}aa^D a^{m+1} = a^D aa^D a^{m+1} = a^m.$$

Thus, $x$ is the pseudo core inverse of $a$.

$(2) \Leftrightarrow (3)$. By Lemma 6.4.1, $aa^D \in R^{\{1,3\}}$ if and only if $a^\pi \in R^{\{1,4\}}$. In this case, $1 - (a^\pi)^{(1,4)}a^\pi \in (aa^D)\{1, 3\}$ and $1 - aa^D(aa^D)^{(1,3)} \in a^\pi\{1, 4\}$. From the proof of previous part, we know that $a^\circledcirc = a^D(aa^D)^{(1,3)}$ and $aa^\circledcirc \in (aa^D)\{1, 3\}$. So $a^\circledcirc = a^D(1 - (a^\pi)^{(1,4)}a^\pi)$ and $1 - aa^D aa^\circledcirc \overset{(6.4.1)}{=} 1 - aa^\circledcirc \in a^\pi\{1, 4\}$.   □

Dually, we have corresponding result for dual pseudo core inverse.

**Theorem 6.4.3 ([111])** *If $a \in R^D$, then the following conditions are equivalent:*

(1) $a \in R^\circledcirc$.
(2) $aa^D \in R^{\{1,4\}}$.
(3) $a^\pi \in R^{\{1,3\}}$.

*In this case, $a^\circledcirc a \in (aa^D)\{1, 4\}$, $1 - a^\circledcirc a \in a^\pi\{1, 3\}$ and*

$$a_\circledcirc = (aa^D)^{(1,4)}a^D = (1 - a^\pi(a^\pi)^{(1,3)})a^D,$$

*for any $(aa^D)^{(1,4)} \in (aa^D)\{1, 4\}$ and $(a^\pi)^{(1,3)} \in a^\pi\{1, 3\}$.*

As an application of Theorem 6.4.2, Jacobson's lemma for pseudo core inverse is explored in the rest of this section.

First, in view of Sect. 4.4, if $\alpha = 1 - ab \in R^D$, then so is $\beta = 1 - ba$; moreover,

$$\beta^D = (1 + ba^D a)(1 - ba^\pi ra), \tag{6.4.3}$$

where $r = 1 + \alpha + \cdots + \alpha^{ind\alpha - 1}$. For $\beta^\pi$, we see

**Lemma 6.4.4 ([78])** *If $\alpha = 1 - ab \in R^D$ with $ind(\alpha) = k$, then $\beta = 1 - ba \in R^D$ with $ind(\alpha) = k$. Moreover, $\beta^\pi = ba^\pi ra$, where $r = 1 + \alpha + \cdots + \alpha^{k-1}$.*

Generally, Jacobson's lemma for pseudo core inverses is not true. There exists a $*$-ring $R$ such that $1 - ab$ is pseudo core invertible while $1 - ba$ is not, for some $a, b \in R$.

**Example 6.4.5 ([111])**  Let $R$ be the ring of all $2 \times 2$ matrices over the complex field $\mathbb{C}$, with transpose as the involution. Suppose that $a = \begin{bmatrix} 1 & -i \\ i & -1 \end{bmatrix}$ and $b = \begin{bmatrix} 0 & 0 \\ 0 & -1 \end{bmatrix}$.

Then we have $1 - ab = \begin{bmatrix} 1 & -i \\ 0 & 0 \end{bmatrix}$ and $1 - ba = \begin{bmatrix} 1 & 0 \\ i & 0 \end{bmatrix}$. It is easy to verify that $1 - ab \in R^{\tiny\textcircled{D}}$ with $(1 - ab)^{\tiny\textcircled{D}} = \begin{bmatrix} 1 & 0 \\ 0 & 0 \end{bmatrix}$ while $1 - ba \notin R^{\tiny\textcircled{D}}$.

A natural question arising here is under what conditions $1 - ab \in R^{\tiny\textcircled{D}}$ implies $1 - ba \in R^{\tiny\textcircled{D}}$. To deal with this question, we need to make some preparations first.

**Lemma 6.4.6**  Let $a_1, a_2, d \in R$. If $a_1, a_2 \in R^D$ and $da_1 = a_2 d$, then $a_2^D d = d a_1^D$.

**Proof**  It follows by Lemma 6.3.1.                                                                $\square$

**Theorem 6.4.7 ([111])**  Let $p, a, q \in R$ with $p'pa = a = aqq'$ for some $p', q' \in R$. If $a \in R^{\{1,4\}}$, then the following conditions are equivalent:

(1) $paq \in R^{\{1,4\}}$.
(2) $v = a^{(1,4)}aqq^* + 1 - a^{(1,4)}a \in R^{-1}$.

In this case, $q^*v^{-1}a^{(1,4)}p' \in (paq)\{1, 4\}$.

**Proof**  (1) $\Rightarrow$ (2). Suppose that $x$ is a $\{1, 4\}$-inverse of $paq$. Then $paq(paq)^*x^* = paq$ by Lemma 2.2.3. We have

$$(a^{(1,4)}aqq^*a^{(1,4)}a + 1 - a^{(1,4)}a)(a^*p^*x^*q' + 1 - a^{(1,4)}a)$$
$$= a^{(1,4)}aqq^*a^{(1,4)}aa^*p^*x^*q' + 1 - a^{(1,4)}a$$
$$= a^{(1,4)}aqq^*a^*p^*x^*q' + 1 - a^{(1,4)}a$$
$$= a^{(1,4)}p'paqq^*a^*p^*x^*q' + 1 - a^{(1,4)}a$$
$$= a^{(1,4)}p'paqq' + 1 - a^{(1,4)}a$$
$$= a^{(1,4)}aqq' + 1 - a^{(1,4)}a$$
$$= 1.$$

Noting that $a^{(1,4)}aqq^*a^{(1,4)}a + 1 - a^{(1,4)}a$ is Hermitian, it is invertible. Let $x = a^{(1,4)}aqq^* - 1$ and $y = a^{(1,4)}a$. Then $1 + xy = a^{(1,4)}aqq^*a^{(1,4)}a + 1 - a^{(1,4)}a$. As $1 + xy \in R^{-1}$, we have $1 + yx = a^{(1,4)}aqq^* + 1 - a^{(1,4)}a = v \in R^{-1}$ by Jacobson's lemma.

(2) $\Rightarrow$ (1). According to the fact that $v \in R^{-1}$, it follows that $v^* = qq^*a^{(1,4)}a + 1 - a^{(1,4)}a \in R^{-1}$. Multiplying by $a$ on the left of $v^* = qq^*a^{(1,4)}a + 1 - a^{(1,4)}a$ yields that $av^* = aqq^*a^{(1,4)}a$, and consequently

$$a = aqq^*a^{(1,4)}a(v^{-1})^*.$$

Then we have

$$
\begin{aligned}
paq &= paqq^*a^{(1,4)}a(v^{-1})^*q \\
&= paqq^*a^*(a^{(1,4)})^*(v^{-1})^*q \\
&= paqq^*a^*p^*(p')^*(a^{(1,4)})^*(v^{-1})^*q \in paq(paq)^*R.
\end{aligned}
$$

Thus, by Lemma 2.2.3, $paq \in R^{\{1,4\}}$ with $q^*v^{-1}a^{(1,4)}p' \in (paq)\{1,4\}$.                    □

Taking $p = 1$ in Theorem 6.4.7, we have the following corollary.

**Corollary 6.4.8 ([111])** *Let $a, q \in R$ with $a = aqq'$ for some $q' \in R$. If $a \in R^{\{1,4\}}$, then the following conditions are equivalent:*

(1) $aq \in R^{\{1,4\}}$.
(2) $v = a^{(1,4)}aqq^* + 1 - a^{(1,4)}a \in R^{-1}$.

*In this case, $q^*v^{-1}a^{(1,4)} \in (aq)\{1,4\}$.*

**Theorem 6.4.9 ([111])** *Let $a, b \in R$. If $\alpha = 1 - ab \in R^{\circledD}$, then the following conditions are equivalent:*

(1) $\beta = 1 - ba \in R^{\circledD}$.
(2) $ba^\pi ra \in R^{\{1,4\}}$.
(3) $v = (1 - \alpha\alpha^{\circledD})ra(ra)^* + \alpha\alpha^{\circledD} \in R^{-1}$,

*where $r = 1 + \alpha + \cdots + \alpha^{k-1}$ and $k = ind(\alpha)$. In this case,*

$$
\beta^{\circledD} = (1 + ba^D a)[1 - (ra)^*v^{-1}(1 - \alpha\alpha^{\circledD})ra].
$$

**Proof** Since $\alpha = 1 - ab \in R^{\circledD}$, suppose that $ind(\alpha) = k$, it follows that $\alpha \in R^D$ with $ind(\alpha) = k$ by Lemma 6.2.2 and $\alpha^\pi \in R^{\{1,4\}}$ with $1 - \alpha\alpha^{\circledD} \in (\alpha^\pi)\{1,4\}$ by Theorem 6.4.2.

(1) $\Leftrightarrow$ (2). By Lemma 6.4.4, $\beta = 1 - ba \in R^D$ with $\beta^\pi = ba^\pi ra$, where $r = 1 + \alpha + \cdots + \alpha^{k-1}$. Hence, by Theorem 6.4.2, $\beta \in R^{\circledD}$ if and only if $ba^\pi ra \in R^{\{1,4\}}$.

(2) $\Leftrightarrow$ (3). First we have

$$
ra(ba^\pi) = r(1-\alpha)(1-\alpha\alpha^D) = (1-\alpha^k)(1-\alpha\alpha^D) = 1-\alpha\alpha^D = \alpha^\pi, \qquad (6.4.4)
$$

and

$$
(\alpha^\pi ra)b = (1-\alpha\alpha^D)r(1-\alpha) = (1-\alpha\alpha^D)(1-\alpha^k) = 1-\alpha\alpha^D = \alpha^\pi. \qquad (6.4.5)
$$

Since $\alpha^\pi \in R^{\{1,4\}}$, $ba^\pi ra \in R^{\{1,4\}}$ if and only if $v = (\alpha^\pi)^{(1,4)}\alpha^\pi ra(ra)^* + 1 - (\alpha^\pi)^{(1,4)}\alpha^\pi \in R^{-1}$ by Theorem 6.4.7. Noting that $1 - \alpha\alpha^{\circledD} \in (\alpha^\pi)\{1,4\}$, we have $(\alpha^\pi)^{(1,4)}\alpha^\pi = (1-\alpha\alpha^{\circledD})(1-\alpha\alpha^D) \overset{(6.4.2)}{=} 1-\alpha\alpha^{\circledD}$. Thus, $v = (1-\alpha\alpha^{\circledD})ra(ra)^* + \alpha\alpha^{\circledD}$.

At last, we give a formula for $\beta^{\circledD}$. If (2) holds, for any $(\beta^\pi)^{(1,4)} \in \beta^\pi\{1,4\}$, we know that $\beta^{\circledD} = \beta^D(1 - (\beta^\pi)^{(1,4)}\beta^\pi)$ by Theorem 6.4.2. Then we have

$$
\begin{aligned}
\beta^{\circledD} &= \beta^D(1 - (\beta^\pi)^{(1,4)}\beta^\pi) \\
&\overset{(6.4.3)}{=} (1 + ba^Da)(1 - ba^\pi ra)[1 - (ba^\pi ra)^{(1,4)}ba^\pi ra] \\
&= (1 + ba^Da)[1 - (ba^\pi ra)^{(1,4)}ba^\pi ra].
\end{aligned}
$$

If (3) holds, for any $(\alpha^\pi)^{(1,4)} \in \alpha^\pi\{1,4\}$, we have $(ra)^*v^{-1}(\alpha^\pi)^{(1,4)}ra \in (ba^\pi ra)\{1,4\}$ by Theorem 6.4.7. Combining this with $1 - \alpha\alpha^{\circledD} \in (\alpha^\pi)\{1,4\}$, we get $(ra)^*v^{-1}(1 - \alpha\alpha^{\circledD})ra \in (ba^\pi ra)\{1,4\}$. It follows that

$$
\begin{aligned}
\beta^{\circledD} &= (1 + ba^Da)[1 - (ba^\pi ra)^{(1,4)}ba^\pi ra] \\
&= (1 + ba^Da)[1 - (ra)^*v^{-1}(1 - \alpha\alpha^{\circledD})raba^\pi ra] \\
&\overset{(6.4.4)}{=} (1 + ba^Da)[1 - (ra)^*v^{-1}(1 - \alpha\alpha^{\circledD})\alpha^\pi ra] \\
&\overset{(6.4.2)}{=} (1 + ba^Da)[1 - (ra)^*v^{-1}(1 - \alpha\alpha^{\circledD})ra].
\end{aligned}
$$

$\square$

**Remark 6.4.10 ([111])** If $\alpha \in R^{\circledD}$, then $\alpha \in R^D$ with $\alpha^D = (\alpha^{\circledD})^{k+1}\alpha^k$ by Lemma 6.2.2, where $k = ind(\alpha)$. Hence we can also write the formula of $\beta^{\circledD}$ in Theorem 6.4.9 as

$$
\beta^{\circledD} = (1 + b(\alpha^{\circledD})^{k+1}\alpha^k a)[1 - (ra)^*v^{-1}(1 - \alpha\alpha^{\circledD})ra].
$$

If $1 - ab, 1 - ba \in R^{\circledD}$, then we have $ind(1 - ab) = ind(1 - ba)$ by Lemma 6.4.4. Taking $ind(1 - ab) = 1$ in Theorem 6.4.9, we have the following corollary.

**Corollary 6.4.11 ([111])** Let $a, b \in R$. If $\alpha = 1 - ab \in R^{\circledcirc}$, then the following conditions are equivalent:

(1) $\beta = 1 - ba \in R^{\circledcirc}$;
(2) $ba^\pi a \in R^{\{1,4\}}$;
(3) $v = (1 - \alpha\alpha^{\circledcirc})aa^* + \alpha\alpha^{\circledcirc} \in R^{-1}$.

In this case,

$$
\beta^{\circledcirc} = (1 + ba^\# a)[1 - a^*v^{-1}(1 - \alpha\alpha^{\circledcirc})a].
$$

In Example 6.4.5, $1 - ba$ is not pseudo core invertible while $1 - ab$ is. However, we find that $1 - ba$ in Example 6.4.5 is dual pseudo core invertible with $(1 - ba)_{\circledD} = \begin{bmatrix} 1 & 0 \\ 0 & 0 \end{bmatrix}$. This inspires us to explore equivalent conditions under which $1 - ba$ is dual pseudo core invertible when $1 - ab$ is pseudo core invertible.

**Lemma 6.4.12 ([16, Theorem 3.2])** *Let* $p, a, q \in R$ *with* $p'pa = a = aqq'$ *for some* $p', q' \in R$. *If* $a \in R^{\{1,4\}}$, *then* $paq \in R^{\{1,3\}}$ *if and only if* $t = (pa)^*pa + 1 - a^{(1,4)}a \in R^{-1}$. *In this case,*

$$q'a^{(1,4)}at^{-1}(pa)^* \in (paq)\{1, 3\}.$$

By using Theorem 6.4.3 and Lemma 6.4.12, we can get the following theorem by a similar discussion as in Theorem 6.4.9.

**Theorem 6.4.13 ([111])** *Let* $a, b \in R$. *If* $\alpha = 1 - ab \in R^{\text{\textcircled{D}}}$, *then the following conditions are equivalent:*

(1) $\beta = 1 - ba \in R_{\text{\textcircled{D}}}$.
(2) $ba^\pi ra \in R^{\{1,3\}}$.
(3) $t = (\alpha^\pi)^*b^*ba^\pi + \alpha\alpha^{\text{\textcircled{D}}} \in R^{-1}$.

*In this case,*

$$\beta_{\text{\textcircled{D}}} = [1 - ba^\pi t^{-1}(\alpha^\pi)^*b^*](1 + ba^D a).$$

## 6.5  Pseudo Core Inverses of $ab$ and $ba$

Let $a, b \in R$. Cline's formula states that if $\alpha = ab \in R^D$, then $\beta = ba \in R^D$ with $\beta^D = b(\alpha^D)^2a$. But, in general, Cline's formula for pseudo core inverse does not hold. Actually, there exist a $*$-ring $R$ and elements $a, b \in R$ such that $ab$ is pseudo core invertible while $ba$ is not.

**Example 6.5.1 ([111])** Let $R = \mathbb{C}^{2\times2}$ with transpose as the involution. Suppose that $a = \begin{bmatrix} 1 & i \\ -i & -1 \end{bmatrix}$ and $b = \begin{bmatrix} \frac{1}{2} & \frac{i}{2} \\ -\frac{i}{2} & \frac{1}{2} \end{bmatrix}$. Then we have $ab = \begin{bmatrix} 1 & i \\ 0 & 0 \end{bmatrix}$ and $ba = \begin{bmatrix} 1 & 0 \\ -i & 0 \end{bmatrix}$. It is easy to verify that $ab \in R^{\text{\textcircled{D}}}$ with $(ab)^{\text{\textcircled{D}}} = \begin{bmatrix} 1 & 0 \\ 0 & 0 \end{bmatrix}$ while $ba \notin R^{\text{\textcircled{D}}}$.

It is natural to ask under what conditions $ab \in R^{\text{\textcircled{D}}}$ implies $ba \in R^{\text{\textcircled{D}}}$. To answer this question, we also need some preparations.

**Theorem 6.5.2 ([111])** *Let* $p, a, q \in R$ *with* $p'pa = a = aqq'$ *for some* $p', q' \in R$. *If* $a \in R^{\{1,3\}}$, *then the following conditions are equivalent:*

(1) $paq \in R^{\{1,3\}}$.
(2) $u = p^*paa^{(1,3)} + 1 - aa^{(1,3)} \in R^{-1}$.

*In this case,* $q'a^{(1,3)}u^{-1}p^* \in (paq)\{1, 3\}$.

*Proof* The proof of this theorem is dual to the one of Theorem 6.4.7 and so is omitted.                                                                                            $\square$

Taking $q = 1$ in Theorem 6.5.2, we have the following corollary.

**Corollary 6.5.3 ([111])** *Let* $p, a \in R$ *with* $p'pa = a$ *for some* $p' \in R$. *If* $a \in R^{\{1,3\}}$, *then the following conditions are equivalent:*

(1) $pa \in R^{\{1,3\}}$.
(2) $u = p^* paa^{(1,3)} + 1 - aa^{(1,3)} \in R^{-1}$.

*In this case,* $a^{(1,3)}u^{-1}p^* \in (pa)\{1, 3\}$.

**Theorem 6.5.4 ([111])** *Let* $a, b \in R$. *If* $\alpha = ab \in R^{\oplus}$, *then the following conditions are equivalent:*

(1) $\beta = ba \in R^{\oplus}$.
(2) $ba^D a \in R^{\{1,3\}}$.
(3) $u = (ba^D)^* ba^{\oplus} + 1 - \alpha\alpha^{\oplus} \in R^{-1}$.

*In this case,*

$$\beta^{\oplus} = b(\alpha^{\oplus})^2 u^{-1}(ba^D)^*.$$

**Proof** Since $\alpha = ab \in R^{\oplus}$, it follows that $\alpha \in R^D$ by Lemma 6.2.2 and $\alpha\alpha^D \in R^{\{1,3\}}$ with $\alpha\alpha^{\oplus} \in (\alpha\alpha^D)\{1, 3\}$ by Theorem 6.4.2.

(1) $\Leftrightarrow$ (2). By Cline's formula, $\beta \in R^D$. Noting that $\alpha a = a\beta$, we have $\alpha^D a = a\beta^D$ by Lemma 6.4.6. It follows that $\beta\beta^D = ba\beta^D = ba^D a$. Thus, by Theorem 6.4.2, $\beta \in R^{\oplus}$ if and only if $ba^D a \in R^{\{1,3\}}$.

(2) $\Leftrightarrow$ (3). First we have

$$(\alpha\alpha^D a)ba^D = \alpha\alpha^D \alpha\alpha^D = \alpha\alpha^D,$$

and

$$a(ba^D \alpha\alpha^D) = \alpha\alpha^D \alpha\alpha^D = \alpha\alpha^D.$$

Since $\alpha\alpha^D \in R^{\{1,3\}}$, it follows that $ba^D a = ba^D \alpha\alpha^D a \in R^{\{1,3\}}$ if and only if $u = (ba^D)^* ba^D \alpha\alpha^D (\alpha\alpha^D)^{(1,3)} + 1 - \alpha\alpha^D (\alpha\alpha^D)^{(1,3)} \in R^{-1}$ by Theorem 6.5.2. Noting that $\alpha\alpha^{\oplus} \in (\alpha\alpha^D)\{1, 3\}$, we have $\alpha\alpha^D (\alpha\alpha^D)^{(1,3)} \overset{(6.4.1)}{=} \alpha\alpha^D (\alpha\alpha^{\oplus})$ $\alpha\alpha^{\oplus}$. Thus, $u = (ba^D)^* ba^{\oplus} + 1 - \alpha\alpha^{\oplus}$.

At last, we give a formula for $\beta^{\oplus}$. If (2) holds, for any $(ba^D a)^{(1,3)} \in (ba^D a)\{1, 3\}$, we know that $\beta^{\oplus} = \beta^D (\beta\beta^D)^{(1,3)} = b(\alpha^D)^2 a(\beta\beta^D)^{(1,3)} = b(\alpha^D)^2 a(ba^D a)^{(1,3)}$ by Theorem 6.4.2 and Cline's formula.

If (3) holds, for any $(\alpha\alpha^D)^{(1,3)} \in (\alpha\alpha^D)\{1, 3\}$, we have

$$ba^D (\alpha\alpha^D)^{(1,3)} u^{-1}(ba^D)^* \in (ba^D \alpha\alpha^D a)\{1, 3\} = (ba^D a)\{1, 3\}$$

by Theorem 6.5.2. Combining this with $\alpha\alpha^{\circledD} \in (\alpha\alpha^D)\{1,3\}$, we get

$$ba^D\alpha\alpha^{\circledD}u^{-1}(ba^D)^* \overset{(6.4.1)}{=} ba^{\circledD}u^{-1}(ba^D)^* \in (ba^Da)\{1,3\}.$$

It follows that

$$
\begin{aligned}
\beta^{\circledD} &= b(\alpha^D)^2 a(ba^Da)^{(1,3)}\\
&= b(\alpha^D)^2 ab\alpha^{\circledD}u^{-1}(ba^D)^*\\
&= ba^D\alpha^{\circledD}u^{-1}(ba^D)^*\\
&= ba^D\alpha(\alpha^{\circledD})^2u^{-1}(ba^D)^*\\
&= b(\alpha^D\alpha\alpha^{\circledD})\alpha^{\circledD}u^{-1}(ba^D)^*\\
&\overset{(6.4.1)}{=} b(\alpha^{\circledD})^2u^{-1}(ba^D)^*.
\end{aligned}
$$

□

**Remark 6.5.5**

(1) If $\alpha \in R^{\circledD}$, then $\alpha \in R^D$ with $\alpha^D = (\alpha^{\circledD})^{k+1}\alpha^k$ by Lemma 6.2.2, where $k = ind(\alpha)$. Hence we can also write the formula of $\beta^{\circledD}$ in Theorem 6.5.4 as

$$\beta^{\circledD} = b(\alpha^{\circledD})^2u^{-1}[b(\alpha^{\circledD})^{k+1}\alpha^k]^*.$$

(2) If $ab \in R^D$, we know that $ba \in R^D$ with $ind(ab)-1 \le ind(ba) \le ind(ab)+1$ by Theorem 4.4.7. Thus, we can not derive that $ab \in R^{\oplus}$ implies $ba \in R^{\oplus}$ by taking $ind(ab) = 1$ in Theorem 6.5.4. For example, let $R = \mathbb{C}^{2\times2}$ with transpose as the involution. Suppose that $a = \begin{bmatrix} 0 & 0 \\ 0 & 1 \end{bmatrix}$ and $b = \begin{bmatrix} 0 & 1 \\ 0 & 0 \end{bmatrix}$. Then we have $ab = 0$ and $ba = \begin{bmatrix} 0 & 1 \\ 0 & 0 \end{bmatrix}$. It is clear that $ab \in R^{\oplus}$ with $(ab)^{\oplus} = 0$ while $ba \notin R^{\oplus}$. However, $ba \in R^{\circledD}$ with $ind(ba) = 2$ and $(ba)^{\circledD} = 0$.

In Example 6.5.1, $ba$ is not pseudo core invertible while $ab$ is. However, we find that $ba$ in Example 6.5.1 is dual pseudo core invertible with $(ba)_{\circledD} = \begin{bmatrix} 1 & 0 \\ 0 & 0 \end{bmatrix}$. This motivates us to explore necessary and sufficient conditions under which $ab$ being pseudo core invertible implies that $ba$ is dual pseudo core invertible.

**Lemma 6.5.6 ([16, Theorem 3.2])** *Let* $p, a, q \in R$ *with* $p'pa = a = aqq'$ *for some* $p', q' \in R$. *If* $a \in R^{\{1,3\}}$, *then* $paq \in R^{\{1,4\}}$ *if and only if* $s = aq(aq)^* + 1 - aa^{(1,3)} \in R^{-1}$. *In this case,*

$$(aq)^*s^{-1}aa^{(1,3)}p' \in (paq)\{1,4\}.$$

With Theorem 6.4.3 and Lemma 6.5.6, we can get the following theorem through an analogous discussion as in Theorem 6.5.4.

**Theorem 6.5.7 ([111])** *Let* $a, b \in R$. *If* $\alpha = ab \in R^{\textcircled{D}}$, *then the following conditions are equivalent:*

(1) $\beta = ba \in R_{\textcircled{D}}$.
(2) $ba^D a \in R^{\{1,4\}}$.
(3) $s = \alpha\alpha^D a(\alpha\alpha^D a)^* + 1 - \alpha\alpha^{\textcircled{D}} \in R^{-1}$.

*In this case,*

$$\beta_{\textcircled{D}} = (\alpha\alpha^D a)^* s^{-1} \alpha^D a.$$

## 6.6   The Pseudo Core Inverse of a Sum of Morphisms

In this section, we present the pseudo core invertibility of $\varphi + \eta - \varepsilon$ and give expression of $(\varphi + \eta - \varepsilon)^{\textcircled{D}}$. Firstly, we give an auxiliary lemma.

**Lemma 6.6.1 ([23])** *Let* $\mathscr{C}$ *be an additive category with an involution* $*$ *and* $\varphi :$ $X \to X$ *be a morphism of* $\mathscr{C}$. *If* $\varphi^{\textcircled{D}}$ *exists with* $\mathrm{ind}(\varphi) = k$, *then the following statements hold:*

(1) $\varphi\varphi^{\textcircled{D}} = \varphi^s(\varphi^{\textcircled{D}})^s$, *for arbitrary positive integer s.*
(2) $\varphi^{\textcircled{D}}\varphi\varphi^{\textcircled{D}} = \varphi^{\textcircled{D}}$.
(3) $\varphi^k = \varphi\varphi^{\textcircled{D}}\varphi^k$.
(4) $\varphi\varphi^{\textcircled{D}} = \varphi^{\textcircled{D}}\varphi^2\varphi^{\textcircled{D}}$.

***Proof*** (1) and (2) are clear.

(3)

$$\varphi^k = \varphi^{\textcircled{D}}\varphi^{k+1} = \varphi(\varphi^{\textcircled{D}})^2\varphi^{k+1}$$
$$= \varphi\varphi^{\textcircled{D}}(\varphi^{\textcircled{D}}\varphi^{k+1}) = \varphi\varphi^{\textcircled{D}}\varphi^k.$$

(4)

$$\varphi\varphi^{\textcircled{D}} = \varphi^k(\varphi^{\textcircled{D}})^k = \varphi^{\textcircled{D}}\varphi^{k+1}(\varphi^{\textcircled{D}})^k$$
$$= \varphi^{\textcircled{D}}\varphi(\varphi^k(\varphi^{\textcircled{D}})^k) = \varphi^{\textcircled{D}}\varphi\varphi\varphi^{\textcircled{D}} = \varphi^{\textcircled{D}}\varphi^2\varphi^{\textcircled{D}}.$$

$\square$

Now, we present a sufficient condition when the pseudo core inverse of $f$ exists.

**Theorem 6.6.2 ([23])** *Let* $\mathscr{C}$ *be an additive category with an involution* $*$. *Suppose that both* $\varphi : X \to X$ *is a morphism of* $\mathscr{C}$ *with pseudo core inverse* $\varphi^{\textcircled{D}}$ *and* $\mathrm{ind}(\varphi) =$

$k$ and $\eta : X \to X$ is a morphism of $\mathscr{C}$ such that $1 + \varphi^\oplus \eta$ is invertible. Let $f = \varphi + \eta - \varepsilon$,

$$\alpha = (1 + \varphi^\oplus \eta)^{-1},$$

$$\beta = (1 + \eta \varphi^\oplus)^{-1},$$

$$\varepsilon = (1 - \varphi \varphi^\oplus) \eta \alpha (1 - \varphi^\oplus \varphi),$$

$$\gamma = \alpha (1 - \varphi^\oplus \varphi) \eta \varphi^\oplus \beta,$$

$$\sigma = \alpha \varphi^\oplus \varphi \alpha^{-1} (1 - \varphi \varphi^\oplus) \beta,$$

$$\delta = \beta^* (\varphi^\oplus)^* \eta^* (1 - \varphi \varphi^\oplus) \beta.$$

If $1 - \gamma$, $1 - \sigma$ and $1 - \delta$ are invertible,

$$\eta(\varphi^\oplus \varphi - 1)\varphi = 0 \tag{6.6.1}$$

and

$$\varphi(\varphi^\oplus \varphi - 1)\eta = 0, \tag{6.6.2}$$

then $f^\oplus$ exists and

$$f^\oplus = (1 - \gamma)^{-1} \alpha \varphi^\oplus (1 - \delta)^{-1}, \quad \text{ind}(f) \le k.$$

**Proof** First of all, we obtain some necessary equalities by direct computation.

(1) Since $\varphi^\oplus (1 - \varphi \varphi^\oplus) = 0$ and $(1 - \varphi^\oplus \varphi)\varphi^\oplus = 0$, $\varphi^\oplus \varepsilon = \varepsilon \varphi^\oplus = 0$.

(2) By (6.6.2) and Lemma 6.6.1 (4), we have

$$(\varphi \varphi^\oplus \varphi - \varphi)(1 - \varphi \varphi^\oplus)\eta = (\varphi \varphi^\oplus \varphi - \varphi \varphi^\oplus \varphi^2 \varphi^\oplus - \varphi + \varphi^2 \varphi^\oplus)\eta$$
$$= (\varphi \varphi^\oplus \varphi - \varphi)\eta = \varphi(\varphi^\oplus \varphi - 1)\eta = 0.$$

So, $(\varphi \varphi^\oplus \varphi - \varphi)\varepsilon = 0$.

(3) Since

$$\alpha^{-1} \varphi^\oplus = (1 + \varphi^\oplus \eta)\varphi^\oplus = \varphi^\oplus + \varphi^\oplus \eta \varphi^\oplus$$
$$= \varphi^\oplus (1 + \eta \varphi^\oplus) = \varphi^\oplus \beta^{-1},$$

we get $\alpha\varphi^{\circledD} = \varphi^{\circledD}\beta$. From

$$\beta^{-1}\eta = (1 + \eta\varphi^{\circledD})\eta = \eta + \eta\varphi^{\circledD}\eta = \eta(1 + \varphi^{\circledD}\eta) = \eta\alpha^{-1},$$

we have $\beta\eta = \eta\alpha$.

(4)

$$\varphi^{\circledD}f = \varphi^{\circledD}(\varphi + \eta - \varepsilon) = \varphi^{\circledD}\varphi + \varphi^{\circledD}\eta - \varphi^{\circledD}\varepsilon$$
$$\overset{(1)}{=} \varphi^{\circledD}\varphi + \varphi^{\circledD}\eta = \varphi^{\circledD}\varphi(1 + \varphi^{\circledD}\eta) = \varphi^{\circledD}\varphi\alpha^{-1}$$

and

$$f\varphi^{\circledD} = (\varphi + \eta - \varepsilon)\varphi^{\circledD} \overset{(1)}{=} \varphi\varphi^{\circledD} + \eta\varphi^{\circledD}$$
$$= (1 + \eta\varphi^{\circledD})\varphi\varphi^{\circledD} = \beta^{-1}\varphi\varphi^{\circledD}.$$

(5) Let $f_0 = \alpha\varphi^{\circledD}$. By the equality in (3), we have $f_0 = \varphi^{\circledD}\beta$. Then

$$f_0f = \alpha\varphi^{\circledD}f \overset{(4)}{=} \alpha\varphi^{\circledD}\varphi\alpha^{-1},$$

$$ff_0 = f\varphi^{\circledD}\beta \overset{(4)}{=} \beta^{-1}\varphi\varphi^{\circledD}\beta,$$

$$1 - f_0f = 1 - \alpha\varphi^{\circledD}\varphi\alpha^{-1} = \alpha(1 - \varphi^{\circledD}\varphi)\alpha^{-1}$$
$$= \alpha(1 - \varphi^{\circledD}\varphi)(1 + \varphi^{\circledD}\eta) = \alpha(1 - \varphi^{\circledD}\varphi)$$

and

$$1 - ff_0 = 1 - \beta^{-1}\varphi\varphi^{\circledD}\beta = \beta^{-1}(1 - \varphi\varphi^{\circledD})\beta$$
$$= (1 + \eta\varphi^{\circledD})(1 - \varphi\varphi^{\circledD})\beta = (1 - \varphi\varphi^{\circledD})\beta.$$

(6)

$$(1 - f_0f)ff_0 \overset{(5)}{=} \alpha(1 - \varphi^{\circledD}\varphi)\beta^{-1}\varphi\varphi^{\circledD}\beta$$
$$= \alpha(1 - \varphi^{\circledD}\varphi)(1 + \eta\varphi^{\circledD})\varphi\varphi^{\circledD}\beta$$
$$= \alpha(\varphi\varphi^{\circledD} - \varphi^{\circledD}\varphi^2\varphi^{\circledD})\beta + \alpha(1 - \varphi^{\circledD}\varphi)\eta\varphi^{\circledD}\varphi\varphi^{\circledD}\beta$$
$$= \alpha(1 - \varphi^{\circledD}\varphi)\eta\varphi^{\circledD}\beta = \gamma.$$

(7) $f_0f(1 - ff_0) \overset{(5)}{=} \alpha\varphi^{\circledD}\varphi\alpha^{-1}(1 - \varphi\varphi^{\circledD})\beta = \sigma.$

(8)

$$(ff_0)^*(1 - ff_0) \overset{(5)}{=} (\beta^{-1}\varphi\varphi^{\circledD}\beta)^*(1 - \varphi\varphi^{\circledD})\beta$$

$$= \beta^*(\beta^{-1}\varphi\varphi^{\circledD})^*(1 - \varphi\varphi^{\circledD})^*\beta$$

$$= \beta^*((1 - \varphi\varphi^{\circledD})\beta^{-1}\varphi\varphi^{\circledD})^*\beta$$

$$= \beta^*((1 - \varphi\varphi^{\circledD})(1 + \eta\varphi^{\circledD})\varphi\varphi^{\circledD})^*\beta$$

$$= \beta^*((1 - \varphi\varphi^{\circledD})\eta\varphi^{\circledD})^*\beta$$

$$= \beta^*(\eta\varphi^{\circledD})^*(1 - \varphi\varphi^{\circledD})\beta = \delta.$$

(9)

$$f_0 f f_0 = f_0(ff_0) = f_0\beta^{-1}\varphi\varphi^{\circledD}\beta$$

$$= \varphi^{\circledD}\beta\beta^{-1}\varphi\varphi^{\circledD}\beta = \varphi^{\circledD}\beta = f_0.$$

(10) By direct computation, we have

$$ff_0 f = (\varphi + \eta - \varepsilon)\alpha\varphi^{\circledD}(\varphi + \eta - \varepsilon)$$

$$\overset{(1)}{=} (\varphi + \eta - \varepsilon)\alpha\varphi^{\circledD}(\varphi + \eta) \overset{(3)}{=} (\varphi + \eta - \varepsilon)\varphi^{\circledD}\beta(\varphi + \eta)$$

$$= (\varphi + \eta)\varphi^{\circledD}\beta(\varphi + \eta) = (\varphi + \eta)\alpha\varphi^{\circledD}(\varphi + \eta)$$

$$= \varphi\alpha\varphi^{\circledD}(\varphi + \eta) + \eta\alpha\varphi^{\circledD}(\varphi + \eta).$$

Since $\alpha\varphi^{\circledD} + \alpha\varphi^{\circledD}\eta\varphi^{\circledD} = \alpha(1 + \varphi^{\circledD}\eta)\varphi^{\circledD} = \varphi^{\circledD}$, we get

$$\alpha\varphi^{\circledD} = \varphi^{\circledD} - \alpha\varphi^{\circledD}\eta\varphi^{\circledD} = (1 - \alpha\varphi^{\circledD}\eta)\varphi^{\circledD}.$$

Thus

$$ff_0 f = \varphi(1 - \alpha\varphi^{\circledD}\eta)\varphi^{\circledD}(\varphi + \eta) + \eta\alpha\varphi^{\circledD}(\varphi + \eta)$$

$$= \varphi\varphi^{\circledD}(\varphi + \eta) - \varphi\alpha\varphi^{\circledD}\eta\varphi^{\circledD}(\varphi + \eta) + \eta\alpha\varphi^{\circledD}(\varphi + \eta)$$

$$\overset{(3)}{=} \varphi\varphi^{\circledD}\varphi + \eta - \eta + \varphi\varphi^{\circledD}\eta - \varphi\varphi^{\circledD}\beta\eta\varphi^{\circledD}(\varphi + \eta) + \eta\alpha\varphi^{\circledD}(\varphi + \eta)$$

$$\overset{(3)}{=} \varphi\varphi^{\circledD}\varphi + \eta - \eta + \varphi\varphi^{\circledD}\eta - \varphi\varphi^{\circledD}\eta\alpha\varphi^{\circledD}(\varphi + \eta) + \eta\alpha\varphi^{\circledD}(\varphi + \eta)$$

$$= \varphi\varphi^{\circledD}\varphi + \eta - (1 - \varphi\varphi^{\circledD})\eta(1 - \alpha\varphi^{\circledD}(\varphi + \eta))$$

$$= \varphi\varphi^{\circledD}\varphi + \eta - (1 - \varphi\varphi^{\circledD})\eta(1 - \alpha\varphi^{\circledD}\varphi - \alpha(\alpha^{-1} - 1))$$

$$= \varphi\varphi^{\circledD}\varphi + \eta - (1 - \varphi\varphi^{\circledD})\eta\alpha(1 - \varphi^{\circledD}\varphi)$$

$$= \varphi\varphi^{\circledD}\varphi + \eta - \varepsilon$$

$$= f + (\varphi\varphi^{\circledD}\varphi - \varphi).$$

(11) $(\varphi\varphi^{\scriptsize\textcircled{D}} - 1)\varphi^m\eta = 0$, for arbitrary positive integer $m$.

In fact, by (6.6.2), $\varphi(\varphi^{\scriptsize\textcircled{D}}\varphi - 1)\eta = 0$, we have $(\varphi\varphi^{\scriptsize\textcircled{D}} - 1)\varphi\eta = 0$, that is, $m = 1$ is true. If $m = t$ is true, i.e., $\varphi^t\eta = \varphi\varphi^{\scriptsize\textcircled{D}}\varphi^t\eta$, then by Lemma 6.6.1 (iv), we have

$$
\begin{aligned}
(\varphi\varphi^{\scriptsize\textcircled{D}} - 1)\varphi^{t+1}\eta &= (\varphi\varphi^{\scriptsize\textcircled{D}} - 1)\varphi(\varphi^t\eta) = (\varphi\varphi^{\scriptsize\textcircled{D}} - 1)\varphi^2\varphi^{\scriptsize\textcircled{D}}\varphi^t\eta \\
&= \varphi\varphi^{\scriptsize\textcircled{D}}\varphi^2\varphi^{\scriptsize\textcircled{D}}\varphi^t\eta - \varphi^2\varphi^{\scriptsize\textcircled{D}}\varphi^t\eta \\
&= \varphi^2\varphi^{\scriptsize\textcircled{D}}\varphi^t\eta - \varphi^2\varphi^{\scriptsize\textcircled{D}}\varphi^t\eta \\
&= 0.
\end{aligned}
$$

(12) $(\varphi\varphi^{\scriptsize\textcircled{D}} - 1)\varphi^m\varepsilon = 0$, for arbitrary positive integer $m$.

In fact, when $m = 1$, by (2), it is true. If $m = t$ is true, i.e.,

$$
\varphi^t\varepsilon = \varphi\varphi^{\scriptsize\textcircled{D}}\varphi^t\varepsilon.
$$

Then

$$
\begin{aligned}
(\varphi\varphi^{\scriptsize\textcircled{D}} - 1)\varphi^{t+1}\varepsilon &= (\varphi\varphi^{\scriptsize\textcircled{D}} - 1)\varphi(\varphi^t\varepsilon) = (\varphi\varphi^{\scriptsize\textcircled{D}} - 1)\varphi^2\varphi^{\scriptsize\textcircled{D}}\varphi^t\varepsilon \\
&= (\varphi\varphi^{\scriptsize\textcircled{D}}\varphi^2\varphi^{\scriptsize\textcircled{D}} - \varphi^2\varphi^{\scriptsize\textcircled{D}})\varphi^t\varepsilon \\
&= (\varphi^2\varphi^{\scriptsize\textcircled{D}} - \varphi^2\varphi^{\scriptsize\textcircled{D}})\varphi^t\varepsilon \\
&= 0.
\end{aligned}
$$

(13) $f f_0 f^s = f^s + (\varphi\varphi^{\scriptsize\textcircled{D}} - 1)\varphi^s$, for arbitrary positive integer $s$.

In fact, when $s = 1$, by (10), it is true. If $s = m$ is true, i.e.,

$$
f f_0 f^m = f^m + (\varphi\varphi^{\scriptsize\textcircled{D}} - 1)\varphi^m.
$$

Then

$$
\begin{aligned}
f f_0 f^{m+1} &= (f f_0 f^m)f = (f^m + (\varphi\varphi^{\scriptsize\textcircled{D}} - 1)\varphi^m)f \\
&= f^{m+1} + (\varphi\varphi^{\scriptsize\textcircled{D}} - 1)\varphi^m(\varphi + \eta - \varepsilon) \\
&\overset{(11)}{=} f^{m+1} + (\varphi\varphi^{\scriptsize\textcircled{D}} - 1)\varphi^{m+1} - (\varphi\varphi^{\scriptsize\textcircled{D}} - 1)\varphi^m\varepsilon \\
&\overset{(12)}{=} f^{m+1} + (\varphi\varphi^{\scriptsize\textcircled{D}} - 1)\varphi^{m+1}.
\end{aligned}
$$

(14) Since $\operatorname{ind}(\varphi) = k$, by Lemma 6.6.1, we have $(\varphi\varphi^{\scriptsize\textcircled{D}} - 1)\varphi^k = 0$. By (13), we get

$$
f f_0 f^k = f^k + (\varphi\varphi^{\scriptsize\textcircled{D}} - 1)\varphi^k = f^k.
$$

(15) Since $1 - \gamma$ is invertible and

$$
\begin{aligned}
f_0 f^{k+1} &= f_0 f f^k = f^k - (1 - f_0 f) f^k \\
&\stackrel{(14)}{=} f^k - (1 - f_0 f) f f_0 f^k \\
&= (1 - (1 - f_0 f) f f_0) f^k \\
&\stackrel{(6)}{=} (1 - \gamma) f^k,
\end{aligned}
$$

we have $f^k = (1 - \gamma)^{-1} f_0 f^{k+1}$.

(16) Since $1 - \delta$ is invertible and

$$
\begin{aligned}
(f^k)^*(1 - \delta) &\stackrel{(8)}{=} (f^k)^*(1 - (f f_0)^*(1 - f f_0)) \\
&= (f^k)^* - (f f_0 f^k)^*(1 - f f_0) \\
&\stackrel{(14)}{=} (f^k)^* - (f^k)^*(1 - f f_0) \\
&= (f^k)^* f f_0,
\end{aligned}
$$

we get $(f^k)^* = (f^k)^* f f_0 (1 - \delta)^{-1}$.

(17)

$$
\begin{aligned}
f f_0 (\varphi \varphi^{\mathbb{D}} \varphi - \varphi) &\stackrel{(10)}{=} f f_0 (f f_0 f - f) = f f_0 f f_0 f - f f_0 f \\
&\stackrel{(9)}{=} f f_0 f - f f_0 f = 0,
\end{aligned}
$$

$$
\begin{aligned}
(\varphi \varphi^{\mathbb{D}} \varphi - \varphi) f f_0 &\stackrel{(5)}{=} (\varphi \varphi^{\mathbb{D}} \varphi - \varphi) \beta^{-1} \varphi \varphi^{\mathbb{D}} \beta \\
&= (\varphi \varphi^{\mathbb{D}} \varphi - \varphi)(1 + \eta \varphi^{\mathbb{D}}) \varphi \varphi^{\mathbb{D}} \beta \\
&= (\varphi \varphi^{\mathbb{D}} \varphi - \varphi) \varphi \varphi^{\mathbb{D}} \beta + \varphi(\varphi^{\mathbb{D}} \varphi - 1) \eta \varphi^{\mathbb{D}} \varphi \varphi^{\mathbb{D}} \beta \\
&= (\varphi \varphi^{\mathbb{D}} \varphi^2 \varphi^{\mathbb{D}} - \varphi^2 \varphi^{\mathbb{D}}) \beta + \varphi(\varphi^{\mathbb{D}} \varphi - 1) \eta \varphi^{\mathbb{D}} \beta \\
&\stackrel{(6.6.2)}{=} 0, \text{ by Lemma 6.6.1.}
\end{aligned}
$$

(18)

$$
\begin{aligned}
f(1 - \sigma) &\stackrel{(7)}{=} f(1 - f_0 f(1 - f f_0)) = f - f f_0 f(1 - f f_0) \\
&\stackrel{(10)}{=} f - (f + (\varphi \varphi^{\mathbb{D}} \varphi - \varphi))(1 - f f_0) \\
&= f - f(1 - f f_0) - (\varphi \varphi^{\mathbb{D}} \varphi - \varphi) + (\varphi \varphi^{\mathbb{D}} \varphi - \varphi) f f_0 \\
&\stackrel{(17)}{=} f^2 f_0 - (\varphi \varphi^{\mathbb{D}} \varphi - \varphi).
\end{aligned}
$$

(19) By (18), we have

$$ff_0f(1-\sigma) = ff_0f^2f_0 - ff_0(\varphi\varphi^\circledR\varphi - \varphi) \overset{(17)}{=} ff_0f^2f_0.$$

Since

$$ff_0f^2f_0 - f^2f_0 = (ff_0f - f)ff_0 \overset{(10)}{=} (\varphi\varphi^\circledR\varphi - \varphi)ff_0 \overset{(17)}{=} 0,$$

we get $ff_0f^2f_0 = f^2f_0$. So, $ff_0f(1-\sigma) = f^2f_0$. Since $1-\sigma$ is invertible, we have $ff_0f = f^2f_0(1-\sigma)^{-1}$. Thus

$$\begin{aligned}
ff_0 &= ff_0ff_0 = f^2f_0(1-\sigma)^{-1}f_0 = f^2f_0t \\
&= f(ff_0)t = f(f^2f_0t)t = f^3f_0t^2 = \cdots \\
&= f^{k+1}f_0t^k,
\end{aligned}$$

where $t = (1-\sigma)^{-1}f_0$.

(20) By (16) and (19), we get

$$\begin{aligned}
(f^k)^* &= (f^k)^*ff_0(1-\delta)^{-1} \\
&= (f^k)^*(f^{k+1}f_0t^k)(1-\delta)^{-1} \\
&= (f^k)^*f^{k+1}v,
\end{aligned}$$

where $v = f_0t^k(1-\delta)^{-1}$.

(21) : By (20), we have $f^k = (fv)^*(f^k)^*f^k$. From [102], it is easy to see that $fv$ is a $\{1,3\}$-inverse of $f^k$. Let $(f^k)^{(1,3)} = fv$. Then

$$f^k = f^kfvf^k = f^{k+1}vf^k.$$

So, $f$ is Drazin invertible and $\mathrm{ind}(f) \le k$. Since $f$ is Drazin invertible and $f^k$ has $\{1,3\}$-inverse, by Theorem 6.2.4, $f^\circledR$ exists. Thus

$$\begin{aligned}
f^\circledR &= f^Df^k(f^k)^{(1,3)} = f^Df^kfv = f^Df^{k+1}v = f^kv \\
&\overset{(15)}{=} (1-\gamma)^{-1}f_0f^{k+1}v = (1-\gamma)^{-1}f_0f^{k+1}f_0t^k(1-\delta)^{-1} \\
&= (1-\gamma)^{-1}f_0(f^{k+1}f_0t^k)(1-\delta)^{-1} \\
&\overset{(19)}{=} (1-\gamma)^{-1}f_0(ff_0)(1-\delta)^{-1} \\
&\overset{(9)}{=} (1-\gamma)^{-1}f_0(1-\delta)^{-1} \\
&= (1-\gamma)^{-1}\alpha\varphi^\circledR(1-\delta)^{-1}.
\end{aligned}$$

$\square$

**Remark 6.6.3** ([23]) By the proof of Theorem 6.6.2, the conditions in Theorem 6.6.2 only need that $1 - \gamma$ is left invertible and both $1 - \delta$ and $1 - \sigma$ are right invertible. Also, we can get dual results of the dual pseudo core inverse. Here, we omit it.

If $k = 1$, then we have the following corollary.

**Corollary 6.6.4** ([23]) *Let $\mathscr{C}$ be an additive category with an involution $*$. Suppose that both $\varphi \cdot X \to X$ is a morphism of $\mathscr{C}$ with core inverse $\varphi^{\scriptsize\textcircled{\tiny\#}}$ and $\eta : X \to X$ is a morphism of $\mathscr{C}$ such that $1 + \varphi^{\scriptsize\textcircled{\tiny\#}}\eta$ is invertible. Let $f = \varphi + \eta - \varepsilon$,*

$$\alpha = (1 + \varphi^{\scriptsize\textcircled{\tiny\#}}\eta)^{-1},$$

$$\beta = (1 + \eta\varphi^{\scriptsize\textcircled{\tiny\#}})^{-1},$$

$$\varepsilon = (1 - \varphi\varphi^{\scriptsize\textcircled{\tiny\#}})\eta\alpha(1 - \varphi^{\scriptsize\textcircled{\tiny\#}}\varphi),$$

$$\gamma = \alpha(1 - \varphi^{\scriptsize\textcircled{\tiny\#}}\varphi)\eta\varphi^{\scriptsize\textcircled{\tiny\#}}\beta,$$

$$\sigma = \alpha\varphi^{\scriptsize\textcircled{\tiny\#}}\varphi\alpha^{-1}(1 - \varphi\varphi^{\scriptsize\textcircled{\tiny\#}})\beta,$$

$$\delta = \beta^*(\varphi^{\scriptsize\textcircled{\tiny\#}})^*\eta^*(1 - \varphi\varphi^{\scriptsize\textcircled{\tiny\#}})\beta.$$

*If $1 - \gamma$, $1 - \sigma$ and $1 - \delta$ are invertible, then $f^{\scriptsize\textcircled{\tiny\#}}$ exists and*

$$f^{\scriptsize\textcircled{\tiny\#}} = (1 - \gamma)^{-1}\alpha\varphi^{\scriptsize\textcircled{\tiny\#}}(1 - \delta)^{-1}.$$

***Proof*** Since $\varphi^{\scriptsize\textcircled{\tiny\#}}$ exists, $(\varphi^{\scriptsize\textcircled{\tiny\#}}\varphi - 1)\varphi = 0$ and $\varphi(\varphi^{\scriptsize\textcircled{\tiny\#}}\varphi - 1) = 0$, equalities (6.6.1) and (6.6.2) hold. Hence, the corollary is right.                                        □

Let $J(R)$ be the Jacobson radical of a ring $R$. If $a \in R$ with $a^{\scriptsize\textcircled{\tiny D}}$ exists and $j \in J(R)$, then we use the following notations for the simplicity of the result :

$$\alpha = (1 + a^{\scriptsize\textcircled{\tiny D}}j)^{-1},$$

$$\beta = (1 + ja^{\scriptsize\textcircled{\tiny D}})^{-1},$$

$$\varepsilon = (1 - aa^{\scriptsize\textcircled{\tiny D}})j\alpha(1 - a^{\scriptsize\textcircled{\tiny D}}a),$$

$$\gamma = \alpha(1 - a^{\scriptsize\textcircled{\tiny D}}a)ja^{\scriptsize\textcircled{\tiny D}}\beta,$$

$$\sigma = \alpha a^{\scriptsize\textcircled{\tiny D}}a\alpha^{-1}(1 - aa^{\scriptsize\textcircled{\tiny D}})\beta,$$

$$\delta = \beta^*(a^{\scriptsize\textcircled{\tiny D}})^*j^*(1 - aa^{\scriptsize\textcircled{\tiny D}})\beta.$$

Firstly, we give a necessary lemma.

**Lemma 6.6.5** ([23]) *Let $R$ be a unital $*$-ring and $J(R)$ its Jacobson radical. If $a \in R$ with $a^\circledcirc$ exists and $j \in J(R)$, then $1 - \sigma$ is invertible.*

**Proof** Since

$$
\begin{aligned}
\sigma &= \alpha a^\circledcirc a \alpha^{-1}(1 - aa^\circledcirc)\beta \\
&= \alpha a^\circledcirc a(1 + a^\circledcirc j)(1 - aa^\circledcirc)\beta \\
&= \alpha a^\circledcirc a(1 - aa^\circledcirc)\beta + \alpha a^\circledcirc aa^\circledcirc j(1 - aa^\circledcirc)\beta \\
&= \alpha a^\circledcirc a(1 - aa^\circledcirc)\beta + \alpha a^\circledcirc j(1 - aa^\circledcirc)\beta,
\end{aligned}
$$

let $n = a^\circledcirc a(1 - aa^\circledcirc)$ and $k_1 = \alpha a^\circledcirc j(1 - aa^\circledcirc)\beta$, we have $\sigma = \alpha n\beta + k_1$, $k_1 \in J(R)$. Since

$$
\begin{aligned}
n^2 &= a^\circledcirc a(1 - aa^\circledcirc)a^\circledcirc a(1 - aa^\circledcirc) \\
&= a^\circledcirc a(a^\circledcirc a - a^\circledcirc a)(1 - aa^\circledcirc) = 0,
\end{aligned}
$$

we get that $1 - n$ is invertible. Set $u_1 = \alpha(1 - n)\beta$. Then $u_1$ is invertible. Let

$$
\begin{aligned}
k_2 &= \alpha^{-1}\beta^{-1} - 1 = (1 + a^\circledcirc j)(1 + ja^\circledcirc) - 1 \\
&= a^\circledcirc j + ja^\circledcirc + a^\circledcirc j^2 a^\circledcirc \in J(R),
\end{aligned}
$$

$$
k_3 = 1 - \alpha\beta = \alpha(\alpha^{-1}\beta^{-1} - 1)\beta = \alpha k_2\beta \in J(R).
$$

Set $u_2 = \alpha\beta - \sigma$. Then $u_2 = \alpha(1 - n)\beta - k_1 = u_1(1 - u_1^{-1}k_1)$. Thus $u_2$ is invertible since $u_1^{-1}k_1 \in J(R)$. By direct computation,

$$
1 - \sigma = \alpha\beta - \sigma + 1 - \alpha\beta = u_2 + k_3 = u_2(1 + u_2^{-1}k_3).
$$

Since $u_2^{-1}k_3 \in J(R)$, $1 + u_2^{-1}k_3$ is invertible. So, $1 - \sigma$ is invertible. $\qquad\square$

Now, we show that pseudo core inverses of a sum of elements in a unital $*$-ring with a radical element.

**Theorem 6.6.6** ([23]) *Let $R$ be a unital $*$-ring and $J(R)$ its Jacobson radical. Suppose that $a \in R$ with $a^\circledcirc$ exists and $j \in J(R)$. Let $f = a + j - \varepsilon$. If $j(a^\circledcirc a - 1)a = 0$ and $a(a^\circledcirc a - 1)j = 0$, then $f^\circledcirc$ exists and*

$$
f^\circledcirc = (1 - \gamma)^{-1}(1 + a^\circledcirc j)^{-1}a^\circledcirc(1 - \delta)^{-1}, \quad \mathrm{ind}(f) \leq \mathrm{ind}(a).
$$

**Proof** By Lemma 6.6.5, we know that $1 - \sigma$ is invertible. Since $j \in J(R)$, we obtain that $\gamma = \alpha(1 - a^\circledcirc a)ja^\circledcirc\beta \in J(R)$. So, we have $1 - \gamma$ is invertible. Since

$\beta^*(1 - aa^{\textcircled{D}})ja^{\textcircled{D}}\beta \in J(R)$ and

$$1 - \delta = 1 - \beta^*(a^{\textcircled{D}})^*j^*(1 - aa^{\textcircled{D}})\beta$$
$$= (1 - \beta^*(1 - aa^{\textcircled{D}})ja^{\textcircled{D}}\beta)^*,$$

we get $1 - \beta^*(1 - aa^{\textcircled{D}})ja^{\textcircled{D}}\beta$ is invertible. Hence, $1 - \delta$ is invertible. By Theorem 6.6.2, we know that $f^{\textcircled{D}}$ exists and

$$f^{\textcircled{D}} = (1 - \gamma)^{-1}(1 + a^{\textcircled{D}}j)^{-1}a^{\textcircled{D}}(1 - \delta)^{-1}, \quad \text{ind}(f) \leq \text{ind}(a).$$

$\square$

Using the same method, we can get a dual theorem of Theorem 6.6.6 for dual pseudo core inverses.

## 6.7   The Pseudo Core Inverse of a Product

In this section, we consider matrices over $R$ unless otherwise indicated. Our main goal is to give necessary and sufficient conditions for a product to have pseudo core inverse, as well as to derive its expression. $I$ denotes the identity matrix. We begin with some auxiliary lemmas.

**Lemma 6.7.1 ([16, Theorem 2.1])** *Let $A$ be regular, and $T$ a square matrix with $T^k = PAQ$ for some positive integer $k$ and matrices $P$ and $Q$. Then the following are equivalent:*

(1) *$T^D$ exists with $\text{ind}(T) \leq k$, and $P'PA = A = AQQ'$ for some matrices $P', Q'$.*
(2) *$A^-AQTPA + I - A^-A$ is invertible.*

   *In this case,*

$$T^D = PA(A^-AQTPA + I - A^-A)^{-1}Q.$$

**Lemma 6.7.2 ([16, Theorem 3.2])** *Let $T, P, A, Q$ be matrices such that $T = PAQ$. Suppose $P'PA = A = AQQ'$ for some matrices $P', Q'$. Then we have the following facts:*

(1) *If $A\{1, 4\} \neq \emptyset$, then $T\{1, 3\} \neq \emptyset$ if and only if $(PA)^*PA + I - A^{(1,4)}A$ is invertible.*
   *In this case, $Q'A^{(1,4)}A[(PA)^*PA + I - A^{(1,4)}A]^{-1}(PA)^* \in T\{1, 3\}$.*
(2) *If $A\{1, 3\} \neq \emptyset$, then $T\{1, 4\} \neq \emptyset$ if and only if $AQ(AQ)^* + I - AA^{(1,3)}$ is invertible.*

*In this case, $(AQ)^*[AQ(AQ)^* + I - AA^{\textcircled{D}}]^{-1}AA^{(1,3)}P' \in T\{1, 4\}$.*

**Lemma 6.7.3 ([44, Theorem 2.3])** *Let $T \in R^{n \times n}$. Then we have the following facts:*

(1) $T^{\circledcirc}$ *exists if and only if* $T^D$ *exists and* $T^k\{1, 3\} \neq \emptyset$*, where* $k \geq \mathrm{ind}(A)$*. In this case,* $T^{\circledcirc} = T^D T^k (T^k)^{(1,3)}$*.*

(2) $T_{\circledcirc}$ *exists if and only if* $T^D$ *exists and* $T^k\{1, 4\} \neq \emptyset$*, where* $k \geq \mathrm{ind}(A)$*. In this case,* $T_{\circledcirc} = (T^k)^{(1,4)} T^k T^D$*.*

(3) $T^{\circledcirc}$ *and* $T_{\circledcirc}$ *exist if and only if* $T^D$ *and* $(T^k)^\dagger$ *exist, where* $k \geq \mathrm{ind}(A)$*. In this case,* $T^{\circledcirc} = T^D T^k (T^k)^\dagger$ *and* $T_{\circledcirc} = (T^k)^\dagger T^k T^D$*.*

Let $A$ be a matrix over $R$ and $T$ a square matrix with $T^k = PAQ$ for some positive integer $k$ and matrices $P$ and $Q$. Suppose $P'PA = A = AQQ'$ for some matrices $P'$, $Q'$. The following result gives a characterization for $T$ to have pseudo core inverse, under the assumption that $A\{1, 4\} \neq \emptyset$.

**Proposition 6.7.4 ([45])** *Let $A \in R^{m \times n}$ with $A\{1, 4\} \neq \emptyset$, and $T$ a square matrix with $T^k = PAQ$ for some positive integer $k$ and matrices $P$ and $Q$. Suppose $P'PA = A = AQQ'$ for some matrices $P'$, $Q'$. Then the following are equivalent:*

(1) $T^{\circledcirc}$ *exists with* $\mathrm{ind}(T) \leq k$*.*

(2) $A^{(1,4)} AQTPA + I - A^{(1,4)} A$ *and* $(PA)^* PA + I - A^{(1,4)} A$ *are invertible.*

*In this case,* $T^{\circledcirc} = PA[A^{(1,4)} AQTPA + I - A^{(1,4)} A]^{-1} QPA[(PA)^* PA + I - A^{(1,4)} A]^{-1} (PA)^*$*.*

***Proof*** Applying Lemma 6.7.1, $T^D$ exists if and only if $A^{(1,4)} AQTPA + I - A^{(1,4)} A$ is invertible, in which case, $T^D = PA[A^{(1,4)} AQTPA + I - A^{(1,4)} A]^{-1} Q$.

If $A\{1, 4\} \neq \emptyset$, then by Lemma 6.7.2, $T^k\{1, 3\} \neq \emptyset$ if and only if $(PA)^* PA + I - A^{(1,4)} A$ is invertible, in which case, $Q' A^{(1,4)} A[(PA)^* PA + I - A^{(1,4)} A]^{-1} (PA)^* \in T^k\{1, 3\}$.

From Lemma 6.7.3, $T^{\circledcirc}$ exists if and only if $T^D$ exists and $T^k\{1, 3\} \neq \emptyset$, where $k \geq \mathrm{ind}(T)$. Thus, (1) is equivalent to (2), moreover, $T^{\circledcirc} = T^D T^k (T^k)^{(1,3)} = PA(A^{(1,4)} AQTPA + I - A^{(1,4)} A)^{-1} QPAQQ' A^{(1,4)} A[(PA)^* PA + I - A^{(1,4)} A]^{-1} (PA)^* = PA[A^{(1,4)} AQTPA + I - A^{(1,4)} A]^{-1} QPA[(PA)^* PA + I - A^{(1,4)} A]^{-1} (PA)^*$.  □

Dually, we have the following result, which characterizes the dual pseudo core invertibility of $T$, under the assumption that $A\{1, 3\} \neq \emptyset$.

**Proposition 6.7.5 ([45])** *Let $A \in R^{m \times n}$ with $A\{1, 3\} \neq \emptyset$, and $T$ a square matrix with $T^k = PAQ$ for some positive integer $k$ and matrices $P$ and $Q$. Suppose $P'PA = A = AQQ'$ for some matrices $P'$, $Q'$. Then the following are equivalent:*

(1) $T_{\circledcirc}$ *exists with* $\mathrm{ind}(T) \leq k$*.*

(2) $AQTPAA^{(1,3)} + I - AA^{(1,3)}$ *and* $AQ(AQ)^* + I - AA^{(1,3)}$ *are invertible.*

*In this case,* $T_{\text{\textcircled{D}}} = (AQ)^*[AQ(AQ)^* + I - AA^{(1,3)}]^{-1}AQP[AQTPAA^{(1,3)} + I - AA^{(1,3)}]^{-1}AQ.$

If $m = 1$, then we get a characterization for the existence of the core inverse (resp. dual core inverse) of a product $PAQ$.

**Corollary 6.7.6 ([45])** *Let* $T, P, A, Q$ *be matrices such that* $T = PAQ$. *Suppose* $P'PA = A = AQQ'$ *for some matrices* $P'$, $Q'$. *Then we have the following facts:*

(1) *If* $A\{1, 4\} \neq \emptyset$, *then* $T^{\text{\textcircled{*}}}$ *exists if and only if both* $A^{(1,4)}AQPA + I - A^{(1,4)}A$ *and* $(PA)^*PA + I - A^{(1,4)}A$ *are invertible. In this case,* $T^{\text{\textcircled{*}}} = PA[A^{(1,4)}AQPA + I - A^{(1,4)}A]^{-1}[(PA)^*PA + I - A^{(1,4)}A]^{-1}(PA)^*.$

(2) *If* $A\{1, 3\} \neq \emptyset$, *then* $T_{\text{\textcircled{*}}}$ *exists if and only if both* $AQPAA^{(1,3)} + I - AA^{(1,3)}$ *and* $AQ(AQ)^* + I - AA^{(1,3)}$ *are invertible. In this case,* $T_{\text{\textcircled{*}}} = (AQ)^*[AQ(AQ)^* + I - AA^{(1,3)}]^{-1}[AQPAA^{(1,3)} + I - AA^{(1,3)}]^{-1}AQ.$

(3) *If* $A^\dagger$ *exists, then* $T^{\text{\textcircled{*}}}$ *and* $T_{\text{\textcircled{*}}}$ *exist if and only if* $A^\dagger AQPA + I - A^\dagger A$, $(PA)^*PA + I - A^\dagger A$ *and* $AQ(AQ)^* + I - AA^\dagger$ *are invertible. In this case,*

$$T^{\text{\textcircled{*}}} = PA[A^\dagger AQPA + I - A^\dagger A]^{-1}[(PA)^*PA + I - A^\dagger A]^{-1}(PA)^*$$

*and*

$$T_{\text{\textcircled{*}}} = (AQ)^*[AQ(AQ)^* + I - AA^\dagger]^{-1}[AQPAA^\dagger + I - AA^\dagger]^{-1}AQ.$$

In what follows, we consider the existence criteria and formulae of the pseudo core inverse (resp. dual pseudo core inverse) of $T$, under the assumption that $A$ is regular. We begin with an auxiliary lemma.

**Lemma 6.7.7 ([45])** *Let* $A$ *be a matrix. If there exists a matrix* $X$ *such that* $XA^*A = A$, *then* $X^* \in A\{1, 3\}$.

**Theorem 6.7.8 ([45])** *Let* $A$ *be regular, and* $T$ *a square matrix with* $T^k = PAQ$ *for some positive integer* $k$ *and matrices* $P$, $Q$. *Suppose*

$$U = A^- AQTPA + I - A^-A, \quad V = AQ(PAQ)^*PAA^- + I - AA^-.$$

*Then the following are equivalent:*

(1) $T^D$ *exists with* $\text{ind}(T) \leq k$, $A = P'PA = AQQ'$ *for some matrices* $P'$ *and* $Q'$, *and* $Z'AQ(PAQ)^*PAQ = PAQ$ *for some matrix* $Z'$.

(2) $U$ *is invertible and* $V$ *is left invertible.*

(3) $XAQTPA = A = AQTPAY$, $ZAQ(PAQ)^*PA = A$ *for some matrices* $X, Y, Z$.

*In this case,* $T^{\text{\textcircled{D}}}$ *exists with*

$$T^{\text{\textcircled{D}}} = PAU^{-1}QPAQQ^*A^*Z^*P^* = PAU^{-1}QPAQQ^*A^*Z'^*.$$

***Proof*** By Lemma 6.7.1, $T^D$ exists with $\text{ind}(T) \leq k$, $A = P'PA = AQQ'$ for some matrices $P'$ and $Q'$ if and only if $U$ is invertible. Next, we prove that $U$ is invertible if and only if $XAQTPA = A = AQTPAY$ for some matrices $X$, $Y$.

In fact, applying Jacobson's lemma, $U = A^-AQTPA + I - A^-A$ is invertible, if and only if,

$$U' = AQTPAA^- + I - AA^- \text{ is left invertible and}$$

$$U = A^-AQTPA + I - A^-A \text{ is right invertible.}$$

From $U' = AQTPAA^- + I - AA^-$ is left invertible, it follows that there exists $X$ such that $XU' = X(AQTPAA^- + I - AA^-) = I$. Thus, $XAQTPA = A$.

From $U = A^-AQTPA + I - A^-A$ is right invertible, it follows that there exists $Y$ such that $UY = (A^-AQTPA + I - A^-A)Y = I$. Hence $AQTPAY = A$.

If $XAQTPA = A = AQTPAY$ for some matrices $X$ and $Y$, then

$$(XAA^- + I - AA^-)(AQTPAA^- + I - AA^-) = I$$

$$= (A^-AQTPA + I - A^-A)(A^-AY + I - A^-A),$$

namely, $U' = AQTPAA^- + I - AA^-$ is left invertible and $U = A^-AQTPA + I - A^-A$ is right invertible.

In the following, we prove that $V$ is left invertible if and only if there exists $Z$ such that $ZAQ(PAQ)^*PA = A$ if and only if there exists $Z'$ such that $Z'AQ(PAQ)^*PAQ = PAQ$.

If $V = AQ(PAQ)^*PAA^- + I - AA^-$ is left invertible, then there exists $Z$ such that $ZV = Z[AQ(PAQ)^*PAA^- + I - AA^-] = 1$. Thus, $ZAQ(PAQ)^*PA = A$. Conversely, if $ZAQ(PAQ)^*PA = A$, then $(ZAA^- + I - AA^-)V = 1$, namely, $V$ is left invertible.

If $ZAQ(PAQ)^*PA = A$, then $PZAQ(PAQ)^*PAQ = PAQ$. Conversely if $Z'AQ(PAQ)^*$
$PAQ = PAQ$, then $P'Z'AQ(PAQ)^*PA = A$ because of $P'PA = A = AQQ'$.

From the above, we derive the equivalences of (1)-(3). In this case, $T^D = PAU^{-1}Q$ follows from Lemma 6.7.1, and $T^kT^{k(1,3)} = PAQQ^*A^*Z^*P^* = PAQQ^*A^*Z'^*$ follows from Lemma 6.7.7. Hence by Lemma 6.7.3, $T^\circledR$ exists with

$$T^\circledR = T^DT^kT^{k(1,3)} = PAU^{-1}QPAQQ^*A^*Z^*P^* = PAU^{-1}QPAQQ^*A^*Z'^*.$$

$\square$

Dually, we have the following result.

**Theorem 6.7.9** ([45]) *Let $A$ be regular, and $T$ a square matrix with $T^k = PAQ$ for some positive integer $k$ and matrices $P$, $Q$. Suppose*

$$U = A^-AQTPA + I - A^-A, \quad V = AQ(PAQ)^*PAA^- + I - AA^-.$$

*Then the following are equivalent:*

(1)  $T^D$ *exists with* $\text{ind}(T) \leq k$, $P'PA = A = AQQ'$ *for some matrices $P'$ and $Q'$, and* $PAQ(PAQ)^*PAZ' = PAQ$ *for some matrix $Z'$.*
(2)  $U$ *is invertible and $V$ is right invertible.*
(3)  $XAQTPA = A = AQTPAY$, $AQ(PAQ)^*PAZ = A$ *for some matrices* $X, Y, Z$.

In this case, $T_{\circledD}$ exists with $T_{\circledD} = Q^*Z^*A^*P^*PAQPAU^{-1}Q = Z'^*A^*P^*PAQPAU^{-1}Q$.

Next, we consider the pseudo core invertibility of $T$. It turns to be a useful tool to characterize the pseudo core invertibility of a lower triangular matrix.

**Theorem 6.7.10 ([45])** *Let $A$ be a matrix with $A\{1, 3\} \neq \emptyset$ and $T$ a square matrix with $T^k = PAQ$ for some positive integer $k$, matrices $P$ and $Q$. Suppose $P(I - AA^{(1,3)}) = I - AA^{(1,3)}$, $P'PA = A = AQQ'$ for some matrix $P'$ and $Q'$, and*

$$U = AQTPAA^{(1,3)} + I - AA^{(1,3)}, \quad V = P^*PAA^{(1,3)} + I - AA^{(1,3)}.$$

*Then the following are equivalent:*

(1)  $T^{\circledD}$ *exists with* $\text{ind}(T) \leq k$.
(2)  $U$ *is invertible and $T^k\{1, 3\} \neq \emptyset$.*
(3)  $U$ *is invertible and $V$ is left invertible.*
(4)  $U$ *is invertible and $V$ is right invertible.*

In this case, $T^{\circledD} = PU^{-1}AQPAA^{(1,3)}V^{-1}P^*$.

**Proof**  Applying Lemma 6.7.3, $T^{\circledD}$ exists if and only if $T^D$ exists and $T^k\{1, 3\} \neq \emptyset$.

Since $P'PA = A = AQQ'$, it follows that $T^D$ exists with $\text{ind}(T) \leq k$ if and only if $A^{(1,3)}AQTPA + I - A^{(1,3)}A$ is invertible by Lemma 6.7.1, which is equivalent to the fact that $U$ is invertible by Jacobson's lemma. In this case, $T^D = PU^{-1}AQ$. We thus get (1) $\Leftrightarrow$ (2).

By Corollary 2.3.4, we obtain $PA\{1, 3\} \neq \emptyset$ if and only if $V$ is left invertible if and only if $V$ is right invertible, in which case, $A^{(1,3)}V^{-1}P^* \in PA\{1, 3\}$. And notice that $T^k\{1, 3\} \neq \emptyset$ if and only if $PA\{1, 3\} \neq \emptyset$ whenever $A = AQQ'$. Thus, the equivalences of (2)-(4) follow. In this case,

$$T^{\circledD} = T^DT^k(T^k)^{(1,3)} = PU^{-1}AQPAQQ'A^{(1,3)}V^{-1}P^*$$

$$= PU^{-1}AQPAA^{(1,3)}V^{-1}P^*.$$

$\square$

**Corollary 6.7.11** *Let A be a matrix with $A\{1, 3\} \neq \emptyset$ and T a square matrix with $T = PA$ for some matrix P. Suppose $P(I - AA^{(1,3)}) = I - AA^{(1,3)}$, $P'PA = A$ for some matrix P'. Then the following are equivalent:*

(1)   $T^{\circledast}$ *exists.*
(2)   *U is invertible and V is left invertible.*
(3)   *U is invertible and V is right invertible.*

*Where $U = APAA^{(1,3)} + I - AA^{(1,3)}$, $V = P^{*}PAA^{(1,3)} + I - AA^{(1,3)}$. In this case, $T^{\circledast} = PU^{-1}AA^{(1,3)}V^{-1}P^{*}$.*

The following result characterizes the dual pseudo core invertibility of $T$.

**Theorem 6.7.12 ([45])** *Let A be a matrix with $A\{1, 4\} \neq \emptyset$ and T a square matrix with $T^{k} = PAQ$ for some positive integer k, matrices P and Q. Suppose $(I - A^{(1,4)}A)Q = I - A^{(1,4)}A$, $P'PA = A = AQQ'$ for some matrices P' and Q', and*

$$U = A^{(1,4)}AQTPA + I - A^{(1,4)}A, \quad V = A^{(1,4)}AQ^{*}Q + I - A^{(1,4)}A.$$

*Then the following are equivalent:*

(1)   $T_{\circledD}$ *exists with $\mathrm{ind}(T) \leq k$.*
(2)   *U is invertible and $T^{k}\{1, 4\} \neq \emptyset$.*
(3)   *U is invertible and V is left invertible.*
(4)   *U is invertible and V is right invertible.*

*In this case, $T_{\circledD} = Q^{*}V^{-1}A^{(1,4)}AQPAU^{-1}Q$.*

**Corollary 6.7.13** *Let A be a matrix with $A\{1, 4\} \neq \emptyset$ and T a square matrix with $T = AQ$ for some matrix Q. Suppose $(I - A^{(1,4)}A)Q = I - A^{(1,4)}A$, $AQQ' = A$ for some matrix Q'. Then the following are equivalent:*

(1)   $T_{\circledR}$ *exists.*
(2)   *U is invertible and V is left invertible.*
(3)   *U is invertible and V is right invertible.*

*Where $U = A^{(1,4)}AQA + I - A^{(1,4)}A$, $V = A^{(1,4)}AQQ^{*} + I - A^{(1,4)}A$. In this case, $T_{\circledR} = Q^{*}V^{-1}A^{(1,4)}AU^{-1}Q$.*

## 6.8   The Pseudo Core Inverse of a Low Triangular Matrix

Many authors have derived a number of results on various generalized inverses of lower triangular matrices. In this section, we consider the pseudo core invertibility of a lower triangular matrix $T$. We begin with a special case of a lower triangular matrix $T$.

**Theorem 6.8.1 ([45])** *Let* $T = \begin{bmatrix} 0 & 0 \\ b & d \end{bmatrix} \in R^{2\times2}$ *such that* $d_{\circledD}$ *exists with* $ind(d) = k$. *Then the following are equivalent:*

(1) $T^{\circledD}$ *exists with* $\mathrm{ind}(T) \le k+1$.
(2) $(d^{k+1})^*d^{k+1} + 1 - d_{\circledD}d$ *is invertible.*
(3) $d^{\circledD}$ *exists.*

*In this case,* $T^{\circledD} = \begin{bmatrix} 0 & 0 \\ 0 & d^k u^{-1}(d^{k+1})^* \end{bmatrix}$, *where* $u = (d^{k+1})^*d^{k+1} + 1 - d_{\circledD}d$ *and*

$$\mathrm{ind}(T) = \begin{cases} 1, & \text{if } d \text{ is invertible} \\ k, & \text{if } (1 - dd^D)d^{k-1}b = 0 \\ k+1, & \text{if } (1 - dd^D)d^{k-1}b \ne 0 \end{cases}.$$

**Proof** Since $d_{\circledD}$ exists with $\mathrm{ind}(d) = k$, it follows that $d^D$ exists and $d^{k+1}\{1, 4\} \ne \emptyset$. Thus, $T^D$ exists with $T^D = \begin{bmatrix} 0 & 0 \\ (d^D)^2 b & d^D \end{bmatrix}$, as well as

$$\mathrm{ind}(T) = \begin{cases} 1, & \text{if } d \text{ is invertible} \\ k, & \text{if } (1 - dd^D)d^{k-1}b = 0 \\ k+1, & \text{if } (1 - dd^D)d^{k-1}b \ne 0 \end{cases}$$

Hence $T^{\circledD}$ exists if and only if $T^{k+1}\{1, 3\} \ne \emptyset$ by Lemma 6.7.3.
Observe that

$$T^{k+1} = \begin{bmatrix} 0 & 0 \\ b & d \end{bmatrix}^{k+1} = \begin{bmatrix} 0 & 0 \\ d^k b & d^{k+1} \end{bmatrix} = \begin{bmatrix} 1 & 0 \\ 0 & 1 \end{bmatrix}\begin{bmatrix} 0 & 0 \\ 0 & d^{k+1} \end{bmatrix}\begin{bmatrix} 1 & 0 \\ d^D b & 1 \end{bmatrix} =: PAQ,$$

and

$$\begin{bmatrix} 0 & 0 \\ 0 & d^{k+1}_{\circledD} \end{bmatrix} \in A\{1, 4\}.$$

So we have

$$(PA)^* PA + I - A^{(1,4)} A = \begin{bmatrix} 1 & 0 \\ 0 & u \end{bmatrix},$$

where $u = (d^{k+1})^*d^{k+1} + 1 - d_{\circledD}d$.

(1) $\Leftrightarrow$ (2). Applying Lemma 6.7.2, $T^{k+1}\{1, 3\} \neq \emptyset$ if and only if $(PA)^* PA + I - A^{(1,4)}A$ is invertible if and only if $(d^{k+1})^* d^{k+1} + 1 - d_{\circledD}d$ is invertible. Hence $T^{\circledD}$ exists if and only if $(d^{k+1})^* d^{k+1} + 1 - d_{\circledD}d$ is invertible. In this case,

$$T^{\circledD} = T^D T^{k+1}(T^{k+1})^{(1,3)} = T^D PA[(PA)^* PA + I - A^{(1,4)}A]^{-1}(PA)^*$$

$$= \begin{bmatrix} 0 & 0 \\ (d^D)^2 b & d^D \end{bmatrix} \begin{bmatrix} 0 & 0 \\ 0 & d^{k+1} \end{bmatrix} \begin{bmatrix} 1 & 0 \\ 0 & [(d^{k+1})^* d^{k+1} + 1 - d_{\circledD}d]^{-1} \end{bmatrix} \begin{bmatrix} 0 & 0 \\ 0 & (d^{k+1})^* \end{bmatrix}$$

$$= \begin{bmatrix} 0 & 0 \\ 0 & d^k[(d^{k+1})^* d^{k+1} + 1 - d_{\circledD}d]^{-1}(d^{k+1})^* \end{bmatrix}.$$

(2) $\Leftrightarrow$ (3). Since $d^D$ exists with $ind(d) = k$ and $d^{k+1}\{1, 4\} \neq \emptyset$, it follows by Lemma 6.7.3 that $d^{\circledD}$ exists with $\text{ind}(d) = k$ if and only if $(d^{k+1})^\dagger$ exists, which is equivalent to the fact that $(d^{k+1})^* d^{k+1} + 1 - d_{\circledD}d$ is invertible.           $\square$

From the above result, we can see that $d \in R^{\circledD}$ is an indispensable condition when investigating a general lower triangular matrix case. Hence, in the following, let us assume $d \in R^{\circledD} \cap R_{\circledD}$.

**Theorem 6.8.2 ([45])** *Let* $T = \begin{bmatrix} a & 0 \\ b & d \end{bmatrix} \in R^{2 \times 2}$ *such that* $d \in R^{\circledD} \cap R_{\circledD}$ *with* $\text{ind}(d) = k$, $a^k\{1, 4\} \neq \emptyset$, $c = \sum_{i+j=k-1} d^i ba^j$, $w = [1 - d^k(d^{\circledD})^k]c[1 - (a^k)^{(1,4)}a^k]$ *with* $w\{1, 4\} \neq \emptyset$. *Then the following are equivalent:*

(1) $T^{\circledD}$ *exists with* $\text{ind}(T) \leq k$.
(2) $\xi$ *and* $\eta$ *are invertible, where*

$$\xi = (a^k)^{(1,4)}a^{2k+1} + w^{(1,4)}wa^{k+1} + 1 - (a^k)^{(1,4)}a^k - w^{(1,4)}w,$$

$$\eta = (a^k)^* a^k + c^*[1 - d^k(d^k)^\dagger]c + 1 - (a^k)^{(1,4)}a^k - w^{(1,4)}w.$$

*In this case,* $T^{\circledD} = \begin{bmatrix} \mu & \nu \\ \delta & \omega \end{bmatrix}$, *where*

$$\mu = a^k \xi^{-1} a^k \eta^{-1}(a^k)^*, \quad \nu = a^k \xi^{-1} a^k \eta^{-1} c^*[1 - d^k(d^k)^\dagger],$$

$$\delta = [1 - d^k(d^k)^\dagger]c\xi^{-1}a^k \eta^{-1}(a^k)^* - d^D \gamma \xi^{-1} a^k \eta^{-1}(a^k)^* +$$
$$d^D(d^k)^\dagger ca^k \eta^{-1}(a^k)^* + d^D[1 - d^k(d^k)^\dagger]c\eta^{-1}(a^k)^*,$$

$$\omega = [1 - d^k(d^k)^\dagger]c\xi^{-1}a^k \eta^{-1} c^*[1 - d^k(d^k)^\dagger] - d^D \gamma \xi^{-1} a^k \eta^{-1} c^*[1 - d^k(d^k)^\dagger]$$
$$+ d^D(d^k)^\dagger ca^k \eta^{-1} c^*[1 - d^k(d^k)^\dagger] + d^D[1 - d^k(d^k)^\dagger]c\eta^{-1} c^*[1 - d^k(d^k)^\dagger]$$
$$+ d^{\circledD}.$$

**Proof** Applying Lemma 6.7.3, $d \in R^{\oplus} \cap R_{\oplus}$ with $\mathrm{ind}(d) = k$ if and only if $d^D$ and $(d^k)^{\dagger}$ exist. In this case, $d^k(d^k)^{\dagger} = d^k(d^k)^{\oplus}$, $(d^{k+1})_{\oplus}d^{k+1} = (d^k)_{\oplus}d^k = (d^k)^{\dagger}d^k$.

Since $T = \begin{bmatrix} a & 0 \\ b & d \end{bmatrix}$, then

$$T^k = \begin{bmatrix} a & 0 \\ b & d \end{bmatrix}^k = \begin{bmatrix} a^k & 0 \\ c & d^k \end{bmatrix} = \begin{bmatrix} 1 & 0 \\ [1 - d^k(d^k)^{\dagger}]c(a^k)^{(1,4)} & 1 \end{bmatrix} \begin{bmatrix} a^k & 0 \\ w & d^k \end{bmatrix} \begin{bmatrix} 1 & 0 \\ (d^k)^{\dagger}c & 1 \end{bmatrix}$$

$$:= PAQ.$$

By computation,

$$\begin{bmatrix} (a^k)^{(1,4)} & w^{(1,4)}[1 - d^k(d^k)^{\dagger}] \\ 0 & (d^k)^{\dagger} \end{bmatrix} \in A\{1,4\},$$

$$A^{(1,4)}A = \begin{bmatrix} (a^k)^{(1,4)} & w^{(1,4)}[1 - d^k(d^k)^{\dagger}] \\ 0 & (d^k)^{\dagger} \end{bmatrix} \begin{bmatrix} a^k & 0 \\ w & d^k \end{bmatrix}$$

$$= \begin{bmatrix} (a^k)^{(1,4)}a^k + w^{(1,4)}w & 0 \\ 0 & (d^k)^{\dagger}d^k \end{bmatrix},$$

$$A^{(1,4)}AQTPA + I - A^{(1,4)}A = \begin{bmatrix} (a^k)^{(1,4)}a^{2k+1} + w^{(1,4)}wa^{k+1} & 0 \\ \gamma & (d^k)^{\dagger}d^{2k+1} \end{bmatrix} +$$

$$\begin{bmatrix} 1 - (a^k)^{(1,4)}a^k - w^{(1,4)}w & 0 \\ 0 & 1 - (d^k)^{\dagger}d^k \end{bmatrix}$$

$$= \begin{bmatrix} \xi & 0 \\ \gamma & (d^k)^{\dagger}d^{2k+1} + 1 - (d^k)^{\dagger}d^k \end{bmatrix},$$

$$(PA)^*PA + I - A^{(1,4)}A = \begin{bmatrix} (a^k)^*a^k + c^*[1 - d^k(d^k)^{\dagger}]c & 0 \\ 0 & (d^k)^*d^k \end{bmatrix} +$$

$$\begin{bmatrix} 1 - (a^k)^{(1,4)}a^k - w^{(1,4)}w & 0 \\ 0 & 1 - (d^k)^{\dagger}d^k \end{bmatrix}$$

$$= \begin{bmatrix} \eta & 0 \\ 0 & (d^k)^*d^k + 1 - (d^k)^{\dagger}d^k \end{bmatrix},$$

where

$$\xi = (a^k)^{(1,4)}a^{2k+1} + w^{(1,4)}wa^{k+1} + 1 - (a^k)^{(1,4)}a^k - w^{(1,4)}w,$$

$$\gamma = (d^k)^{\dagger}d^k ca^{k+1} + (d^k)^{\dagger}d^k ba^k + (d^k)^{\dagger}d^{k+1}[1 - d^k(d^k)^{\dagger}]c,$$

$$\eta = (a^k)^*a^k + c^*[1 - d^k(d^k)^{\dagger}]c + 1 - (a^k)^{(1,4)}a^k - w^{(1,4)}w.$$

Since $(d^{k+1})^{\#}$ exists, it follows that $d^{k+1} + 1 - (d^k)^{\dagger} d^k = d^{k+1} + 1 - (d_{\circledo})^k d^k = d^{k+1} + 1 - (d^{k+1})_{\circledo} d^{k+1}$ is invertible. Applying Jacobson's lemma, $(d^k)^{\dagger} d^{2k+1} + 1 - (d^k)^{\dagger} d^k$ is invertible. Since $(d^k)^{\dagger}$ exists, it follows that $(d^k)^* d^k + 1 - (d^k)^{\dagger} d^k$ is invertible.

According to Proposition 6.7.4, $T^{\circledo}$ exists with $\mathrm{ind}(T) \leq k$ if and only if

$$A^{(1,4)} AQTPA + I - A^{(1,4)} A \text{ and } (PA)^* PA + I - A^{(1,4)} A \text{ are invertible,}$$

which is equivalent to $\xi$ and $\eta$ being invertible. Moreover,

$$T^{\circledo} = PA[A^{(1,4)} AQTPA + I - A^{(1,4)} A]^{-1}$$

$$QPA[(PA)^* PA + I - A^{(1,4)} A]^{-1} (PA)^*$$

$$= \begin{bmatrix} a^k & 0 \\ [1 - d^k (d^k)^{\dagger}] c \ d^k & d^k \end{bmatrix} \begin{bmatrix} \xi^{-1} & 0 \\ -d_{\circledo}^{k+1} \gamma \xi^{-1} \ d_{\circledo}^{k+1} + 1 - (d^k)^{\dagger} d^k \end{bmatrix}$$

$$\begin{bmatrix} a^k & 0 \\ (d^k)^{\dagger} c a^k + [1 - d^k (d^k)^{\dagger}] c \ d^k \end{bmatrix}$$

$$\cdot \begin{bmatrix} \eta^{-1} & 0 \\ 0 & (d^k)^{\dagger} (d^k)^{\dagger *} + 1 - (d^k)^{\dagger} d^k \end{bmatrix} \begin{bmatrix} (a^k)^* & c^* [1 - d^k (d^k)^{\dagger}] \\ 0 & (d^k)^* \end{bmatrix}$$

$$= \begin{bmatrix} \mu & \nu \\ \delta & \omega \end{bmatrix},$$

where

$$\mu = a^k \xi^{-1} a^k \eta^{-1} (a^k)^*, \quad \nu = a^k \xi^{-1} a^k \eta^{-1} c^* [1 - d^k (d^k)^{\dagger}],$$

$$\delta = [1 - d^k (d^k)^{\dagger}] c \xi^{-1} a^k \eta^{-1} (a^k)^* - d^D \gamma \xi^{-1} a^k \eta^{-1} (a^k)^* + d^D (d^k)^{\dagger} c a^k \eta^{-1} (a^k)^*$$

$$+ d^D [1 - d^k (d^k)^{\dagger}] c \eta^{-1} (a^k)^*,$$

$$\omega = [1 - d^k (d^k)^{\dagger}] c \xi^{-1} a^k \eta^{-1} c^* [1 - d^k (d^k)^{\dagger}] - d^D \gamma \xi^{-1} a^k \eta^{-1} c^* [1 - d^k (d^k)^{\dagger}]$$

$$+ d^D (d^k)^{\dagger} c a^k \eta^{-1} c^* [1 - d^k (d^k)^{\dagger}] + d^D [1 - d^k (d^k)^{\dagger}] c \eta^{-1} c^* [1 - d^k (d^k)^{\dagger}]$$

$$+ d^{\circledo}.$$

$\square$

With an application of Theorem 6.7.10, we obtain the following result.

**Theorem 6.8.3 ([45])** *Let* $T = \begin{bmatrix} a & 0 \\ b & d \end{bmatrix} \in R^{2 \times 2}$ *such that* $a \in R^{\tiny\textcircled{D}}$ *with* $\mathrm{ind}(a) = k$, $d^k\{1, 3\} \neq \emptyset$, *and let* $c = \sum\limits_{i+j=k-1} d^i b a^j$, $w = [1 - d^k(d^k)^{1,3}]c[1 - (a^{\tiny\textcircled{D}})^k a^k]$ *with* $w\{1, 3\} \neq \emptyset$. *Then the following are equivalent:*

(1)  $T^{\tiny\textcircled{D}}$ *exists with* $\mathrm{ind}(T) \leq k$.
(2)  $\alpha$ *and* $\beta$ *are invertible, where*

$$\alpha = 1 + [c(a^{\tiny\textcircled{D}})^k]^*[1 - d^k(d^k)^{(1,3)} - ww^{(1,3)}]c(a^{\tiny\textcircled{D}})^k,$$

$$\beta = d^{k+1}ww^{(1,3)} + d^{2k+1}(d^k)^{(1,3)} + 1 - ww^{(1,3)} - d^k(d^k)^{(1,3)}.$$

*In this case,* $T^{\tiny\textcircled{D}} = \begin{bmatrix} t_1 & t_2 \\ t_3 & t_4 \end{bmatrix}$, *where*

$$t_1 = (a^{\tiny\textcircled{D}})^{k+1}a^k x_1, \quad t_2 = (a^{\tiny\textcircled{D}})^{k+1}a^k x_2,$$

$$t_3 = \epsilon x_1 + [ww^{(1,3)} + d^k(d^k)^{(1,3)}]\beta^{-1}d^k x_3,$$

$$t_4 = \epsilon x_2 + [ww^{(1,3)} + d^k(d^k)^{(1,3)}]\beta^{-1}d^k x_4,$$

$$x_1 = aa^{\tiny\textcircled{D}}\alpha^{-1}; \quad x_2 = aa^{\tiny\textcircled{D}}\alpha^{-1}[c(a^{\tiny\textcircled{D}})^k]^*[1 - d^k(d^k)^{(1,3)} - ww^{(1,3)}],$$

$$x_3 = [1 - d^k(d^k)^{(1,3)} - ww^{(1,3)}]c(a^{\tiny\textcircled{D}})^k\alpha^{-1},$$

$$x_4 = [1 - d^k(d^k)^{(1,3)} - ww^{(1,3)}]c(a^{\tiny\textcircled{D}})^k\alpha^{-1}[c(a^{\tiny\textcircled{D}})^k]^*[1 - d^k(d^k)^{(1,3)} - ww^{(1,3)}]$$
$$+ ww^{(1,3)} + d^k(d^k)^{(1,3)},$$

$$\theta = d^k(d^k)^{(1,3)}ca^2a^{\tiny\textcircled{D}} + d^k baa^{\tiny\textcircled{D}} + d^{k+1}[1 - d^k(d^k)^{(1,3)}]c(a^{\tiny\textcircled{D}})^k,$$

$$\epsilon = [1 - d^k(d^k)^{(1,3)} - ww^{(1,3)}]c(a^{\tiny\textcircled{D}})^{2k+1}a^k + [ww^{(1,3)} + d^k(d^k)^{(1,3)}]\beta^{-1} \times$$
$$[w + d^k(d^k)^{(1,3)}c - \theta(a^{\tiny\textcircled{D}})^{k+1}a^k].$$

***Proof*** Since $T = \begin{bmatrix} a & 0 \\ b & d \end{bmatrix}$, then

$$T^k = \begin{bmatrix} a & 0 \\ b & d \end{bmatrix}^k = \begin{bmatrix} a^k & 0 \\ c & d^k \end{bmatrix}$$

$$= \begin{bmatrix} 1 & 0 \\ [1 - d^k(d^k)^{1,3}]c(a^{\tiny\textcircled{D}})^k & 1 \end{bmatrix}\begin{bmatrix} a^k & 0 \\ w & d^k \end{bmatrix}\begin{bmatrix} 1 & 0 \\ (d^k)^{(1,3)}c & 1 \end{bmatrix} := PAQ.$$

It is easy to check that $\begin{bmatrix} (a^{\oslash})^k & [1 - (a^{\oslash})^k a^k] w^{(1,3)} \\ 0 & (d^k)^{(1,3)} \end{bmatrix} \in A\{1,3\}$, and

$$P[I - AA^{(1,3)}] = \begin{bmatrix} 1 - aa^{\oslash} & 0 \\ 0 & 1 - ww^{(1,3)} - d^k(d^k)^{(1,3)} \end{bmatrix} = I - AA^{(1,3)},$$

$$P^*PAA^{(1,3)} + I - AA^{(1,3)}$$

$$= \begin{bmatrix} [c(a^{\oslash})^k]^*[1 - d^k(d^k)^{(1,3)}]c(a^{\oslash})^k + 1 & [c(a^{\oslash})^k]^* ww^{(1,3)} \\ [1 - d^k(d^k)^{(1,3)}]c(a^{\oslash})^k & 1 \end{bmatrix}$$

$$= \begin{bmatrix} 1 & [c(a^{\oslash})^k]^* ww^{(1,3)} \\ 0 & 1 \end{bmatrix} \begin{bmatrix} \alpha & 0 \\ 0 & 1 \end{bmatrix} \begin{bmatrix} 1 & 0 \\ [1 - d^k(d^k)^{(1,3)}]c(a^{\oslash})^k & 1 \end{bmatrix},$$

$$AQTPAA^{(1,3)} + I - AA^{(1,3)} = \begin{bmatrix} a^{k+2}a^{\oslash} + 1 - aa^{\oslash} & 0 \\ \theta & \beta \end{bmatrix}.$$

where

$$\alpha = 1 + [c(a^{\oslash})^k]^*[1 - d^k(d^k)^{(1,3)} - ww^{(1,3)}]c(a^{\oslash})^k,$$

$$\beta = d^{k+1}ww^{(1,3)} + d^{2k+1}(d^k)^{(1,3)} + 1 - ww^{(1,3)} - d^k(d^k)^{(1,3)},$$

$$\theta = d^k(d^k)^{(1,3)}ca^2a^{\oslash} + d^k baa^{\oslash} + d^{k+1}[1 - d^k(d^k)^{(1,3)}]c(a^{\oslash})^k.$$

Applying Theorem 6.7.10, $T^{\oslash}$ exists if and only if $U = A^{(1,3)}AQTPA + I - A^{(1,3)}A$ and $V = P^*PAA^{(1,3)} + I - AA^{(1,3)}$ are invertible, by Jacobson's lemma, equivalent to $AQTPAA^{(1,3)} + I - AA^{(1,3)}$ and $V = P^*PAA^{(1,3)} + I - AA^{(1,3)}$ are invertible.

Since $a^D$ exists with $ind(a) = k$, $(a^{k+1})^{\#}$ exists. Thus, $a^{k+1} + 1 - aa^{\oslash}$ is invertible. Again, applying Jacobson's lemma, $a^{k+2}a^{\oslash} + 1 - aa^{\oslash}$ is invertible. Hence $T^{\oslash}$ exists if and only if $\alpha$ and $\beta$ are invertible. Moreover,

$$T^k(T^k)^{(1,3)} = PA(PA)^{(1,3)}$$

$$= PAA^{(1,3)}[P^*PAA^{(1,3)} + I - AA^{(1,3)}]^{-1}P^* = \begin{bmatrix} x_1 & x_2 \\ x_3 & x_4 \end{bmatrix},$$

$$T^D = PAA^{(1,3)}(AQTPAA^{(1,3)} + I - AA^{(1,3)})^{-1}AQ$$

$$= \begin{bmatrix} (a^{\oslash})^{k+1}a^k & 0 \\ \epsilon & [ww^{(1,3)} + d^k(d^k)^{(1,3)}]\beta^{-1}d^k \end{bmatrix},$$

where

$$x_1 = aa^{\circledD}\alpha^{-1}, \quad x_2 = aa^{\circledD}\alpha^{-1}[c(a^{\circledD})^k]^*[1 - d^k(d^k)^{(1,3)} - ww^{(1,3)}],$$

$$x_3 = [1 - d^k(d^k)^{(1,3)} - ww^{(1,3)}]c(a^{\circledD})^k\alpha^{-1},$$

$$x_4 = [1 - d^k(d^k)^{(1,3)} - ww^{(1,3)}]c(a^{\circledD})^k\alpha^{-1}[c(a^{\circledD})^k]^*[1 - d^k(d^k)^{(1,3)} - ww^{(1,3)}]$$
$$+ ww^{(1,3)} + d^k(d^k)^{(1,3)},$$

$$\epsilon = [1 - d^k(d^k)^{(1,3)} - ww^{(1,3)}]c(a^{\circledD})^{2k+1}a^k + [ww^{(1,3)} + d^k(d^k)^{(1,3)}]\beta^{-1} \times$$
$$[w + d^k(d^k)^{(1,3)}c - \theta(a^{\circledD})^{k+1}a^k].$$

Hence $T^{\circledD} = T^D T^k (T^k)^{(1,3)} = \begin{bmatrix} t_1 & t_2 \\ t_3 & t_4 \end{bmatrix}$, where

$$t_1 = (a^{\circledD})^{k+1}a^k x_1, \quad t_2 = (a^{\circledD})^{k+1}a^k x_2,$$

$$t_3 = \epsilon x_1 + [ww^{(1,3)} + d^k(d^k)^{(1,3)}]\beta^{-1}d^k x_3,$$

$$t_4 = \epsilon x_2 + [ww^{(1,3)} + d^k(d^k)^{(1,3)}]\beta^{-1}d^k x_4.$$

□

**Remark 6.8.4 ([45])**

(1) Both $a$ and $d$ are pseudo core invertible may not imply that $T = \begin{bmatrix} a & 0 \\ b & d \end{bmatrix}$ is pseudo core invertible. For example, let $T = \begin{bmatrix} 1 & 0 \\ i & 0 \end{bmatrix} \in \mathbb{C}^{2\times 2}$ with transpose as involution. Then both 1 and 0 are pseudo core invertible, but for any positive integer $k$, $T^k = T$ has no $\{1, 3\}$-inverse. Hence $T$ is not pseudo core invertible.

(2) If $d$ is pseudo core invertible, then $T = \begin{bmatrix} 0 & 0 \\ b & d \end{bmatrix} \in R^{2\times 2}$ is pseudo core invertible with $T^{\circledD} = \begin{bmatrix} 0 & 0 \\ 0 & d^{\circledD} \end{bmatrix}$.

# References

1. G. Azumaya, Strongly $\pi$-regular rings. J. Fac. Sci. Hokkaido Univ. **13**, 34–39 (1954)
2. O.M. Baksalary, G. Trenkler, Core inverse of matrices. Linear Multilinear Algebra **58**(6), 681–697 (2010)
3. B.A. Barnes, Common operator properties of the linear operators $RS$ and $SR$. Proc. Amer. Math. Soc. **126**(4), 1055–1061 (1998)
4. A. Ben-Israel, T.N.E. Greville, Generalized Inverses: Theory and Applications, 2nd edn. (Springer, New York, 2003)
5. J. Benítez, Moore-Penrose inverses and commuting elements of C*-algebras. J. Math. Anal. Appl. **345**, 766–770 (2008)
6. J. Benítez, D.S. Cvetković-Ilić, Equalities of ideals associated with two projections in rings with involution. Linear Multilinear Algebra **61**, 1419–1435 (2012)
7. S.K. Berberian, Baer *-Rings (Springer, New York, 1972)
8. K.P.S. Bhaskara Rao, Theory of Generalized Inverses over Commutative Rings (Taylor and Francis, London, 2002)
9. V.P. Camillo, F.J. Costa-Cano, J.J. Simón, Relating properties of a ring and its ring of row and column finite matrices. J. Algebra **244**(2), 435–449 (2001)
10. C.G. Cao, Generalized inverses of matrices over rings (in Chinese). Acta. Math. Sin. **1**, 131–133 (1988)
11. N. Castro-González, J.J. Koliha, Y.M. Wei, Perturbation of the Drazin inverse for matrices with equal eigenprojections at zero. Linear Algebra Appl. **312**, 181–189 (2000)
12. N. Castro-González, J.J. Koliha, I. Straškraba, Perturbation on the Drazin inverse. Soochow J. Math. **27**, 201–211 (2001)
13. N. Castro-González, C. Mendes-Araújo, P. Patrício, Generalized inverses of a sum in rings. Bull. Austral. Math. Soc. **82**, 156–164 (2010)
14. N. Castro-González, J. Robles, J.Y. Vélez-Cerrada, The group inverse of $2 \times 2$ matrices over a ring. Linear Algebra Appl. **483**, 3600–3609 (2013)
15. N. Castro-González, J.L. Chen, L. Wang, Further results on generalized inverses in rings with involution. Electron. J. Linear Algebra **30**, 118–134 (2015)
16. J.L. Chen, A note on generalized inverses of a product. Northeast Math. J. **12**, 431–440 (1996)
17. J.L. Chen, J. Cui, Two questions of L. Vǎs on *-clean rings. Bull. Austral. Math. Soc. **88**, 499–505 (2013)
18. J.L. Chen, X.X. Zhang, Coherent Rings and FP-injective Rings (Science Press, Beijing, 2014)
19. J.L. Chen, X.X. Zhang, Decomposition and generalized inverses of matrices. (Chinese) College Math. **36**(5), 57–66 (2020)

© The Author(s), under exclusive license to Springer Nature Singapore Pte Ltd. 2024
J. Chen, X. Zhang, *Algebraic Theory of Generalized Inverses*,
https://doi.org/10.1007/978-981-99-8285-1

20. J.L. Chen, H.H. Zhu, Drazin invertibility of product and difference of idempotents in a ring. Filomat **28**(6), 1133–1137 (2014)
21. J.L. Chen, G.F. Zhuang, Y.M. Wei, The Drazin inverse of a sum of morphisms. Acta Math. Sci. Ser. A **29**(3), 538–552 (2009)
22. J.L. Chen, H.H. Zhu, P. Patrício, Y.L. Zhang, Characterizations and representations of core and dual core inverses. Canad. Math. Bull. **60**(2), 269–282 (2017)
23. J.L. Chen, W.D. Li, M.M. Zhou, Pseudo core inverses of a sum of morphisms. Quaest. Math. **44**(10), 1321–1332 (2021)
24. A.H. Clifford, Semigroups admitting relative inverses. Ann. Math. **42**(2), 1037–1049 (1941)
25. R.E. Cline, An application of representation for the generalized inverse of a matrix. MRC Technical Report 592, 1965
26. R.E. Cline, Inverses of rank invariant powers of a matrix. SIAM J. Numer. Anal. **5**, 182–197 (1968)
27. J. Cui, X.B. Yin, Some characterizations of ∗-regular rings. Comm. Algebra **45**(2), 841–848 (2017)
28. J. Cui, X.B. Yin, A question on ∗-regular rings. Bull. Korean Math. Soc. **55**(5), 1333–1338 (2018)
29. D.S. Cvetković-Ilić, Some results on the (2,2,0) Drazin inverse problem. Linear Algebra Appl. **438**, 4726–4741 (2013)
30. D.S. Cvetković-Ilić, C.Y. Deng, Some results on the Drazin invertibility and idempotents. J. Math. Anal. Appl. **359**, 731–738 (2009)
31. D.S. Cvetković-Ilić, C.Y. Deng, The Drazin invertibility of the difference and the sum of two idempotent operators. J. Comput. Appl. Math. **233**, 1717–1722 (2010)
32. D.S. Cvetković-Ilić, R.E. Harte, On the algebraic closure in rings. Proc. Amer. Math. Soc. **135**(11), 3547–3582 (2007)
33. D.S. Cvetković-Ilić, R.E. Harte, On Jacobson's lemma and Drazin invertibility. Appl. Math. Lett. **23**(4), 417–420 (2010)
34. C.Y. Deng, The Drazin inverses of products and differences of orthogonal projections. J. Math. Anal. Appl. **335**, 64–71 (2007)
35. C.Y. Deng, The Drazin inverses of sum and difference of idempotents. Linear Algebra Appl. **430**, 1282–1291 (2009)
36. C.Y. Deng, Characterizations and representations of group inverse involving idempotents. Linear Algebra Appl. **434**, 1067–1079 (2011)
37. C.Y. Deng, Y.M. Wei, Characterizations and representations of the Drazin inverse involving idempotents. Linear Algebra Appl. **431**, 1526–1538 (2009)
38. C.Y. Deng, Y.M. Wei, Further results on the Moore-Penrose invertibility of projectors and its applications. Linear Multilinear Algebra **60**(1), 109–129 (2012)
39. M.P. Drazin, Pseudo-inverses in associative rings and semigroups. Amer. Math. Mon. **65**, 506–514 (1958)
40. M.P. Drazin, A class of outer generalized inverses. Linear Algebra Appl. **436**, 1909–1923 (2012)
41. M.P. Drazin, Commuting properties of generalized inverses. Linear Multilinear Algebra **61**(12), 1675–1681 (2013)
42. I. Erdélyi, On the matrix equation $Ax = \lambda Bx$. J. Math. Anal. Appl. **17**, 119–132 (1967)
43. I. Fredholm, Sur une classe d'équations fonctionnelles. (French) Acta Math. **27**(1), 365–390 (1903)
44. Y.F. Gao, J.L. Chen, Pseudo core inverses in rings with involution. Comm. Algebra **46**(1), 38–50 (2018)
45. Y.F. Gao, J.L. Chen, The pseudo core inverse of a lower triangular matrix. Rev. R. Acad. Cienc. Exactas Fíc. Nat. Ser. A Mat. RACSAM **113**, 423–434 (2019)
46. Y.F. Gao, J.L. Chen, P. Patrício, D.G. Wang, The pseudo core inverse of a companion matrix. Stud. Sci. Math. Hug. **55**(3), 407–420 (2018)
47. K.R. Goodearl, Von Neumann Regular Rings. Monographs and Studies in Mathematics, 4 (Pitman, London, 1979)

48. M.C. Gouveia, R. Puystjens, About the group inverse and Moore-Penrose inverse of a product. Linear Algebra Appl. **150**, 361–369 (1991)
49. F.J. Hall, R.E. Hartwig, Algebraic properties of governing matrices used in Cesaro-Neumann iterations. Rev. Roumaine Math. Pures Appl. **26**(7), 959–978 (1981)
50. F.J. Hall, R.E. Hartwig, Applications of the Drazin Inverse to Cesaro-Neumann Iterations (Pitman, Boston, 1982)
51. R.Z. Han, J.L. Chen, Generalized inverses of matrices over rings. Chinese Q. J. Math. **7**(4), 40–47 (1992)
52. R.E. Hartwig, Block generalized inverses. Arch. Rational Mech. Anal. **61**, 197–251 (1976)
53. R.E. Hartwig, More on the Souriau-Frame algorithm and the Drazin inverse. SIAM J. Appl. Math. **31**(1), 42–46 (1976)
54. R.E. Hartwig, An application of the Moore-Penrose inverse to antisymmetric relations. Proc. Amer. Math. Soc. **78**(2), 181–186 (1980)
55. R.E. Hartwig, The group-inverse of a block triangular matrix, in *Current Trends in Matrix Theory, Proceedings of the Third Auburn Matrix Theory Conference*, ed. by F. Ulhig, R. Gore (Elsevier, Amsterdam, 1987)
56. R.E. Hartwig, J. Luh, On finite regular rings. Pacific J. Math. **69**(1), 73–95 (1977)
57. R.E. Hartwig, J. Luh, A note on the group structure of unit regular ring elements. Pacific J. Math. **71**(2), 449–461 (1977)
58. R.E. Hartwig, P. Patrício, When does the Moore-Penrose inverse flip? Oper. Matrices **6**, 181–192 (2012)
59. R.E. Hartwig, J. Shoaf, Group inverses and Drazin inverses of bidiagonal and triangular Toeplitzmatrices. J. Austral. Math. Soc. Ser. A. **24**, 10–34 (1977)
60. R.E. Hartwig, K. Spindelböck, Matrices for which $A^*$ and $A^\dagger$ commute. Linear Multilinear Algebra **14**, 241–256 (1984)
61. R.E. Hartwig, G.R. Wang, Y.M. Wei, Some additive results on Drazin inverse. Linear Algebra Appl. **322**, 207–217 (2001)
62. C.H. Hung, T.L. Markham, The Moore-Penrose inverse of a partitioned matrix $M = \begin{bmatrix} A & O \\ B & C \end{bmatrix}$. Czech. Math. J. **25**, 354–361 (1975)
63. D. Huylebrouck, The generalized inverse of a sum with radical element: applications. Linear Algebra Appl. **246**, 159–175 (1996)
64. D. Huylebrouck, R. Puystjens, Generalized inverses of a sum with a radical element. Linear Algebra Appl. **84**, 289–300 (1986)
65. N. Jacobson, Some remarks on one-sided inverses. Proc. Amer. Math. Soc. **1**(3), 352–355 (1950)
66. G. Kantún-Montiel, Outer generalized inverses with prescribed ideals. Linear Multilinear Algebra **62**(9), 1187–1196 (2014)
67. Y.Y. Ke, L. Wang, J.L. Chen, The core inverse of a product and $2 \times 2$ matrices. Bull. Malays. Math. Sci. Soc. **42**(1), 51–66 (2019)
68. Y.Y. Ke, J.L. Chen, P. Stanimirović, M. Ćirić, Characterizations and representations of outer inverse for matrices over a ring. Linear Multilinear Algebra **69**(1), 155–176 (2021)
69. D. Khurana, T.Y. Lam, Commutators and anti-commutators of idempotents in rings. Contemp. Math. **715**, 205–224 (2018)
70. J.J. Koliha, P. Patrício, Elements of rings with equal spectral idempotents. J. Aust. Math. Soc. **72**, 137–152 (2002)
71. J.J. Koliha, V. Rakčević, Invertibility of the sum of idempotents. Linear Multilinear Algebra **50**(4), 285–292 (2002)
72. J.J. Koliha, V. Rakčević, Invertibility of the difference of idempotents. Linear Multilinear Algebra **51**(1), 97–110 (2003)
73. J.J. Koliha, V. Rakočević, Range projections and the Moore-Penrose inverse in rings with involution. Linear Multilinear Algebra **55**(2), 103–112 (2007)

74. J.J. Koliha, D. Djordjević, D. Cvetković, Moore-Penrose inverse in rings with involution. Linear Algebra Appl. **426**, 371–381 (2007)

75. J.J. Koliha, D.S. Cvetković-Ilić, C.Y. Deng, Generalized Drazin invertibility of combinations of idempotents. Linear Algebra Appl. **437**, 2317–2324 (2012)

76. T.Y. Lam, Lectures on Modules and Rings (Springer, New York, 1999)

77. T.Y. Lam, A First Course in Noncommutative Rings, 2nd edn. (Springer, New York, 2001)

78. T.Y. Lam, P.P. Nielsen, Jacobson's lemma for Drazin inverses. Contemp. Math. **609**, 185–195 (2014)

79. Y. Li, The Moore-Penrose inverses of products and differences of projections in a $C^*$-algebra. Linear Algebra Appl **428**, 1169–1177 (2008)

80. Y. Li, The Drazin inverses of products and differences of projections in a $C^*$-algebra. J. Aust. Math. Soc. **86**, 189–198 (2009)

81. T.T. Li, J.L. Chen, Characterizations of core and dual core inverses in rings with involution. Linear Multilinear Algebra **66**(4), 717–730 (2018)

82. T.T. Li, J.L. Chen, The core invertibility of a companion matrix and a Hankel matrix. Linear Multilinear Algebra **68**(3), 550–562 (2020)

83. T.T. Li, J.L. Chen, D.G. Wang, S.Z. Xu, Core and dual core inverses of a sum of morphisms. Filomat **33**(10), 2931–2941 (2019)

84. Y.H. Liao, J.L. Chen, J. Cui, Cline's formula for the generalized Drazin inverse. Bull. Malays. Math. Sci. Soc. **37**(1), 37–42 (2014)

85. X.J. Liu, L. Xu, Y.M. Yu, The representations of the Drazin inverse of differences of two matrices. Appl. Math. Comput. **216**, 3652–3661 (2010)

86. J. Ljubisavljević, D.S. Cvetković-Ilić, Additive results for the Drazin inverse of block matrices and applications. J. Comput. Appl. Math. **235**(12), 3683–3690 (2011)

87. S.B. Malik, N. Thome, On a new generalized inverse for matrices of an arbitrary index. Appl. Math. Comput. **226**, 575–580 (2014)

88. X. Mary, On generalized inverses and Green's relations. Linear Algebra Appl. **434**, 1836–1844 (2011)

89. X. Mary, P. Patrício, The inverse along a lower triangular matrix. Appl. Math. Comput. **219**, 1130–1135 (2012)

90. X. Mary, P. Patrício, Generalized inverses modulo $\mathcal{H}$ in semigroups and rings. Linear Multilinear Algebra **61**, 886–891 (2013)

91. X. Mary, P. Patrício, The group inverse of a product. Linear Multilinear Algebra **64**(9), 1776–1784 (2016)

92. E.H. Moore, On the reciprocal of the general algebraic matrix. Bull. Amer. Math. Soc. **26**, 394–395 (1920)

93. P. Patrício, The Moore-Penrose inverse of von Neumann regular matrices over a ring. Linear Algebra Appl. **332**, 469–483 (2001)

94. P. Patrício, The Moore-Penrose inverse of a factorization. Linear Algebra Appl. **370**, 227–235 (2003)

95. P. Patrício, The Moore-Penrose inverse of a companion matrix. Linear Algebra Appl. **437**, 870–877 (2012)

96. P. Patrício, R.E. Hartwig, The (2,2,0) Group Inverse Problem. Appl. Math. Comput. **217**(2), 516–520 (2010)

97. P. Patrício, R.E. Hartwig, Some regular sums. Linear Multilinear Algebra **63**(1), 185–200 (2015)

98. P. Patrício, C. Mendes-Araújo, Moore-Penrose invertibility in involutory rings: the case $aa^\dagger = bb^\dagger$. Linear Multilinear Algebra **58**(3–4), 445–452 (2010)

99. P. Patrício, R. Puystjens, About the von Neumann regularity of triangular block matrices. Linear Algebra Appl. **332**, 485–502 (2001)

100. P. Patrício, R. Puystjens, Drazin-Moore-Penrose invertibility in rings. Linear ALgebra Appl. **389**, 159–173 (2004)

101. P. Patrício, A. Veloso Da Costa, On the Drazin index of regular elements. Cent. Eur. J. Math. **7**(2), 200–205 (2009)

102. R. Penrose, A generalized inverse for matrices. Proc. Camb. Philos. Soc. **51**, 406–413 (1955)
103. K.M. Prasad, K.S. Mohana, Core-EP inverse. Linear Multilinear Algebra **62**(6), 792–802 (2014)
104. R. Puystjens, M.C. Gouveia, Drazin invertibility for matrices over an arbitrary ring. Linear Algebra Appl. **385**, 105–116 (2004)
105. R. Puystjens, R.E. Hartwig, The group inverse of a companion matrix. Linear Multilinear Algebra **43**(1–3), 137–150 (1997)
106. R. Puystjens, D.W. Robinson, The Moore-Penrose inverse of a morphism with factorization. Linear Algebra Appl. **40**, 129–141 (1981)
107. R. Puystjens, D.W. Robinson, Symmetric morphisms and existence of Moore-Penrose inverses. Linear Algebra Appl. **131**, 51–69 (1990)
108. D.S. Rakić, N.Č. Dinčić, D.S. Djordjević, Group, Moore-Penrose, core and dual core inverse in rings with involution. Linear Algebra Appl. **463**, 115–133 (2014)
109. C.R. Rao, S.K. Mitra, *Generalized Inverse of Matrices and its Applications* (Wiley, New York, 1971)
110. H. Schwerdtfeger, *Introduction to Linear Algebra and the Theory of Matrices* (Noordhoff, Groningen, 1950)
111. G.Q. Shi, J.L. Chen, T.T. Li, M.M. Zhou, Jacobson's lemma and Cline's formula for generalized inverses in a ring with involution. Comm. Algebra **48**(9), 3948–3961 (2020)
112. Y.Y. Tseng, Generalized inverses of unbounded operators between two unitary spaces. Dokl. Akad. Nauk. SSSR. **67**, 431–434 (1949)
113. Y.Y. Tseng, Properties and classifications of generalized inverses of closed operators. Dokl. Akad. Nauk. SSSR. **67**, 607–610 (1949)
114. H.X. Wang, Core-EP decomposition and its applications. Linear Algebra Appl. **508**, 289–300 (2016)
115. H.X. Wang, X.J. Liu, Characterizations of the core inverse and the partial ordering. Linear Multilinear Algebra **63**(9), 1829–1836 (2015)
116. H.X. Wang, X.J. Liu, EP-nilpotent decomposition and its applications. Linear Multilinear Algebra **68**(8), 1682–1694 (2020)
117. L. Wang, X. Zhu, J.L. Chen, Additive property of Drazin invertibility of elements in a ring. Filomat **30**(5), 1185–1193 (2016)
118. L. Wang, N. Castro-González, J.L. Chen, Characterizations of outer generalized inverses. Canad. Math. Bull. **60**(4), 861–871 (2017)
119. Y.M. Wei, Perturbations bound of the Drazin inverse. Appl Math Comput. **125**, 231–244 (2002)
120. Y.M. Wei, G.R. Wang, The perturbation theory for the Drazin inverse and its applications. Linear Algebra Appl. **258**, 179–186 (1997)
121. C. Wu, J.L. Chen, Left core inverses in rings with involution. Rev. R. Acad. Cienc. Exactas Fís. Nat. Ser. A Mat. RACSAM **116**(2), Page No. 67, 15 pp. (2022)
122. S.Z. Xu, J.L. Chen, The Moore-Penrose inverse in rings with involution. Filomat **33**(18), 5791–5802 (2019)
123. S.Z. Xu, J.L. Chen, J. Benítez, EP elements in rings with involution. Bull. Malays. Math. Sci. Soc. **42**, 3409–3426 (2019)
124. S.Z. Xu, J.L. Chen, X.X. Zhang, New characterizations for core inverses in rings with involution. Front. Math. China **12**(1), 231–246 (2017)
125. H. Yao, J.C. Wei, EP elements and *-strongly regular rings. Filomat **32**, 117–125 (2018)
126. H. You, J.L. Chen, Generalized inverses of a sum of morphisms. Linear Algebra Appl. **338**, 261–273 (2001)
127. H. You, J.L. Chen, The Drazin inverse of a morphism in additive category. (Chinese) J. Math. (Wuhan) **22**(3), 359–364 (2002)
128. X.X. Zhang, J.L. Chen, *Counterexamples in Ring Theory* (Science Press, Beijing, 2019)
129. X.X. Zhang, S.S. Zhang, J.L. Chen, Moore-Penrose invertibility of differences and products of projections in rings with involution. Linear Algebra Appl. **439**, 4101–4109 (2013)

130. X.X. Zhang, J.L. Chen, L. Wang, Generalized symmetric ∗-rings and Jacobson's lemma for Moore-Penrose inverse. Publ. Math. Debrecen **91**(3–4), 321–329 (2017)

131. M.M. Zhou, J.L. Chen, Characterizations and maximal classes of elements related to pseudo core inverses. Rev. R. Acad. Cienc. Exactas Fíc. Nat. Ser. A Mat. RACSAM **114**, 104 (2020)

132. M.M. Zhou, J.L. Chen, X. Zhu, The group inverse and core inverse of sum of two elements in a ring. Comm. Algebra **48**(2), 676–690 (2020)

133. M.M. Zhou, J.L. Chen, D.G. Wang, The core inverses of linear combinations of two core invertible matrices. Linear Multilinear Algebra **69**(4), 702–718 (2021)

134. H.H. Zhu, X.X. Zhang, J.L. Chen, Generalized inverses of a factorization in a ring with involution. Linear Algebra Appl. **477**, 142–150 (2015)

135. H.H. Zhu, J.L. Chen, P. Patrício, Further results on the inverse along an element in semigroups and rings. Linear Multilinear Algebra **64**(3), 393–403 (2016)

136. H.H. Zhu, P. Patrício, J.L. Chen, Y.L. Zhang, The inverse along a product and its applications. Linear Multilinear Algebra **64**(5), 834–841 (2016)

137. H.H. Zhu, J.L. Chen, P. Patrício, Reverse order law for the inverse along an element. Linear Multilinear Algebra **65**(1), 166–177 (2017)

138. H.H. Zhu, J.L. Chen, P. Patrício, X. Mary, Centralizer's applications to the inverse along an element. Appl. Math. Comput. **315**, 27–33 (2017)

139. G.F. Zhuang, J.L. Chen, D.S. Cvetković-Ilić, Y.M. Wei, Additive property of Drazin invertibility of elements in a ring. Linear Multilinear Algebra **60**(8), 903–910 (2012)

# Index

© The Author(s), under exclusive license to Springer Nature Singapore Pte Ltd. 2024     321
J. Chen, X. Zhang, *Algebraic Theory of Generalized Inverses*,
https://doi.org/10.1007/978-981-99-8285-1

# Mathematics Monograph Series

1.  Finite Element Methods:Accuracy and Improvement (有限元方法：精度及其改善) 2006.4 Qun Lin, Jiafu Lin

2.  Spectral Analysis of Large Dimensional Random Matrices (大维随机矩阵的谱分析) 2006.9 Zhidong Bai

3.  Spectral and High-Order Methods with Applications (谱方法和高精度算法及其应用) 2006.12 Jie Shen, Tao Tang

4.  Functional Inequalities, Markov Semigroups and Spectral Theory (泛函不等式，马尔可夫半群和谱理论) 2005.3　Fengyu Wang

5.  Kac-Moody Algebras and Their Representations (卡茨–穆迪代数及其表示) 2007.3 Xiaoping Xu

6.  Adaptive Computations：Theory and Algorithms (自适应计算：理论与算法) 2007.3　Tao Tang, Jinchao Xu

7.  On the Study of Singular Nonlinear Traveling Wave Equations: Dynamical System Approach (奇非线性行波方程研究的动力系统方法) 2007.5　Jibin Li, Huihui Dai

8.  Growth Curve Models and Statistical Diagnostics (生长曲线模型及其统计诊断) 2007.8 Jianxin Pan, Kaitai Fang

9.  Theory of Polyhedra (多面形理论) 2008.2　Yanpei Liu

10.  Herz Type Spaces and Their Applications (赫兹型空间及其应用) 2008.2　Shangzhen Lu, Dachun Yang, Guoen Hu

11.  Some Topics on Value Distribution and Differentiability in Complex and P-adic Analysis (复与 P 进位分析中有关值分布及微分性的一些论题) 2008. 5　Escassut, W.Tutschke, C.C.Yang

12.  Nonlinear Complex Analysis and Its Applications (非线性复分析及其应用) 2008.5 Guochun Wen, Dechang Chen, Zuoliang Xu

13.  Introduction to Mathematical Logic and Resolution Principle (数理逻辑引论与归结原理) 2009.4　Guojun Wang, Hongjun Zhou

14.  General Theory of Map Census (地图计数通论) 2009.5　Yanpei Liu

15.  Spectral Analysis of Large Dimensional Random Matrices(Second Edition) (大维随机矩阵的谱分析) (二版) 2010.2　Zhidong Bai

16.  Association Schemes of Matrices (矩阵结合方案) 2010.4　Yangxian Wang, Yuanji Huo and Changli

17.  Empirical Likelihood in Nonparametric and Semiparametric Models (非参数和半参数模型中的经验似然) 2010. 6　Liu Xue and Lixing Zhu

18.  Mathematical Theory of Elasticity of Quasicrystals and Its Applications (准晶数学的弹性理论及应用) 2010.9　Tianyou Fan

19.  Economic Operation of Electricity Market and Its Mathematical Methods (电力市场的经济运